Mycotoxins in Food and Beverages

Innovations and Advances

Part I

Books Published in *Food Biology* series

Mycotoxins in Food and Beverages
Innovations and Advances
Part I

Editors

Didier Montet

Researcher and Expert in Food Safety
UMR Qualisud, CIRAD, Montpellier, France

Catherine Brabet

Senior Researcher, UMR Qualisud, CIRAD,
Montpellier, France

Sabine Schorr-Galindo

Professor, Senior Researcher, UMR Qualisud,
Univ. Montpellier, Montpellier, France

Ramesh C. Ray

Retired Senior Principal Scientist (Microbiology)
ICAR - Central Tuber Crops Research Institute,
Regional Centre, Bhubaneswar, India

CRC Press
Taylor & Francis Group
Boca Raton London New York

CRC Press is an imprint of the
Taylor & Francis Group, an **informa** business

A SCIENCE PUBLISHERS BOOK

First edition published 2021
by CRC Press
6000 Broken Sound Parkway NW, Suite 300, Boca Raton, FL 33487-2742

and by CRC Press
2 Park Square, Milton Park, Abingdon, Oxon, OX14 4RN

© 2021 Taylor & Francis Group, LLC

CRC Press is an imprint of Taylor & Francis Group, LLC

Library of Congress Cataloging-in-Publication Data

Names: Montet, Didier, editor. | Brabet, Catherine, editor. | Galindo,
  Sabine, 1967- editor. | Ray, Ramesh C. editor
Title: Mycotoxins in food and beverages : innovations and advances /
  editors, Didier Montet, Catherine Brabet, Sabine Galindo, Ramesh C. Ray.
Other titles: Food biology series.
Description: Boca Raton : CRC Press, 2021- | Series: Food biology series |
  Includes bibliographical references and index. | Summary: "Mycotoxins
  are secondary metabolites produced by fungi in a wide range of foods
  (cereals, peanut, tree nuts, dried fruits, coffee, cocoa, grapes,
  spices...) both in the field and after harvest, particularly during
  storage. They can also be found in processed foods of plant origin, or
  by transfer, in food products of animal (milk, eggs, meat and offal).
  Mycotoxins are of major concern since they can cause acute or chronic
  intoxications in both humans and animals which are sometimes fatal. Many
  countries, particularly in Europe, have set maximum acceptable levels
  for mycotoxins in food and feed"-- Provided by publisher.
Identifiers: LCCN 2020049316 | ISBN 9780367422097 (hardback ; alk. paper)
Subjects: MESH: Mycotoxins | Food Microbiology | Food
  Contamination--prevention & control
Classification: LCC RA1242.M94 | NLM QW 630.5.M9 | DDC 615.9/5295--dc23
LC record available at https://lccn.loc.gov/2020049316

ISBN: 978-0-367-42209-7 (hbk)
ISBN: 978-0-367-68280-4 (pbk)
ISBN: 978-1-003-03581-7 (ebk)

Typeset in Palatino Roman
by Innovative Processors

Foreword

"Quo vadis" is a historically well-known term, applicable for many disciplines across the past. In fact, questioning the direction of the future has never been as challenging and so important. One anchor point is the past and the second is the *status quo*.

This book addresses specific topics in mycotoxin research regarding food safety. It presents the reader a cross section of recent developments, ranging from lesser-known mycotoxins with potentially relevant toxicology over the influence of actual developments with impact on the generation of mycotoxins (*e.g. climate change, agricultural practices, biotechnological production changes*) spanning to management plans and the scientific-analytical tools needed to answer open questions.

Addressing such specific aspects in an all-inclusive perspective is more important than ever, as global human population continues to grow and the impact of seizing of limited resources relevant for (intensive) agriculture progresses (*e.g. phosphates*) correlates with population growth. The resulting demand of agricultural produce in the future – in sufficient quantity and sustainable quality – will largely effected by plant health and mycotoxin management along the food and feed production chain.

Since the discovery of aflatoxins in the 1960s as result of the *turkey-X disease* incident in the United Kingdom, mycotoxin research has made vast progress. As a result, many countries in the world have regulations in place to protect citizens and animal welfare from unfit produce contaminated with mycotoxins. Europe alone regulated until 2020 nine groups of mycotoxins in more than 60 different food commodity groups based on previous risk assessments made by the European Food Safety Authority (EFSA). This regulatory development is remarkable, as it resulted in agricultural imports (to the EU) based on sound risk evaluation and risk management, while in return generated profitable market prices for the exporting countries.

The *turkey-X-disease* incident in the 1960s linked to the identification of aflatoxin B1 as causal agent marked the beginning of today's mycotoxin research. Aflatoxin B1 is the most potent natural occurring carcinogen and the search for other threads from fungal origin led to the discovery of ochratoxin A, the trichothecenes, zearalenone, patulin and finally the fumonisins in the mid 1980s. With reference to the Food and Agricultural Organisation (FAO), it is estimated that more than 20% of all agricultural produces is affected by mycotoxins, though not all above levels of acute health concern.

In early days after discovery, the available analytical methods relied predominantly on thin-layer chromatography in combination with optical detection methods, like fluorescence or UV absorption, with native fluorescence being the favourable feature resulting in a relatively specific and sufficiently visible/measurable result even for low amounts present. Therefore, it is worth mentioning that three of the above-mentioned mycotoxins (groups) that have been studied in the earlier days (aflatoxins, ochratoxin A and zearalenone) possess such a natural fluorescence. Patulin levels in apple juice (from *Penicillium expansum*) can reach up to 1 g per litre. While patulin does not possess any fluorescence, the high concentrations facilitated its isolation and characterisation, resulting initially in attempts to utilise it as antibiotic.

A more challenging effort was the identification of fumonisins as the causal agent responsible for what is known as Equine LEM (leukoencephalomalacia) to be developed; a deadly disease affecting horses in South Africa. Once identified, their danger to human health was consequently spotted. The fumonisins initially escaped the scientists in absence of any optically useful absorbance or fluorescence and other challenging chemo-physical properties. Only with great effort, they were finally identified in the 1980s in South Africa as mycotoxins of global relevance.

Over the last 25 years the scientific development, namely in instrumental analysis and immune-chemistry, led to the rather universally availability of liquid-chromatography coupled to mass selective detectors and to methods using particularly tailored (fragmented) antibodies and aptamers that recognise mycotoxins (or structurally similar mycotoxin groups) rather specifically. These two technologies were game-changers that put numerous other mycotoxins on the radar for public health risks, which so far escaped scientists due to the analytical limitations until then. It must be mentioned however that numerous mycotoxins were already identified and structurally identified (in fungal cultures) in the 1970s, but these were not necessary identified in food, too.

As our world became a vastly changing place the words of Heraklit: πάντα ῥέí (panta rhei, meaning "everything flows") appear more relevant for scientists than ever.

Climate change, a growing population and the limits of natural resources require humankind to manage progress under these conditions. As a result, novel agricultural practices (e.g. vertical agriculture or hydroponic agriculture) or even new food production strategies (e.g. industrial utilisation of insects for nutrition or precision fermentation/biology) seem to offer solutions for global nutrition and are promoted by international think tanks with enormous financial leverage as "disruptive technologies" of the future.

Such developments will offer fungi new opportunities to compete for survival and eventually bear the potential of new mycotoxin scenarios in the food chain.

As a result sound risk evaluation and risk management is key and it should target the identification of possible mycotoxin risks in a changing food system. The contributions in this book reflect on the current knowledge and strategies regards these challenges ahead. Further, they give the reader an option to verify the assumption based on Aristotle *"the whole thing* (multidisciplinary reduction of mycotoxins in food and feed) *is more than the sum of each single strategic approach"*.

Dr Joerg Stroka
European Commission Joint Research Centre (JRC)
Geel, Retieseweg 111, 2440 Geel, Belgium

Preface to the Series

Food is the essential source of nutrients (such as carbohydrates, proteins, fats, vitamins, and minerals) for all living organisms to sustain life. A large part of daily human efforts is concentrated on food production, processing, packaging and marketing, product development, preservation, storage, and ensuring food safety and quality. It is obvious therefore, our food supply chain can contain microorganisms that interact with the food, thereby interfering in the ecology of food substrates. The microbe-food interaction can be mostly beneficial (as in the case of many fermented foods such as cheese, butter, sausage, etc.) or in some cases, it is detrimental (spoilage of food, mycotoxin, etc.). The *Food Biology* series aims at bringing all these aspects of microbe-food interactions in form of topical volumes, covering food microbiology, food mycology, biochemistry, microbial ecology, food biotechnology and bio-processing, new food product developments with microbial interventions, food nutrification with nutraceuticals, food authenticity, food origin traceability, and food science and technology. Special emphasis is laid on new molecular techniques relevant to food biology research or to monitoring and assessing food safety and quality, multiple hurdle food preservation techniques, as well as new interventions in biotechnological applications in food processing and development.

The series is broadly broken up into food fermentation, food safety and hygiene, food authenticity and traceability, microbial interventions in food bio-processing and food additive development, sensory science, molecular diagnostic methods in detecting food borne pathogens and food policy, etc. Leading international authorities with background in academia, research, industry and government have been drawn into the series either as authors or as editors. The series will be a useful reference resource base in food microbiology, biochemistry, biotechnology, food science and technology for researchers, teachers, students and food science and technology practitioners.

Ramesh C Ray
Series Editor

Preface

This first of the two volumes of an important book on mycotoxins entitled "Mycotoxins in Food and Beverages: Innovations and Advances", was finalized in the midst of the coronavirus crisis that has affected almost the entire world. We hope that this crisis will come to an end very soon and by the time this book is published, we – our families, friends and colleagues would have forgotten this painful episode of COVID 19.

Such crises pass but the problem of food hazards remain with us. We would like to draw attention now to defining the food crisis which in fact is a food hazard and disseminate the findings or research worldwide.

If we return to the mycotoxins, the object of this book, we are on a science that can be described as new since the first serious analyzes date from the 1980s period when some of the coordinators of this book were only simple students.

The coordinators of the first volume of this book wish to thank Dr Joerg Stroka from the Joint Research Center (JRC) of the European Commission located at Geel (Belgium) for his nice preface. His opinion on the history of mycotoxins and his knowledge of the analysis of the problem of mycotoxins by world experts are of the greatest importance.

We have experienced spectacular progress on mycotoxin knowledge in recent years partly due to the progress in mycotoxins analysis. The major innovations are taken up in the chapter by Professor Ray Coker (United Kingdom), the inventor of the ToxiMet system, an automatic multi mycotoxins analyzer. It gives us an overview of all the mycotoxin purification and analysis systems currently in use around the world.

Even if current modern analytical devices allow analyzing more than 400 mycotoxins, the analysis of mycotoxins remains difficult and long and requires sophisticated devices. Two teams have offered rapid and innovative methods in this book, one at the marketing stage and the other at the advanced research stage. Thus María Luz Rodríguez et al. (United Kingdom) describe the method developed by the Randox company which relates to Biochip Array Technology capable of carrying out multi-analyzes of mycotoxins from a food sample. Dr Moez El Saadani et al. (Egypt) take stock of aptameric sensors, of which they have developed an electronic prototype capable of analyzing ochratoxin A in a real environment in a quantitative and instantaneous way. A 3 D piece of DNA permitted to specifically capture a mycotoxin.

The fate of mycotoxins during storage in food is described by Professor Philippe Dantigny et al. (France). They tell us the importance of studying mycotoxins migration in foods by quantifying their concentration and by correlating these concentrations to simple observations made by consumers (i.e. visual aspect of fungal growth). Safe recommendations about moldy foods are provided by these scientists.

Developing countries are the most affected by the presence and toxicity of mycotoxins as described by Dr William Stafstrom et al. from Cornell University (USA). They show the importance of improving mycotoxin surveillance and response capacity in developing countries. Their chapter explores and describes the current state of monitoring and documenting mycotoxins in under-regulated contexts, how multi-disciplinary approaches can enhance more efficient and effective monitoring. They also propose a mycotoxin surveillance framework for use in low-resource settings.

As an example of the biodiversity of mycotoxin-producing fungi, Dr Amaranta Carvajal-Campos et al. from CIRAD and INRAE (France) and University Nangui Abrogoua (Côte d'Ivoire) focus on the biodiversity of aflatoxigenic *Aspergillus* section *Flavi* species according to food matrices and geographic areas.

It is rare to have knowledge on this subject from developing countries collated in a book. Some of these countries have trained specialists and some have set up research and/or monitoring laboratories. Among these experts, Dr Pauline Mounjouenpou from the Institute of Agricultural Research for Development (IRAD) at Yaoundé (Cameroon) reports on mycotoxins and their regulation in Cameroon. Professor Amina Bouseta et al. from Fez University discuss the occurrence of mycotoxins in foods and feeds in Morocco and their sources of contamination. They also give an overview on their control by official laboratories and the national regulation and finish by sharing some ideas of prevention.

Ms Ruth Nyagah, a food expert from Kenya, gives rare information on the link between mycotoxins and cancer in Kenya. She also provides an overview of the aflatoxin levels in farm products and explains the reactions of the Kenyan government by the creation of the National Food Safety Coordinating Committee.

Dr Larissa Yacine Waré et al. makes us aware of the presence of mycotoxins in milk and infant formula in Burkina Faso. They show the importance of chronic exposures to toxins that have a negative impact on the health of infants and young children. They give important and appropriate recommendations to producers of infant and young child feeding.

The second volume of this book will focus on prevention and decontamination, biological fight and the uses of biocides. We will give to the readers new information about toxicology and how molds will react to climate change and adapt their production of mycotoxins. We will also take stock of what genetically modified plants can bring to consumers.

We wish all of our readers an excellent reading time.

Didier Montet
Catherine Brabet
Sabine Schorr-Galindo
Ramesh C. Ray

Contents

Mycotoxin Surveillance for Low-resource Settings

William Stafstrom, Anthony Wenndt and Rebecca Nelson[*]

School of Integrative Plant Science, Cornell University, Ithaca, New York, 14853 USA

1. Introduction

Mycotoxins are a global health and economic challenge, estimated to affect a quarter of the global food supply. This 25% figure represents the frequency of samples contaminated at levels that exceed European Union and Codex Alimentarius regulatory standards, and it is estimated that 60–80% of samples contain detectable mycotoxins (Eskola et al. 2019). Since the identification of aflatoxins (AF) as the cause of Turkey-X disease nearly 60 years ago, highly-resourced countries have instituted costly food safety monitoring and regulatory mechanisms that largely insulate consumers from mycotoxins' negative health effects (Labuza 1983, Wu 2004). In tropical and subtropical regions, crops and foods are frequently contaminated by mycotoxins produced by fungi of the genera *Aspergillus*, *Fusarium*, and *Penicillium* such as AF, fumonisins (FUM), ochratoxin A (OTA), tricothecenes such as deoxynevalenol (DON), and zearalenone (ZEA) (Shephard 2008).

In contrast to farmers in high-income countries, where the burden of mycotoxins is experienced in the form of economic losses, many smallholder farmers and consumers in the global south encounter mycotoxins directly as a threat to health and nutrition. Governments in many developing countries lack the resources and capacity required to comprehensively monitor and regulate their presence in the food system (Shephard 2008). Smallholder farming systems figure prominently in developing countries' agricultural sectors and contribute to both formal and informal markets. Both formal and informal sectors in these countries are inadequately regulated and are sources of mycotoxin risk (Grace et al. 2015).

The negative health effects of consuming different mycotoxins can generally be classified as resulting from chronic or acute toxicity. The threat

*Corresponding author: rjn7@cornell.edu

of chronic mycotoxin exposure is often overlooked, whereas human disease outbreaks caused by acute mycotoxin exposure garner international attention. Trichothecene (e.g. DON) contamination of wheat caused an outbreak of gastrointestinal illness in India in 1987 (Bhat et al. 1989). In 1995, a foodborne disease outbreak in several Indian villages was linked to FUM contamination of household maize and sorghum stores (Bhat et al. 1997). Most prominently, the consumption of AF-contaminated homegrown maize in eastern Kenya in 2004 caused aflatoxicosis and resulted in 125 deaths (Azziz-Baumgartner et al. 2005). In contrast, the relatively rare acute outbreaks in developed countries have mainly affected pets or livestock (Leung et al. 2006, Morgavi and Riley 2007). These and other acute mycotoxin outbreaks predominantly affect localities that share several important features: small farm sizes, low farm incomes, and limited regulatory systems.

Communities that depend on smallholder agricultural systems are also at risk of mycotoxins' less obvious chronic effects (Shirima et al. 2015). The agricultural sectors of many low-income countries rely significantly upon smallholder farming systems in which most crops are grown for subsistence or traded in informal and underregulated markets. Smallholder farms disproportionately occupy marginal lands and experience unfavorable conditions that induce plant stress. These systems face myriad challenges including vulnerability to extreme weather conditions (e.g. drought), poor soil quality, limited access to inputs, and various types of market failure. Worryingly, climate change threatens to exacerbate many of these challenges in increasingly unpredictable ways (Morton 2007).

Improving mycotoxin surveillance and response capacity in these settings was one of the key recommendations from a report conducted by the Centers for Disease Control and Prevention and the World Health Organization (Strosnider et al. 2006). Coordinated efforts to improve monitoring of mycotoxins and avoid such events in these settings have made some progress, but mycotoxins remain a prominent public health challenge in developing countries (Ladeira et al. 2017).

Innovations in monitoring mycotoxins in underregulated and low-resource areas are needed but remain a significant challenge. Most validated assays are expensive, require advanced instruments and highly trained personnel, and are not suited for use in the field (Harvey et al. 2013). Small-scale farms are difficult to sample as they are dispersed in remote areas, their households are sensitive to excessive sampling, and their crops rarely enter formal markets. Landscape-wide mycotoxin prediction models are emerging as an option to guide and support on-the-ground surveillance, but successful deployment in resource-limited environments will require substantial technical advancement (Battilani 2016).

Many developing countries have not yet implemented regulatory guidelines for mycotoxins, and in those that have, regulations are often only enforced in the context of international trade with developed countries (Matumba et al. 2017). In these cases, mycotoxins harm not only the health of local communities but also disrupt their access to lucrative markets.

In developing countries, mycotoxin surveys have provided snapshot views of mycotoxin contamination and have often been targeted towards areas known to be at higher risk of exposure. Few studies have assessed mycotoxin dynamics over long periods of time and/or over large areas, which would provide a more complete understanding of mycotoxin contamination within and across diverse food systems. In the future, it will be more possible to establish systems that monitor the complex spatial and temporal dynamics of mycotoxins within these systems and analyze their interactions with biological, environmental, and socio-cultural factors. This is necessary to establish effective mycotoxin surveillance systems and to implement innovative mitigation strategies. Because of resource constraints, the high-income country model for monitoring and regulating mycotoxins is not broadly applicable in these systems, and alternative approaches must be considered. Despite these challenges, recent advances in the understanding of mycotoxins and potential monitoring technologies offer hope for more effective mycotoxin surveillance in low-resource areas.

This chapter aims to explore and describe: (1) the current state of monitoring and documenting mycotoxins in underregulated contexts, (2) how multi-disciplinary approaches enhance mycotoxin surveys and enable more efficient and effective monitoring, (3) a proposed mycotoxin surveillance framework for use in low-resource settings that integrates across scales and is context-dependent and resource-sensitive, and 4) how this framework would be amenable to sustainable and community-motivated interventions.

2. Status of Mycotoxin Surveillance in Low-Resource Settings

Mycotoxin surveillance in low-resource settings is limited in scope and is infrequently institutionalized outside of export markets. Understanding how mycotoxins have been monitored in these contexts in the past can provide the basis for identifying areas for improvement. Doing so requires becoming acquainted with a variety of stakeholders, mycotoxins, crops, sampling methodologies, and assay technologies. In unregulated and informal food systems, mycotoxin surveys along the food value chain offer insights into the levels of contamination faced by consumers. Without systematic surveillance mechanisms, such snapshot surveys hint at the true state of mycotoxin exposure faced by consumers, especially where enforcement of mycotoxin regulations is limited. Though there are rarely comprehensive mycotoxin surveillance programs in developing countries, some isolated efforts to build infrastructure and capacity for such programs have been initiated.

2.1 Monitors of Mycotoxins

The participants in (intermittent) monitoring of mycotoxins in low-resource areas are affiliated with a wide range of governmental agencies, domestic and international academic institutions, and research centers. These actors represent diverse fields such as plant pathology, analytical chemistry, food science, nutrition, agronomy, and others.

The role of governments in mycotoxin surveillance in low-resource settings varies significantly among countries and is typically limited to the formal sector. Generally, national food safety programs monitor the presence of mycotoxins in commercially produced goods. For instance, the Kenya Bureau of Standards (KEBS) recently banned five domestic brands of maize flour after finding them to exceed Kenya's 10 ppb regulatory limit for AF (BBC 2019). KEBS previously banned domestic peanut butters that exceeded AF standards (The Citizen 2019). These actions had regional repercussions, as Tanzania and Rwanda also instituted bans on the identified brands. In India, the Food Safety and Standards Authority of India (FSSAI) measured AF (M1 form) for the first time in their milk quality survey and found that 5.7% of samples were above the limit of 0.5 ppb (FSSAI 2018). Encouragingly, this FSSAI survey was conducted using mobile laboratories and real time data analysis, but only cities of >50,000 people were visited, and India does not regulate mycotoxins in the livestock feeds that are the source of the milk contamination. These actions by domestic authorities speak to the growing awareness of the threat of AF contamination, but also underline that such efforts are focused on commercially available products contaminated by a single mycotoxin. As governments progress in their surveillance of formal markets, there must be alternative strategies to address the mycotoxin exposure of citizens in subsistence farming systems and those who access informal markets.

Domestic research institutions are active and important contributors in surveying their countries' foods and beverages for mycotoxins. Domestic groups are uniquely positioned to look beyond the commonly surveyed staple foods and beverages (e.g. maize, groundnuts, wheat, milk, etc.) and explore other less globally prominent crops that are widely consumed in local contexts. For instance, the most comprehensive survey of AF in rice was carried out by a team of Indian plant pathologists (Reddy et al. 2009). A diverse group of researchers from the University of Harare showed that AF was common in multiple legumes grown by smallholder farmers throughout Zimbabwe (Maringe et al. 2017). Traditional South African alcoholic beverages were analyzed for multiple mycotoxins and different mycotoxins were associated with commercial and home-brewed beverages (Odhav and Naicker 2002). However, domestic institutions frequently face resource constraints and their surveys tend to be relatively small-scale.

International and regional research centers like the CGIAR's International Livestock Research Institute (ILRI) have played a key role in the understanding of mycotoxin dynamics in developing countries. CGIAR's agricultural research centers possess the human and technical resources necessary to conduct mycotoxin surveys, and they are mandated to research and improve agriculture in low-resource contexts. Because of their relative resource advantages, they can collect and analyze more samples and conduct multi-year studies. For example, ILRI's regional hub in Vietnam conducted the country's first comprehensive survey of AF in maize for food and livestock feed (Lee et al. 2017). The International Crops Research Institute for the Semi-Arid Tropics (ICRISAT) spearheaded a survey of Zambian peanut butter over three years, showing significant and erratic AF contamination in commercially available brands (Njoroge et al. 2016).

Collaborations among international, regional, and domestic scientists can be effective partnerships for carrying out mycotoxin surveys. Often, this blend leverages international perspectives and resources, regional hubs of excellence with high capacity, and local knowledge of diverse and complex food systems. For example, authors affiliated with ILRI, the University of Rwanda, and international academic institutions collaborated to characterize maize AF contamination in Kigali marketplaces (Nishimwe et al. 2017). Intragovernmental cooperation has also proven beneficial, as in the case of American and Nepalese government scientists assaying FUM and trichothecenes in Nepalese maize and wheat (Desjardins et al. 2000). Collaborations can help to expand the scope of a survey; Nigerian and German institutions collaborated to assess the presence of multiple mycotoxins in Nigerian maize, including the first documentation of ergosterol in this context (Bankole et al. 2010). Surveys conducted by these diverse teams of researchers offer extensive insight into the status of mycotoxin contamination in developing countries. However, this reliance on diverse and informal groups also emphasizes the lack of a consistent and defined system of mycotoxin surveillance in these contexts.

Local communities are notably absent as active monitors in these efforts, as few mycotoxin surveys have significantly engaged with community groups. Within surveys, their role is often limited to being the source of the contaminated products. This is partly driven by the reliance on validated mycotoxin assays that are analyzed at a centralized location far from where samples are collected. Though such validated assays may not be feasible at the local level, communities offer deep knowledge of many of the drivers of mycotoxins and should play an important role in monitoring mycotoxins.

2.2 Food and Beverage Types

Mycotoxin surveys have analyzed a wide range of food and beverages, often focusing on AF contamination in the staple crops maize and groundnuts because of their notable vulnerability to contamination. For instance, mycotoxins were tested in a wide variety of foods and feeds from Bangladeshi markets and the highest levels of contamination were from AF in maize, groundnuts, and poultry feed (Dawlatana et al. 2002). While these crops are among the most vulnerable to AF accumulation, the potential sources of mycotoxins are numerous within varied and diverse diets, and surveillance efforts need to be balanced across the foods at most obvious risk and other possible sources.

Consumption of mycotoxin-contaminated feeds by livestock leads to a reduction in animal health and productivity and results in the presence of mycotoxins in animal products such as milk, eggs, and meat. For example, in Kenyan peri-urban and rural settings, the AF and DON concentrations in livestock feed and AF concentrations in milk were surveyed (Makau et al. 2016). Consumers in peri-urban areas were found to be at higher risk of exposure to AF from milk (M1 form) because the feed in those areas was more highly contaminated by mycotoxins. Other beverages traditionally consumed in low-resource settings, particularly beer and coffee, have been monitored. In Africa, sorghum malts and beers are regularly contaminated with AF, FUM, and DON (Matumba et al. 2011, Roger 2011). In major coffee producing and consuming nations, such as Ethiopia, Cote d'Ivoire, and Colombia, substantial levels of OTA have been observed in coffee for local consumption (Diaz et al. 2004, Geremew et al. 2016, Manda et al. 2016).

Surveys demonstrate that the sources of mycotoxin exposure vary significantly across food systems, and a surveillance framework should account for local variation in both the quantity and types of foods and beverages consumed.

2.3 Sampling Methodologies

Mycotoxins are often distributed highly heterogeneously within a commodity or food product, which makes them difficult to sample and assay accurately. Improvements in sampling and testing methodologies lead to significant improvements in accurately monitoring mycotoxins (Herrman et al. 2020). Sampling methodologies must be locally adapted and context-sensitive for determining exposure levels in developing countries. Food and beverages consumed in these countries are often derived from subsistence farms or local markets, and commercial products can represent a diverse array of farm types and regions of origin.

For rural areas, farm- or household-based sampling methods are typically employed. Procuring samples directly from farmers' fields allows researchers to assess mycotoxin concentrations prior to storage or processing. In Zambia, harvested cobs were sampled from standing maize plants in smallholder farms to investigate pre-harvest AF and FUM contamination (Njeru et al. 2019). Sampling from household stores can include crops at different stages of processing. Matumba et al. (2015) directly sampled from subsistence farmers' bags of shelled maize, while Shephard et al. (2013) also sampled cooked maize products like porridge. Household and farm surveys are effective at determining highly local exposure to mycotoxins, as the crop or food products tested are derived from the farm or local trade. However, visiting hundreds of farms or homes is difficult to scale, as it is expensive and tedious, poses multiple social and cultural constraints, and the ethics of sampling from food insecure households must be considered.

To more comprehensively survey outside of the farm or household setting, alternative approaches have been used. Nabwire et al. (2019) leveraged relationships with schools to procure household maize samples that they tested for AF. Mutiga et al. (2014, 2015) sampled maize at local grain mills in eastern and western Kenya. In developing countries, small-scale mills are common in both rural and urban environments and are patronized by local customers. Sampling at these local processing nodes offers improvements in survey efficiency and scale, and, if sampling flour, representative sample sizes can be smaller because mycotoxin distribution is more homogeneous in milled than in unmilled samples. Local mills are used for cereals like maize, sorghum, and rice, and shelling machines are analogous hubs for groundnuts.

Sampling outside of the rural context might also be necessary, as crops originating from smallholder farms are traded throughout a country. The 2016 Tanzanian aflatoxicosis outbreak demonstrated this phenomenon; though deaths were linked to subsistence farmers' home-grown stores of maize, samples of commercial maize flour during this time period also contained high concentrations of AF (Kamala et al. 2018). Surveys of traded goods have sampled food commodities or processed foods downstream of producers in the value chain. Potential sampling nodes in these areas are local markets, milling and processing facilities, and commercially available products (Makau et al. 2016, Moser et al. 2014, Nishimwe et al. 2017, Njoroge et al. 2016). Depending on the size of the market, the sourcing of the processor, or the distribution of processed goods, these samples can represent mycotoxin contamination from single or multiple origins. This complexity underlines that understanding the flow of crops through the food value chain is necessary for optimizing sampling methods and integrating mycotoxin surveillance at different scales.

In light of this inherent complexity, a sampling methodology that allows for less granular surveys would improve broader surveillance of mycotoxins. For crops that are processed at highly local facilities (e.g. maize, sorghum, groundnuts), sampling at these community nodes might offer an ideal balance of sampling resolution and efficiency. Further down the value chain, sampling local markets, traders, or processors could provide an even broader view but would sacrifice geographical resolution.

2.4 Mycotoxin Assays

A technical challenge to mycotoxin surveillance in low-resource settings is the lack of low-cost, field-based assays (Shephard 2018). Portable options, such as lateral flow assays, are prohibitively expensive for large-scale monitoring. Cheap, portable, and non-destructive methods, like near infrared reflectance (NIR), are promising but have not been validated at regulatory standard levels (Harvey et al. 2013). Aptamer-based assays have not yet come to market, but could be inexpensive, stable, and accurate enough to significantly enhance field-based mycotoxin monitoring (Sharma et al. 2017, Shephard 2018).

Most mycotoxin assays' levels of detection (LOD) are fit for purpose and based on regulatory standards that depend on toxicity and risk exposure assessments in different foods and beverages. For example, the high toxicity of AF and its presence in staple crops, like maize and sorghum, means that their LODs are typically orders of magnitude lower than for mycotoxins such as FUM and trichothecenes. However, in areas at risk of extreme exposure to mycotoxins, qualitative or semi-quantitative assays could still prove useful, especially if there is a substantial savings in cost. In the context of smallholder farming systems, assay costs make large-scale continuous monitoring prohibitively expensive, but a compromise that integrates more advanced techniques with cheaper and less precise methods could prove to be effective.

2.5 Mycotoxin Surveys Summary

Published surveys collectively paint a picture of largely unregulated food systems that are contaminated with highly variable and often dangerously high levels of mycotoxins depending on the year, sample type, and location. Presently, mycotoxin surveillance in developing countries relies on surveys that provide a relatively scattered and likely biased view. The general state of mycotoxin risk documented by these surveys has helped to enhance awareness and mobilize some action at local and global scales, yet awareness of mycotoxins within many at risk communities remains low (Adekoya et al. 2017, James et al. 2007, Johnson et al. 2018, Lee et al. 2017, Mboya and Kolanisi 2014). For mycotoxin surveys to connect the awareness of the problem to actionable solutions, the status of mycotoxins in the food system needs to be further contextualized in a more holistic fashion.

3. Monitoring Mycotoxin Risk Factors in Low-Resource Settings

Mycotoxin surveys of foods and beverages have established a basic understanding of the mycotoxin challenge in most countries' food systems. However, food systems in developing countries are inherently complex and distinctive, with diverse social and cultural contexts and varied agricultural systems. This complexity means that surveys that provide a snapshot view of mycotoxin contamination may not offer insights into the underlying causes. Integrated surveys that analyze mycotoxin concentrations along with their various drivers have been critical to contextualizing the degree to which different risk factors influence contamination. To establish an effective surveillance framework, factors associated with mycotoxin contamination from every step along the food value chain should be monitored to improve risk prediction.

3.1 Pre-Harvest Risk Factors

Mycotoxin susceptibility is largely driven by plant stress, which is influenced by weather and soil conditions (Ferrigo et al. 2014). Pre-harvest abiotic factors thus play a large part in determining the extent of mycotoxin accumulation, and they form the basis of predictive models for pre-harvest AF, DON, and FUM contamination (Battilani et al. 2008, Cao et al. 2014, Cotty and Jaime-Garcia 2007, Kerry et al. 2017, Schaafsma and Hooker 2007). In smallholder farming systems, surveys have documented how certain fungi and their mycotoxins are associated with different agroecological zones, which provides insight into how mycotoxins are affected by factors such as altitude, soil, temperature, humidity, seasonality, and rainfall (Monyo et al. 2012, Mukanga et al. 2010, Mutiga et al. 2015, Ndemera et al. 2018, Sserumaga et al. 2019, Toteja et al. 2006).

There is some geographic overlap of mycotoxins, and general trends about what drives their geographical distributions across growing seasons have been documented. For example, in lower latitudes, hotter and drier lowlands tend to have more AF contamination, while cooler and wetter highlands experience more FUM and DON contamination. Smith et al. (2016) examined environmental effects in Kenyan smallholder farming communities by using remotely sensed datasets of rainfall, soil parameters, and a vegetation index to model the concentrations of maize AF and FUM in Kenyan smallholder farming systems. Crucially, they found evidence for mycotoxin distribution being affected by both across- and within-season environmental variation. From a surveillance perspective, remote sensing datasets provide detailed and, typically, freely available information on smallholder farming systems that could help to predict areas at risk of certain mycotoxins both across and within years.

Agronomic factors during a growing season significantly influence end-of-season mycotoxin contamination, and many of these are controlled by the grower. In smallholder farming systems, numerous studies have focused on how mycotoxin contamination of crops (primarily AF and/or FUM in maize and/or groundnuts) are associated with myriad pre-harvest agronomic practices. These data are usually acquired by conducting questionnaires or surveys with farmers during sample collection. Some of the good agricultural practices that could effectively mitigate mycotoxins are intercropping, crop rotation, residue removal, use of early maturing and resistant varieties, soil amendments, reducing field drying time, and pest management by pesticides or push-pull systems (Atukwase et al. 2009, Degraeve et al. 2015, Kaaya et al. 2005, Kimanya et al. 2009, Mutiga et al. 2014, 2015, Ndemera et al. 2018, Njeru et al. 2019, Phokane et al. 2019). Accounting for these factors in a surveillance framework can help to evaluate risk and to suggest the most effective interventions.

The types and varieties of crops grown also influence toxin accumulation. Information about cropping systems is relevant to risk assessment and targeting of surveillance efforts, although a crop's inherent vulnerability to mycotoxins is difficult to quantify given the large environmental effects on fungal infection, colonization, and mycotoxin production. Within smallholder farming systems, there is often high genetic diversity within crops, as landraces and open pollinated varieties are common. Certain crops are more resistant to mycotoxin contamination because of their innate resistance or their adaptation to an environment. Within a crop, varietal differences explain some variance in mycotoxin contamination. AF contamination of western Kenyan groundnuts was lower in improved varieties than in landraces, and resistant genotypes were identified in a west African groundnut breeding program (Mutegi et al. 2009, Waliyar et al. 1994). In Kenyan maize, varieties with flint kernels were less contaminated with AF than dent kernel varieties (Mutiga et al. 2017).

There is a considerable diversity of mycotoxigenic fungal pathogens both across and within species and understanding this diversity in the context of the food value chain can be relevant to mycotoxin surveillance. Viewing this diversity through the lens of the plant disease triangle, which considers the interactions among the pathogen, host crop, and environment, can help to inform which mycotoxins are likely to contaminate a given food or beverage. Multiple fungal pathogens are present in smallholder farming systems and the distribution of different species across landscapes is largely dictated by environmental factors such as temperature and humidity (Bankole et al. 2006, Giorni et al. 2009, Hove et al. 2016, Mukanga et al. 2010). Within a species, strains can vary in their mycotoxin production potential and different strains can be adapted to different environments (Fandohan et al. 2005, Kpodo et al. 2000, Monyo et al. 2012). For example, in an area of Kenya that is notably affected by aflatoxicosis outbreaks, Probst et al. (2012) determined that the genetically distinct and extremely toxigenic S strain morphotypes of *A. flavus* predominated. For rapid diagnostics in the field, Radhakrishnan et al. (2019) devised a DNA sequencing-based monitoring system that distinguished different strains of wheat yellow rust in Ethiopia. These findings suggest that surveillance of mycotoxins can be enhanced by understanding the ecogeographical distribution of mycotoxigenic fungi species and their strains in crops and soils.

3.2 Post-Harvest Risk Factors

Post-harvest events also contribute to mycotoxin risk and are often considered the most tractable ways to manage mycotoxin risk under resource constraints. Within smallholder farming systems, critical post-harvest actions to reduce mycotoxin levels involve drying, storage, sorting, and processing (Manandhar et al. 2018). Some storage methods that inhibit fungal growth or insect damage include storing in a cool, dry location, using hermetically sealed bags that limit insect damage, using natural deterrents to insects like essential oils or neem leaves, or application of synthetic insecticides (Kamala et al. 2016, Magembe et al. 2016, Sasamalo et al. 2018). Sorting is typically done by hand and involves removing discolored, damaged, or non-uniform grains, which are more likely to be infected by fungal pathogens and contain mycotoxins. Hand sorting has proven effective for reducing FUM in maize and AF in groundnut, but it has not been consistently effective for reducing AF in maize (Afolabi et al. 2006, Galvez et al. 2003, Mutiga et al. 2014, Xu et al. 2017). Optical sorting devices that use spectral reflectance signals can reduce AF and FUM in maize, but they are not validated to regulatory limits and are not commercially available (Harvey et al. 2013, Stasiewicz et al. 2017). Decortication, the removal of the maize germ and seed coat by milling, is common and significantly reduces FUM (Fandohan et al. 2005, Vanara et al. 2009). Another processing technique, nixtamalization, is commonly employed in Central America and can reduce AF and FUM in maize food products (Elias-Orozco et al. 2002, Palencia et al. 2003). Post-harvest mitigation methods vary in their effectiveness, depend on the crop and mycotoxin in question, and have a range of costs. Monitoring their implementation in a food system would help to better model mycotoxin exposure risk and identify optimal interventions.

Dietary composition significantly influences exposure to mycotoxins. Cultural preferences for crops that are more readily contaminated by mycotoxins increase the risk of exposure. Also, certain demographics may be particularly at risk; for example, children's weaning foods can be contaminated with a greater number of mycotoxins than foods consumed by the general population (Matumba et al. 2015, Yacine Ware et al. 2017). This is troubling, as infants and young children are especially sensitive to mycotoxins' negative health effects. Understanding the dietary patterns of consumers is the last step along the food value chain and is critical to informing the mycotoxin risk faced by a community.

3.3 Integrated Monitoring and Mitigation

A diverse array of interconnected factors affects the mycotoxin contamination of a food system. As opposed to traditional surveys of mycotoxins, interdisciplinary surveys elucidate how these factors interact and influence mycotoxin levels. Thus, by monitoring and modeling these associated factors, patterns of mycotoxin contamination across landscapes and communities can be discerned. There remains the challenge of how to scale monitoring techniques of mycotoxins and their drivers efficiently and effectively in low-resource settings. Each combination of crop and mycotoxin has its own unique blend of drivers, which multiplies this challenge (Pitt et al. 2013). Combining mycotoxin surveillance with expertise from diverse fields (e.g. plant pathology, plant genetics, agronomy, animal science, epidemiology, etc.) provides a more comprehensive platform for translatable mitigation strategies that have multiplicative benefits beyond just reducing mycotoxins. For example, sustainable approaches to improving soil quality not only reduce mycotoxins indirectly by lessening stress on the plant, but also improve yields and provide resiliency against drought. If applied correctly, integrative and multi-disciplinary approaches to mycotoxin surveillance could enable more effective and nuanced interventions that are informed by and adapted to local communities.

4. Mycotoxin Surveillance Framework

Mycotoxins are symptoms of food systems under stress. Their presence indicates difficulties in farming, storage, processing, and human and livestock health. In the absence of formal monitoring, surveys of mycotoxins in low-resource settings have described seemingly idiosyncratic and noisy systems in which consumers are exposed to a wide range of mycotoxins and concentrations. Integrated surveys have detailed how this highly variable contamination is influenced by numerous and diverse factors. To date, a variety of barriers, compounded by resource constraints, have prevented long-term and continuous monitoring of mycotoxins in these systems. The presence of extreme levels of mycotoxin contamination is often not detected until after devastating outbreaks occur, and the effects of exposure to moderate levels of mycotoxins are less well understood.

Pitt et al. (2018) proposed the first systematic surveillance approach for constructing an early warning system for AF in maize and groundnuts in African and Asian smallholder farming systems. Their proposed framework emphasized the necessity of linking centralized governance structures and laboratory monitors with community officials based in at-risk areas. This section will leverage insights from past mycotoxin research to inform the structure of an integrated mycotoxin surveillance framework in low-resource settings that aims to function as both an early warning system and a tool for building sustained resilience to mycotoxins within chronically threatened systems.

4.1   Proposed Surveillance Framework

We propose a dual-track framework for monitoring mycotoxins in low-resource contexts that combines regional surveillance by remote sensing with targeted local mycotoxin surveys and community engagement. By combining these sources of information and using them to construct spatiotemporal models of mycotoxin risk, areas in danger of either acute or chronic exposure to a mycotoxin can be targeted for interventions that correspond to the level of risk (Figure 1). Monitoring different types of exposure requires slightly different inputs and enables different responses.

4.2   Acute vs. Chronic Risk

The two tracks aim to identify localities that are at either acute or chronic risk of mycotoxin exposure. Surveillance of acute risk involves predicting where potential mycotoxicosis outbreaks will occur and intervening rapidly in those communities to verify the risk and rapidly implement interventions. This track relies heavily on information on environmental conditions during a growing season, such as the occurrence and timing of drought events.

Monitoring of chronic mycotoxin risk is less time sensitive and would aim to identify areas at risk of mycotoxin exposure over longer time periods. Communities at chronic risk of mycotoxin exposure can be identified by including historical weather or mycotoxin datasets in the model. Modeling chronic risk can also involve future climate scenarios to discern where mycotoxin risk will be elevated in the future. Whereas acute risk monitoring would focus on the potential of an outbreak of a single mycotoxin on a small number of crops, chronic risk monitoring would consider the possibility that a community could be exposed to different mycotoxins in different years.

4.3   Framework Inputs

The key inputs to both framework components are the environmental variables that are associated with the presence of a mycotoxin in a crop or food product. These environmental indicators form the basis for either mechanistic

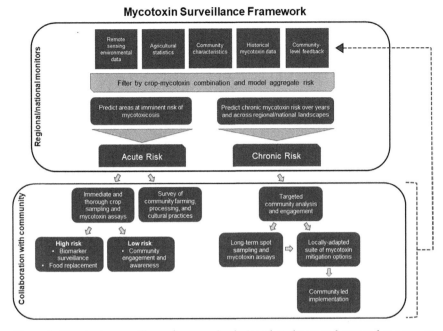

Figure 1: Mycotoxin surveillance frameworks designed to detect and respond to acute or chronic exposure within at-risk communities. Remote regional/national monitors model the aggregate risk of all mycotoxin-crop combinations for the current growing season (acute risk) or across multiple years (chronic risk). Communities identified will be engaged to verify the type of risk present and determine the optimal actions. All data generated from these processes will feed into the models and improve their performance in future years (dashed line).

or empirical models that predict the presence of mycotoxins. To predict the level of mycotoxin contamination, mechanistic models integrate knowledge from *in vitro* work on the optimal mycotoxin-producing conditions for a given pathogen and crop. Mechanistic models have been developed for AF in maize and groundnuts, DON in wheat, and OTA in grapes (Battilani 2016). Empirical models could be developed and validated by using ground-truthed mycotoxin data generated as part of the surveillance framework. One difference in the inputs would be that an early warning system would be sensitive to current year growing season conditions in predicting pre-harvest mycotoxin risk. A model of chronic mycotoxin risk would leverage both historical data and predicted climate scenarios to the cumulative mycotoxin risk of an area across years.

Aside from environmental mycotoxin drivers, these prediction models would be improved by contextualizing them within other known factors such as cropland distribution, farming and agronomic techniques, insect pressure, and cultural practices. Some of these data are accessible within large and established datasets, while others could be acquired via community feedback loops. For example, national agricultural statistics would indicate which regions grow certain crops, and that information can be used to adjust risk estimates. On the other hand, knowing the types of varieties grown by farmers across countries or regions is more difficult to monitor remotely, but this information could be acquired by building communication channels with communities.

4.4 Monitors

In an integrated surveillance framework, communities should play a prominent role as local monitors, and centralized monitors should serve to support them. A centralized national or regional center(s) would be responsible for predicting landscape-wide risk, performing mycotoxin assays, engaging at-risk communities, and managing the diverse streams of data that are generated. These centers require personnel with experience in diverse fields such as mycotoxin assays, geospatial modeling, and community extension. These research teams could be integrated with other institutions, such as those involved in protection of public health, leveraging existing infrastructure and expertise.

Centralized surveillance efforts must be matched with community-level awareness and action. Within communities prone to mycotoxin exposure, relationships between the centralized monitors and local stakeholders must be formed. Depending on the type of challenge identified and the local organizational capacity, different types of local actors could be involved. In many cases a regional- or district-level agricultural extension agent may be the ideal contact for a community, as they would be aware of the agricultural practices within their locality and are tasked with outreach to farmers. In other scenarios, such as a mycotoxin risk identified in a more urban setting, local community health workers might be more effective collaborators. Larger existing local institutions, like farmer organizations or farmer research networks, could also fill this role. These types of local actors would be vital and active participants in raising awareness of mycotoxins, promoting community engagement, identifying and choosing optimal mitigation strategies, and conducting efficient sampling. Without community cooperation and active participation, mycotoxin surveillance will merely serve to document a problem.

4.5 Contextualizing Proposed Frameworks

Surveillance of mycotoxins across space and time depends on several contextual factors (Table 1). These factors define the components of our proposed framework.

4.6 Scale

Different aspects of this framework will operate at different scales. At the broadest level, risk is defined by cropping patterns and landscape-wide environmental conditions, ideally modified by cultural and agronomic factors that also influence mycotoxin risk. Across a landscape there is significant

Table 1: Contextual factors for mycotoxin surveillance across space and time

Contextual factor	Description
Scale	Spatial and temporal extent of surveillance framework
Scope	Combination(s) of mycotoxin and crop/food/beverage targeted by the surveillance framework
Function	Intended use of the framework (e.g. strict enforcement of regulations, predicting mycotoxicosis outbreaks, estimating level of chronic exposure, etc.)
Mycotoxin characteristics	Variable effects of different mycotoxins, livestock vs. human, infant vs. adult, chronic vs. acute effects, interactive effects, etc.
Technical	Technological sophistication (e.g. assay cost/accuracy/ portability, landscape risk model performance, etc.)

variation in the presence of mycotoxins within and across years and landscape-scale surveillance would identify areas at higher risk of mycotoxin exposure either across or within years. Remote sensing-based monitoring helps to predict pre-harvest mycotoxin contamination risk by detecting environmental conditions associated with the health of the crop and the mycotoxigenic fungi.

Once informed by a landscape-wide model, samples would be collected from communities identified as being at risk of mycotoxin contamination. Pitt et al. (2018) noted the importance of grouping nearby samples together to gain a sense of the community's AF load. Similarly, we propose that thorough sampling and mycotoxin quantification of crops at greatest risk (e.g. maize and groundnuts) would occur at important nodes within identified communities. Also, understanding overall mycotoxin levels at these nodes is more important than understanding the concentrations of individual samples. For instance, sampling at local grain mills would provide information on the mycotoxin levels at a given locality more efficiently than visiting multiple homes or farms. The intensity and duration of sample collection would depend on the level of risk identified; if a community is at risk of acute exposure, the sampling plan would accommodate more thorough sampling, whereas a smaller number of samples would be collected in areas not at immediate risk. In this scheme, a large number of samples would be tested with a spectral device, a subset of those samples would be measured with less precise field-based assays, and an even smaller subset could be sent to a laboratory and validated with costly and precise assays (Figure 2). Importantly, information generated by the assays would be georeferenced to the sampling locations and could be used to retrain and improve landscape-wide models in future years.

4.7 Scope

The scope of this framework depends on the technical capabilities of the monitors, the characteristics of the pathosystem(s), and the features of the food system. Ideally, it could monitor a wide range of mycotoxins in a diverse array of crops, foods/feeds, and beverages, but resource constraints would

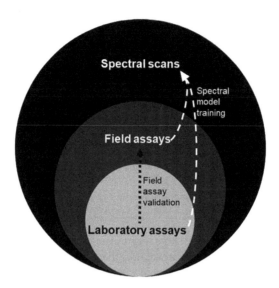

Figure 2: Potential sampling and assay plan for proposed framework. Within an identified "at risk" community, many pooled and georeferenced samples would be non-destructively sampled and scanned by a spectral device. A subset of scanned samples would be assayed with field-based techniques (e.g. lateral flow assay). A subset of field-assayed samples would be sent to a laboratory for further testing (e.g. HPLC or ELISA). The size of each circle is roughly proportional to the number of samples assayed by that method. Red dashed line indicates that laboratory assays would be used to validate field assay measurements. White dashed lines denote that field and laboratory assays will be used to train a model for spectral prediction of mycotoxin(s) in the remaining samples.

likely narrow this extensive list to those that are associated with negative health outcomes and are more easily monitored. Typically, this would entail sampling staple crops further upstream in the food value chain. For instance, rather than trying to sample all processed foods or beverages derived from maize, it would be more efficient to sample the inputs to those processing operations. However, in this example, the potential effects of processing on mycotoxins and the variable quality of maize entering different processing operations could be confounding factors. For example, dehulling maize generally reduces FUM content and lower quality maize is often used to produce local beers.

Landscape-wide prediction models would be developed for specific mycotoxin-crop combinations (Figure 1). These combinations could be integrated to acquire an aggregate mycotoxin risk for a given area. This aggregate risk might be modified by the dietary patterns of a specific area or demographic.

From a technical perspective, this framework is limited by the ability to perform field-based mycotoxin assays which means that, depending on the environment, the mycotoxins monitored would be some combination of AF, DON, and/or FUM. Development of multi-mycotoxin field-based assays could expand the mycotoxins assayed or enable more efficient sampling of mycotoxins that commonly co-contaminate samples like AF and FUM in maize or sorghum.

Most surveillance systems for mycotoxins test food or beverages sampled from various points in the value chain. An alternative or complementary approach to surveillance could be monitoring human exposure levels by testing mycotoxin biomarkers. For example, anonymized urine samples collected at health clinics (those collected for routine health screens) and/or community sanitation systems (especially those that employ urine-diverting separating toilets) could permit testing of urinary mycotoxin biomarkers. Doing this across a city would provide estimates of mycotoxin exposure in different neighborhoods that are traceable over time and could facilitate highly targeted interventions.

4.8 Function

In developed countries and many export markets, the function of mycotoxin surveillance is to strictly enforce established regulations. However, instituting rigorous enforcement of regulations in low-resource settings would be expensive with current technologies and would likely have unintended consequences for food safety and security. For example, in Malawian groundnuts, most of which derive from smallholder farms, highly regulated exported groundnuts do not exceed the 4 ppb limit on AF. Groundnuts from local markets, however, are highly contaminated, in part, because out-sorted groundnuts are consumed domestically, concentrating the toxin in the local food system (Matumba et al. 2014). Without valuing waste streams generated by sorting such crops, it is unlikely that a regulatory system could exist without dangerous and inequitable downstream consequences on food safety and security.

Given these challenges, the proposed framework would not function to strictly regulate mycotoxin concentrations. Instead, it would aim to predict and verify areas where consumers are at risk of acute or chronic exposure to certain mycotoxins, and, in doing so, enable responses that are tailored to the type of risk identified. A primary challenge will be to effectively communicate the landscape-wide information that is generated by a remote centralized monitor to local communities.

Monitoring acute risk would entail modeling based on current year growing conditions that are predictive of mycotoxin contamination. Identifying mycotoxin hotspots prior to or at harvest would enable early verification of dangerous levels of contamination and would be followed by early intervention. In these cases, intervention may be substantial and involve costly measures such as food replacement. Presently, responses to mycotoxicosis outbreaks are reactive and occur after communities experience acute effects of exposure.

Historical data and future predictions could inform which areas and communities are chronically at risk of mycotoxin contamination. Responses to these predictions would be less reactive and emphasize building awareness and capacity to mitigate exposure. For instance, in a rural community, the response to identifying an area frequently at risk of AF contamination in groundnuts might consist of building awareness and capacity within local farmer groups, improving soil health (organic matter and nutrient content), increasing access to resistant varieties, and improving storage conditions. In a peri-urban or urban setting, the response would focus more on actions like improving sorting methods and altering weaning foods to contain less contaminated ingredients while still maintaining nutritional value.

4.9　Mycotoxin Characteristics

The framework must grapple with the variable effects of different mycotoxins on human health as well as their variable effects on more susceptible sub-populations. One way in which this framework confronts the issue of exposure level is by creating tiered responses levels for acute and chronic mycotoxin exposure. The risk of acute exposure, especially in the scenario of aflatoxicosis, should require the most immediate and strongest response. Prolonged exposure to a single mycotoxin or multiple mycotoxins demands a different category of response that focuses more on building awareness and capacity within local communities and implementing sustainable mitigation strategies.

This framework's risk assessments would also incorporate more vulnerable sub-populations such as infants and young children. In addition to being more sensitive to the effects of mycotoxins, children's weaning foods can contain more types and higher concentrations of mycotoxins than adult foods (Matumba et al. 2015, Yacine Ware et al. 2017). Assimilating these considerations within a risk assessment model would mean creating different levels of risk for different sub-populations.

4.10　Technical

Some of the barriers to mycotoxin surveillance in low-resource contexts can be addressed by real and anticipated progress in technologies that can assess indirect indicators of pre-harvest mycotoxin risk across landscapes (e.g. remote sensing mapping) and in cheaper field-based assays (e.g. spectral methods). While there are significant tradeoffs in utilizing these surveillance mechanisms, integrating these techniques with established methods can address some of their shortcomings (Figure 2). Also, applying these methods over long periods of time allows for retraining and improvement of models as newer ground-truthed data is generated. Ultimately, the success of integrating these systems will rely on increased awareness and engagement, especially at the grassroots level.

Landscape-level detection of mycotoxins would rely heavily on a variety of remote sensing datasets. An essential layer of information is the geographical distribution of crops. You et al. (2014) generated a global crop allocation model on a 5 arc minute grid (~10 km) that maps the distribution of 20 major crops and their cultivation system (rainfed vs. irrigation). The same research group proposed improving this model at the country level by incorporating national agricultural statistics (Van Dijk et al. 2018). Prediction of pre-harvest mycotoxin contamination can be generated by integrating this data with remote sensing environmental data such as soil moisture, vegetation indices, temperature, precipitation, and others. These environmental predictors of mycotoxin contamination can be derived from datasets such as the Famine Early Warning Systems Network (FEWS NET), Land Data Assimilation System (FLDAS) and WorldClim2 (Fick and Hijmans 2017, McNally et al. 2017). Thomas et al. (2019) implemented a crop-simulation based model with the Decision Support System for Agrotechnology Transfer (DSSAT) software that is based on WorldClim2 data. They used this model to predict maize and groundnut AF contamination in West Africa, Central America, and South Asia under different climate change scenarios. With refinement, this type of model could be an essential part of an early warning system by predicting AF hotspots. It could also accommodate other mycotoxins like FUM and DON, and other commonly contaminated crops like wheat.

Once developed, historical data from diverse sources (e.g. remote sensing data, agricultural statistics, census information, mycotoxin surveys, etc.) would predict which areas are at chronically high risk of mycotoxin contamination across years, and contemporaneous environmental data from a crop's growing season could be used to predict areas at risk of contamination within a given year. International or regional actors could cooperate to implement such a system at a global or regional scale, or it could be implemented within individual countries. A drawback of this kind of model is that it only accounts for pre-harvest drivers of contamination and does not explicitly consider a crop's path through the food value chain. Feedback from communities could reveal some of this information and improve future models. This makes peri-urban and urban food systems, which receive food products from a variety of areas, more difficult to model without knowing where their foods originated.

Rather than being concerned with the exact concentration of mycotoxins, less precise field-based methods (e.g. qualitative or semi-quantitative assays, spectral methods) would give an immediate sense of the level of risk. We propose measuring the largest possible number of samples with the least precise and most affordable method (e.g. spectral method), then sub-sampling from scanned samples and using field-based assays to receive real-time data on mycotoxin levels. Finally, an even smaller subset of samples assayed by both field-based methods would be sent to a laboratory for more precise and expensive validated assays (Figure 2). Because field-based assays are less precise and less likely to be operated by highly trained personnel, retesting some of them in a laboratory can help to validate field assays. Results from the field-based and laboratory-based assays would be used to train spectral-based models for the mycotoxin(s) assayed. This is critical as spectral reflectance models of mycotoxins perform well within a narrow population of samples but are liable to without adequate model calibration.

4.11 Proposed Framework Summary

The proposed framework for mycotoxin surveillance in low-resource settings strives to be inclusive and adaptable to the diverse systems that it intends to serve. These ambitious will require compromises in assay quality and flexibility in sampling plans. Ultimately, its success will be determined by its ability to integrate local communities' knowledge and capacity for implementing change with insights from centralized monitors.

5. Translating Awareness into Action

The implementation of the proposed surveillance framework would improve understanding of the complex spatiotemporal dynamics of mycotoxins within complex and difficult-to-monitor systems. However, its effectiveness will also be measured by how well its outputs can be utilized. A mycotoxin surveillance framework should be optimized to encourage effective and, in all but the most extreme cases, sustainable interventions. In low-resource settings, where governmental capacity and community resources are often limited, a framework should strive to enable community-led actions that are sensitive to these challenges. These actions do not necessarily need to address the mycotoxin issue directly, but, rather, should embrace the interconnected nature of the problem and work towards solutions that are motivated by associated issues like economics or nutrition.

5.1 Aflasafe Case Study

Mycotoxin surveillance has been fundamental to the success of both large- and small-scale mycotoxin mitigation projects. A prominent example of using large-scale surveillance to implement an intervention is the development and deployment of atoxigenic *A. flavus* strains to reduce AF by competitive inhibition. In sub-Saharan Africa this product, marketed as Aflasafe, is used to mitigate AF contamination on maize and groundnut crops (Bandyopadhyay et al. 2016). The rollout of Aflasafe first needed substantial surveillance to identify areas at risk of AF contamination, isolate local atoxigenic genotypes, assess local awareness and capacity to employ the product, and validate its effectiveness on farms (Agbetiameh et al. 2019, Ayedun et al. 2017, Marechera and Ndwiga 2015, Senghor et al. 2020). The diverse types of surveys undertaken by the developers of Aflasafe underlines the importance of integrating mycotoxin surveys with a more nuanced understanding of affected communities, to inform mitigation efforts.

5.2 Awareness to Action

In the case of Aflasafe, the nature of the associated surveys was dictated by the implementation of a specific technology. However, the surveillance framework proposed here aims to allow communities to adopt intervention options that best suit their capacity and characteristics. For example, a rural community with significant AF contamination in their groundnuts might identify planting improved varieties and improving storage as their primary intervention options. In a more urban setting that receives AF-contaminated groundnuts from that rural community, the focus may be on improved grading and sorting of groundnuts and finding value for out-sorted groundnuts, such as their use in making fuel briquettes (the rural community might also implement training on effective sorting for home consumption and re-purposing of contaminated nuts). The key will be to link these communities with the resources that can help them carry out and sustain their interventions.

To ensure that communities have agency over the decisions they make, a strong relationship needs to be formed between them and the centralized mycotoxin monitors. Initially, the appropriate leaders in that community need to be identified and a process of building awareness of the general mycotoxin challenge should occur. The leaders engaged will depend on the characteristics of the community and the nature of their mycotoxin problem. In raising awareness, the focus should not be solely on mycotoxins, but rather on how they are representative of a larger economic and health issues.

Verifying the community's specific mycotoxin challenge by surveying food products for mycotoxins can be a part of the process of increasing awareness. However, it should be accompanied by broader assessments of the farming, food storage, socioeconomic, and cultural characteristics of that community. Community leaders should play an active role in this process and cooperate closely with experts from the centralized unit (e.g. academic, governmental, or other) to characterize community assets and challenges. This should result in a list of possible long-term interventions, and the community should select those that they believe will be most effective. This intervention list should be inclusive of both direct and indirect mitigation actions. For instance, developing waste-to-value schemes, such as the use of out-sorted AF-contaminated groundnuts and maize as inputs for biofuel, would reduce the economic disincentive for sorting and indirectly help reduce contamination of food products. The recycling of organic wastes to produce organic fertilizers could address the critical issue of soil health: soil organic matter and fertility are critical for ensuring plant health. Given that mycotoxin problems often reflect plant stress caused by climate perturbations and/or depletion of soil fertility and water holding capacity, efforts to improve soil health will benefit both the quality and quantity of food.

To enable the uptake and adoption of a selected intervention, centralized monitors will connect the community with the appropriate resources. For example, if the community opts to increase the use of hermetic storage bags, then centralized monitors could help connect the community with suppliers and engage with the government or an NGO to support the initial rollout of this technology. Optimal interventions would be sustainable and multiplicative in their benefits. Pre-harvest interventions that embody agro-ecological approaches, like crop rotation or integrated pest management, often use fewer costly inputs and address challenges beyond just mycotoxins. Garnering the resources necessary to enable community motivated interventions would be a challenge given the climate of top-down driven interventions. However, the chance of long-term success of such initiatives would be increased by having significant community buy-in.

Though this process is relatively protracted, the benefits of implementing these actions would extend beyond just the reduction of mycotoxins. Also, building the capacity of at-risk communities to be resilient to the effects of mycotoxins would reduce their risk of a future acute outbreak. These long-term and sustainable actions are increasingly necessary as climate change threatens to make mycotoxin contamination even more pervasive and extreme.

References

Adekoya, I., Njobeh, P., Obadina, A., Chilaka, C., Okoth, S., De Boevre, M. et al. 2017. Awareness and prevalence of mycotoxin contamination in selected Nigerian fermented foods. Toxins 9(11): 1-16 (Article 363).

Afolabi, C.G., Bandyopadhyay, R., Leslie, J.F. and Ekpo, E.J.A. 2006. Effect of sorting on incidence and occurrence of fumonisins and *Fusarium verticillioides* on maize from Nigeria. Journal of Food Protection 69(8): 2019–2023.

Agbetiameh, D., Ortega-Beltran, A., Awuah, R.T., Atehnkeng, J., Islam, M.S., Callicott, K.A. et al. 2019. Potential of atoxigenic *Aspergillus flavus* vegetative compatibility groups associated with maize and groundnut in Ghana as biocontrol agents for aflatoxin management. Frontiers in Microbiology 10.

Atukwase, A., Kaaya, A.N. and Muyanja, C. 2009. Factors associated with fumonisin contamination of maize in Uganda. Journal of the Science of Food and Agriculture 89(14): 2393–2398.

Ayedun, B., Okpachu, G., Manyong, V., Atehnkeng, J., Akinola, A., Abu, G.A. et al. 2017. An assessment of willingness to pay by maize and groundnut farmers for aflatoxin biocontrol product in northern Nigeria. Journal of Food Protection 80(9): 1451–1460.

Azziz-Baumgartner, E., Lindblade, K., Gieseker, K., Rogers, H.S., Kieszak, S., Njapau, H. et al. 2005. Case-control study of an acute aflatoxicosis outbreak, Kenya, 2004. Environmental Health Perspectives 113(12): 1779–1783.

Bandyopadhyay, R., Ortega-Beltran, A., Akande, A., Mutegi, C., Atehnkeng, J., Kaptoge, L. et al. 2016. Biological control of aflatoxins in Africa: Current status and potential challenges in the face of climate change. World Mycotoxin Journal 9(5): 771–789.

Bankole, S.A., Schollenberger, M. and Drochner, W. 2010. Survey of ergosterol, zearalenone and trichothecene contamination in maize from Nigeria. Journal of Food Composition and Analysis 23(8): 837–842.

Bankole, S., Schollenberger, M. and Drochner, W. 2006. Mycotoxins in food systems in Sub Saharan Africa: A review. Mycotoxin Research 22(3): 163–169.

Battilani, P. 2016. Recent advances in modeling the risk of mycotoxin contamination in crops. Current Opinion in Food Science 11: 10–15.

Battilani, P., Pietri, A., Barbano, C., Scandolara, A., Bertuzzi, T. and Marocco, A. 2008. Logistic regression modeling of cropping systems to predict fumonisin contamination in maize. Journal of Agricultural and Food Chemistry 56(21): 10433–10438.

BBC. 2019. Kenya's ugali scare: How safe is your maize flour? BBC News.

Bhat, R.V., Ramakrishna, Y., Beedu, S.R. and Munshi, K.L. 1989. Outbreak of trichothecene mycotoxicosis associated with consumption of mould-damaged wheat production in Kashmir Valley, India. The Lancet 333(8628): 35–37.

Bhat, R.V., Shetty, P.H., Amruth, R.P. and Sudershan, R.V. 1997. A foodborne disease outbreak due to the consumption of moldy sorghum and maize containing fumonisin mycotoxins. Journal of Toxicology: Clinical Toxicology 35(3): 249–255.

Cao, A., Santiago, R., Ramos, A.J., Souto, X.C., Aguín, O., Malvar, R.A. et al. 2014. Critical environmental and genotypic factors for *Fusarium verticillioides* infection, fungal growth and fumonisin contamination in maize grown in northwestern Spain. International Journal of Food Microbiology 177: 63–71.

Cotty, P.J. and Jaime-Garcia, R. 2007. Influences of climate on aflatoxin producing fungi and aflatoxin contamination. International Journal of Food Microbiology 119(1–2): 109–115.

Dawlatana, M., Coker, R.D., Nagler, M.J., Wild, C.P., Hassan, M.S. and Blunden, G. 2002. The occurrence of mycotoxins in key commodities in Bangladesh: surveillance results from 1993 to 1995. Journal of Natural Toxins 11(4): 379–386.

Degraeve, S., Madege, R.R., Audenaert, K., Kamala, A., Ortiz, J., Kimanya, M. et al. 2015. Impact of local pre-harvest management practices in maize on the occurrence of *Fusarium* species and associated mycotoxins in two agro-ecosystems in Tanzania. Food Control 59: 225–233.

Desjardins, A.E., Manandhar, G., Plattner, R.D., Maragos, C.M., Shrestha, K. and McCormick, S.P. 2000. Occurrence of *Fusarium* species and mycotoxins in Nepalese maize and wheat and the effect of traditional processing methods on mycotoxin levels. Journal of Agricultural and Food Chemistry 48(4): 1377–1383.

Diaz, G.J., Ariza, D. and Perula, N.S. 2004. Method validation for the determination of ochratoxin A in green and soluble coffee by immunoaffinity column cleanup and liquid chromatography. Mycotoxin Research 20(2): 59–67.

Elias-Orozco, R., Castellanos-Nava, A., Gaytán-Martínez, M., Figueroa-Cárdenas, J.D. and Loarca-Piña, G. (2002). Comparison of nixtamalization and extrusion processes for a reduction in aflatoxin content. Food Additives and Contaminants 19(9): 878–885.

Eskola, M., Kos, G., Elliott, C.T., Hajšlová, J., Mayar, S. and Krska, R. 2019. Worldwide contamination of food-crops with mycotoxins: Validity of the widely cited 'FAO estimate' of 25%. Critical Reviews in Food Science and Nutrition 60(16): 2773–2789.

Fandohan, P., Gnonlonfin, B., Hell, K., Marasas, W.F.O. and Wingfield, M.J. 2005. Natural occurrence of *Fusarium* and subsequent fumonisin contamination in preharvest and stored maize in Benin, West Africa. International Journal of Food Microbiology 99(2): 173–183.

Fandohan, P., Zoumenou, D., Hounhouigan, D.J., Marasas, W.F.O., Wingfield, M.J. and Hell, K. 2005. Fate of aflatoxins and fumonisins during the processing of maize into food products in Benin. International Journal of Food Microbiology 98(3): 249–259.

Ferrigo, D., Raiola, A. and Causin, R. 2014. Plant stress and mycotoxin accumulation in maize. Agrochimica 58: 116–127.

Fick, S.E. and Hijmans, R.J. 2017. WorldClim 2: New 1-km spatial resolution climate surfaces for global land areas. International Journal of Climatology 37(12): 4302–4315.

FSSAI. 2018. Food Safety and Standards Authority of India Annual Report 2017-18.

Galvez, F.C.F., Francisco, M.L., Villarino, B.J., Lustre, A.O. and Resurreccion, A.V. 2003. Manual sorting to eliminate aflatoxin from peanuts. Journal of Food Protection 66(10): 1879–1884.

Geremew, T., Abate, D., Landschoot, S., Haesaert, G. and Audenaert, K. 2016. Occurrence of toxigenic fungi and ochratoxin A in Ethiopian coffee for local consumption. Food Control 69: 65–73.

Giorni, P., Magan, N. and Battilani, P. 2009. Environmental factors modify carbon nutritional patterns and niche overlap between *Aspergillus flavus* and *Fusarium verticillioides* strains from maize. International Journal of Food Microbiology 130(3): 213–218.

Grace, D., Mahuku, G., Hoffmann, V., Atherstone, C., Upadhyaya, H.D. and Bandyopadhyay, R. (2015). International agricultural research to reduce food risks: Case studies on aflatoxins. Food Security 7(3): 569–582.

Harvey, J., Gnonlonfin, B., Fletcher, M., Fox, G., Trowell, S., Berna, A. et al. 2013. Improving diagnostics for aflatoxin detection. *In:* L. Unnevehr and D. Grace (eds.). Aflatoxins: Finding Solutions for Improved Food Safety, IFPRI, Washington, DC.

Herrman, T.J., Hoffman, V., Muiriri, A. and McCormick, C. 2020. Aflatoxin proficiency testing and control in Kenya. Journal of Food Protection 83(1): 142–146.

Hove, M., De Boevre, M., Lachat, C., Jacxsens, L., Nyanga, L.K. and De Saeger, S. 2016. Occurrence and risk assessment of mycotoxins in subsistence farmed maize from Zimbabwe. Food Control 69: 36–44.

James, B., Adda, C., Cardwell, K., Annang, D., Hell, K., Korie, S. et al. 2007. Public information campaign on aflatoxin contamination of maize grains in market stores in Benin, Ghana and Togo. Food Additives & Contaminants 24(11): 1283–1291.

Johnson, A.M., Fulton, J.R., Abdoulaye, T., Ayedun, B., Widmar, N.J.O., Akande, A. et al. 2018. Aflatoxin awareness and Aflasafe adoption potential of Nigerian smallholder maize farmers. World Mycotoxin Journal 11(3): 437–446.

Kaaya, A.N., Warren, H.L., Kyamanywa, S. and Kyamuhangire, W. 2005. The effect of delayed harvest on moisture content, insect damage, moulds and aflatoxin contamination of maize in Mayuge district of Uganda. Journal of the Science of Food and Agriculture 85(15): 2595–2599.

Kamala, A., Shirima, C., Jani, B., Bakari, M., Sillo, H., Rusibamayila, N. et al. 2018. Outbreak of an acute aflatoxicosis in Tanzania during 2016. World Mycotoxin Journal 11(3): 311–320.

Kamala, A., Kimanya, M., Haesaert, G., Tiisekwa, B., Madege, R., Degraeve, S. et al. 2016. Local post-harvest practices associated with aflatoxin and fumonisin contamination of maize in three agro ecological zones of Tanzania. Food Additives & Contaminants: Part A 33(3): 551–559.

Kerry, R., Ortiz, B.V., Ingram, B.R. and Scully, B.T. 2017. A spatio–temporal investigation of risk factors for aflatoxin contamination of corn in southern Georgia, USA using geostatistical methods. Crop Protection 94: 144–158.

Kimanya, M.E., De Meulenaer, B., Tiisekwa, B., Ugullum, C., Devlieghere, F., Van Camp, J. et al. 2009. Fumonisins exposure from freshly harvested and stored maize and its relationship with traditional agronomic practices in Rombo district, Tanzania. Food Additives & Contaminants: Part A 26(8): 1199–1208.

Kpodo, K., Thrane, U. and Hald, B. 2000. *Fusaria* and fumonisins in maize from Ghana and their co-occurrence with aflatoxins. International Journal of Food Microbiology 61(2–3): 147–157.

Labuza, T. 1983. Regulation of mycotoxins in food. Journal of Food Protection 46(3): 260–265.

Ladeira, C., Frazzoli, C. and Orisakwe, O.E. 2017. Engaging one health for non-communicable diseases in Africa: Perspective for mycotoxins. Frontiers in Public Health, 5: 1–15 (Article 266).

Lee, H.S., Nguyen-Viet, H., Lindahl, J., Thanh, H.M., Khanh, T.N., Hien, L.T.T. et al. 2017. A survey of aflatoxin B1 in maize and awareness of aflatoxins in Vietnam. World Mycotoxin Journal 10(2): 195–202.

Leung, M.C., Diaz-Llano, G. and Smith, T.K. 2006. Mycotoxins in pet food: A review on worldwide prevalence and preventative strategies. Journal of Agricultural and Food Chemistry 54(26): 9623–9635.

Magembe, K.S., Mwatawala, M.W. and Mamiro, D.P. 2016. Mycotoxin contamination in stored maize and groundnuts based on storage practices and conditions in subhumid tropical Africa: The case of Kilosa District, Tanzania. Journal of Food Protection 79(12): 2160–2166.

Makau, C.M., Matofari, J.W., Muliro, P.S. and Bebe, B.O. 2016. Aflatoxin B1 and deoxynivalenol contamination of dairy feeds and presence of aflatoxin M1 contamination in milk from smallholder dairy systems in Nakuru, Kenya. International Journal of Food Contamination 3: 1–10 (Article 6).

Manandhar, A., Milindi, P. and Shah, A. 2018. An overview of the post-harvest grain storage practices of smallholder farmers in developing countries. Agriculture 8(4): 57.

Manda, P., Jean, A., Adepo, B., Ngbé, J.V. and Dano, D.S. 2016. Assessment of ochratoxin A intake due to consumption of coffee and cocoa derivatives marketed in Abidjan (Cote d'Ivoire). Journal of Toxicology and Environmental Health Sciences 8(6): 41–45.

Marechera, G. and Ndwiga, J. 2015. Estimation of the potential adoption of Aflasafe among smallholder maize farmers in lower eastern Kenya. African Journal of Agricultural and Resource Economics 10(1): 72–85.

Maringe, D.T., Chidewe, C., Benhura, M.A., Mvumi, B.M., Murashiki, T.C., Dembedza, M.P. et al. 2017. Natural postharvest aflatoxin occurrence in food legumes in the smallholder farming sector of Zimbabwe. Food Additives & Contaminants: Part B 10(1): 21–26.

Matumba, L., Monjerezi, M., Khonga, E.B. and Lakudzala, D.D. 2011. Aflatoxins in sorghum, sorghum malt and traditional opaque beer in southern Malawi. Food Control 22(2): 266–268.

Matumba, L., Sulyok, M., Monjerezi, M., Biswick, T. and Krska, R. 2015. Fungal metabolites diversity in maize and associated human dietary exposures relate to micro-climatic patterns in Malawi. World Mycotoxin Journal 8(3): 269–282.

Matumba, L., Van Poucke, C., Monjerezi, M., Ediage, E.N. and De Saeger, S. 2014. Concentrating aflatoxins on the domestic market through groundnut export: A focus on Malawian groundnut value and supply chain. Food Control 51: 236–239.

Matumba, L., Van Poucke, C., Njumbe Ediage, E. and De Saeger, S. 2017. Keeping mycotoxins away from the food: Does the existence of regulations have any impact in Africa? Critical Reviews in Food Science and Nutrition 57(8): 1584–1592.

Mboya, R.M. and Kolanisi, U. 2014. Subsistence farmers' mycotoxin contamination awareness in the SADC region: Implications on Millennium Development Goals 1, 4 and 6. Journal of Human Ecology 46(1): 21–31.

McNally, A., Arsenault, K., Kumar, S., Shukla, S., Peterson, P., Wang, S. et al. 2017. A land data assimilation system for sub-Saharan Africa food and water security applications. Scientific Data, 4: 1–19 (Article 170012).

Monyo, E.S., Njoroge, S.M.C., Coe, R., Osiru, M., Madinda, F., Waliyar, F. et al. 2012. Occurrence and distribution of aflatoxin contamination in groundnuts (*Arachis hypogaea* L.) and population density of Aflatoxigenic *Aspergilli* in Malawi. Crop Protection 42: 149–155.

Morgavi, D.P. and Riley, R.T. 2007. An historical overview of field disease outbreaks known or suspected to be caused by consumption of feeds contaminated with *Fusarium* toxins. Animal Feed Science and Technology 137(3–4): 201–212.

Morton, J.F. 2007. The impact of climate change on smallholder and subsistence agriculture. Proceedings of the National Academy of Sciences of the United States of America 104(50): 19680–19685.

Moser, C., Hoffmann, V. and Ordonez, R. 2014. Firm heterogeneity in food safety provision: Evidence from aflatoxin tests in Kenya. Agricultural & Applied Economics Association's 2014 AAEA Annual Meeting.

Mukanga, M., Derera, J., Tongoona, P. and Laing, M.D. 2010. A survey of pre-harvest ear rot diseases of maize and associated mycotoxins in south and central Zambia. International Journal of Food Microbiology 141(3): 213–221.

Mutegi, C.K., Ngugi, H.K., Hendriks, S.L. and Jones, R.B. 2009. Prevalence and factors associated with aflatoxin contamination of peanuts from Western Kenya. International Journal of Food Microbiology 130(1): 27–34.

Mutiga, S.K., Hoffmann, V., Harvey, J.W., Milgroom, M.G. and Nelson, R.J. 2015. Assessment of aflatoxin and fumonisin contamination of maize in Western Kenya. Phytopathology 105(9): 1250–1261.

Mutiga, S.K., Were, V., Hoffmann, V., Harvey, J.W., Milgroom, M.G. and Nelson, R.J. 2014. Extent and drivers of mycotoxin contamination: Inferences from a survey of Kenyan maize mills. Phytopathology 104(11): 1221–1231.

Mutiga, S.K., Morales, L., Angwenyi, S., Wainaina, J., Harvey, J., Das, B. et al. 2017. Association between agronomic traits and aflatoxin accumulation in diverse maize lines grown under two soil nitrogen levels in Eastern Kenya. Field Crops Research 205: 124–134.

Nabwire, W.R., Ombaka, J., Dick, C.P., Strickland, C., Tang, L., Xue, K.S. et al. 2019. Aflatoxin in household maize for human consumption in Kenya, East Africa. Food Additives & Contaminants: Part B 13(1): 45–51.

Ndemera, M., Landschoot, S., De Boevre, M., Nyanga, L.K. and De Saeger, S. 2018. Effect of agronomic practices and weather conditions on mycotoxins in maize: A case study of subsistence farming households in Zimbabwe. World Mycotoxin Journal 11(3): 421–436.

Nishimwe, K., Wanjuki, I., Karangwa, C., Darnell, R. and Harvey, J. 2017. An initial characterization of aflatoxin B1 contamination of maize sold in the principal retail markets of Kigali, Rwanda. Food Control 73: 574–580.

Njeru, N.K., Midega, C.A.O., Muthomi, J.W., Wagacha, J.M. and Khan, Z.R. 2019. Influence of socio-economic and agronomic factors on aflatoxin and fumonisin contamination of maize in western Kenya. Food Science & Nutrition 7(7): 2291–2301.

Njoroge, S.M.C., Matumba, L., Kanenga, K., Siambi, M., Waliyar, F., Maruwo, J. et al. 2016. A case for regular aflatoxin monitoring in peanut butter in sub-Saharan Africa: Lessons from a 3-year survey in Zambia. Journal of Food Protection 79(5): 795–800.

Odhav, B. and Naicker, V. 2002. Mycotoxins in South African traditionally brewed beers. Food Additives and Contaminants 19(1): 55–61.

Palencia, E., Torres, O., Hagler, W., Meredith, F.I., Williams, L.D. and Riley, R.T. 2003. Total fumonisins are reduced in tortillas using the traditional nixtamalization method of Mayan communities. The Journal of Nutrition 133(10): 3200–3203.

Phokane, S., Flett, B.C., Ncube, E., Rheeder, J.P. and Rose, L.J. 2019. Agricultural practices and their potential role in mycotoxin contamination of maize and groundnut subsistence farming. South African Journal of Science 115(9/10): 1–6.

Pitt, J.I., Boesch, C., Whitaker, T.B. and Clarke, R. 2018. A systematic approach to monitoring high preharvest aflatoxin levels in maize and peanuts in Africa and Asia. World Mycotoxin Journal 11(4): 485–491.

Pitt, J.I., Taniwaki, M.H. and Cole, M.B. 2013. Mycotoxin production in major crops as influenced by growing, harvesting, storage and processing, with emphasis on the achievement of food safety objectives. Food Control 32(1): 205–215.

Probst, C., Callicott, K.A. and Cotty, P.J. 2012. Deadly strains of Kenyan *Aspergillus* are distinct from other aflatoxin producers. European Journal of Plant Pathology 132(3): 419–429.

Radhakrishnan, G.V., Cook, N.M., Bueno-Sancho, V., Lewis, C.M., Persoons, A., Mitiku, A.D. et al. 2019. MARPLE, a point-of-care, strain-level disease diagnostics and surveillance tool for complex fungal pathogens. BMC Biology 17(65): 1–17.

Reddy, K.R.N., Reddy, C.S. and Muralidharan, K. 2009. Detection of *Aspergillus* spp. and aflatoxin B1 in rice in India. Food Microbiology 26(1): 27–31.

Robertson, C., Nelson, T.A., MacNab, Y.C. and Lawson, A.B. 2010. Review of methods for space–time disease surveillance. Spatial and Spatio-Temporal Epidemiology 1(2–3): 105–116.

Roger, D. 2011. Deoxynivanol (DON) and fumonisins B1 (FB1) in artisanal sorghum opaque beer brewed in north Cameroon. African Journal of Microbiology Research 5(12): 1565–1567.

Sasamalo, M.M., Mugula, J.K. and Nyangi, C.J. 2018. Aflatoxins contamination of maize at harvest and during storage in Dodoma, Tanzania. International Journal of Innovative Research and Development 7(6): 11–15.

Schaafsma, A.W. and Hooker, D.C. 2007. Climatic models to predict occurrence of *Fusarium* toxins in wheat and maize. International Journal of Food Microbiology 119(1–2): 116–125.

Senghor, L.A., Ortega-Beltran, A., Atehnkeng, J., Callicott, K.A., Cotty, P.J. and Bandyopadhyay, R. 2020. The atoxigenic biocontrol product Aflasafe SN01 is a valuable tool to mitigate aflatoxin contamination of both maize and groundnut cultivated in Senegal. Plant Disease 104(2): 510–520.

Sharma, A., Goud, K.Y., Hayat, A., Bhand, S. and Marty, J.L. 2017. Recent advances in electrochemical-based sensing platforms for aflatoxins detection. Chemosensors 5(1): 1–15.

Shephard, G.S. 2008. Impact of mycotoxins on human health in developing countries. Food Additives & Contaminants: Part A 25(2): 146–151.

Shephard, G.S., Burger, H.M., Gambacorta, L., Krska, R., Powers, S.P., Rheeder, J.P. 2013. Mycological analysis and multimycotoxins in maize from rural subsistence farmers in the former Transkei, South Africa. Journal of Agricultural and Food Chemistry 61(34): 8232–8240.

Shephard, G.S. 2018. Mycotoxin crises: Fit-for-purpose analytical responses in the developing world. Journal of AOAC International 101(3): 609–612.

Shirima, C.P., Kimanya, M.E., Routledge, M.N., Srey, C., Kinabo, J.L., Humpf, H.U. et al. 2015. A prospective study of growth and biomarkers of exposure to aflatoxin and fumonisin during early childhood in Tanzania. Environmental Health Perspectives 123(2): 173–178.

Smith, L.E., Stasiewicz, M., Hestrin, R., Morales, L., Mutiga, S. and Nelson, R.J. 2016. Examining environmental drivers of spatial variability of aflatoxin accumulation in Kenyan maize: Potential utility in risk prediction models. African Journal of Food, Agriculture, Nutrition and Development 16(3): 11086–11105.

Sserumaga, J.P., Ortega-Beltran, A., Wagacha, J.M., Mutegi, C.K. and Bandyopadhyay, R. 2019. Aflatoxin-producing fungi associated with pre-harvest maize contamination in Uganda. International Journal of Food Microbiology 313(108376): 1–8.

Stasiewicz, M.J., Falade, T.D.O., Mutuma, M., Mutiga, S.K., Harvey, J.J.W., Fox, G. et al. 2017. Multi-spectral kernel sorting to reduce aflatoxins and fumonisins in Kenyan maize. Food Control 78: 203–214.

Strosnider, H., Azziz-Baumgartner, E., Banziger, M., Bhat, R.V., Breiman, R., Brune, M.N. et al. 2006. Workgroup report: Public health strategies for reducing aflatoxin exposure in developing countries. Environmental Health Perspectives 114(12): 1898–1903.

The Citizen. 2019. Tanzania suspends importation of banned Kenyan peanut butter. The Citizen.

Thomas, T.S., Robertson, R. and Boote, K. 2019. Evaluating risk of aflatoxin field contamination from climate change using new modules inside DSSAT. *In:* IFPRI Discussion Paper 01859. International Food Policy Research Institute.

Toteja, G.S., Mukherjee, A., Diwakar, S., Singh, P., Saxena, B.N., Sinha, K.K. et al. 2006. Aflatoxin B1 contamination in wheat grain samples collected from different geographical regions of India: A multicenter study. Journal of Food Protection 69(6): 1463–1467.

Van Dijk, M., You, L., Havlik, P., Palazzo, A. and Mosnier, A. 2018. Generating high-resolution national crop distribution maps: Combining statistics, gridded data and surveys using an optimization approach. 30th International Conference of Agricultural Economists, August 2018. Vancouver, British Columbia, Canada.

Vanara, F., Reyneri, A. and Blandino, M. 2009. Fate of fumonisin B1 in the processing of whole maize kernels during dry-milling. Food Control 20(3): 235–238.

Waliyar, F., Ba, A., Hassan, H., Bonkoungou, S. and Bosc, J.P. 1994. Sources of resistance to *Aspergillus flavus* and aflatoxin contamination in groundnut genotypes in west Africa. Plant Disease 78(7): 704–708.

Wu, F. 2004. Mycotoxin risk assessment for the purpose of setting international regulatory standards. Environmental Science and Technology 38(15): 4049–4055.

Xu, Y., Doel, A., Watson, S., Routledge, M.N., Elliott, C.T., Moore, S.E. et al. 2017. Study of an educational hand sorting intervention for reducing aflatoxin B1 in groundnuts in rural Gambia. Journal of Food Protection 80(1): 44–49.

Yacine Ware, L., El Durand, N., Nikiema, P.A., Alter, P., Fontana, A., Montet, D. et al. 2017. Occurrence of mycotoxins in commercial infant formulas locally produced in Ouagadougou (Burkina Faso). Food Control 73: 518–523.

You, L., Wood, S., Wood-Sichra, U. and Wu, W. 2014. Generating global crop distribution maps: From census to grid. Agricultural Systems 127: 53–60.

Overview of Mycotoxins and Regulations in Cameroon

Pauline Mounjouenpou Limi

Toxicology, Food Science and Technoloy, Institute of Agricultural Research for Development (IRAD), PO Box 2067, Yaoundé, Cameroon

1. Introduction

Agriculture occupies an important place in the Cameroon's economy, contributing to more than 70% of its economic growth and providing around 25% of its Gross Domestic Product (GDP). It is practiced by about 70% of the active population. Maize, peanuts, cassava, sorghum, cocoa and coffee are the main crops contributing to this agricultural GDP. These crops are highly consumed locally by the population and also exported all over the world.

Cameroon, like many other countries, faces the serious challenge of postharvest losses which is partially due to molds contamination and the presence of mycotoxins in food. The tropical climate (high temperature and humidity) and the poor crop storage conditions facilitate fungal growth (Milićević et al. 2010). This contamination may occur during cultivation, harvesting or storage (Karlovsky et al. 2016, Cheli et al. 2017). In Cameroon, more than 25% of food crops are contaminated with mycotoxins annually.

Under tropical climatic conditions, certain molds species, mainly of the *Aspergillus*, *Penicillium* and *Fusarium* genera, produced mycotoxins. Analysis of food from animal and plant origins in Cameroon during the last 30 years, have highlighted the presence of several mycotoxins (aflatoxins, fumonisins, ochratoxins, zearalenone, deoxynivalenol…) (Nguegwouo et al. 2019). They are of major agro-economic and public health importance (Cardwell 2000, Zain 2011) as they negatively impact human and animal health; hence the necessity to consider mycotoxin contamination as an important quality and safety parameter for foods.

In order to limit the associated health risk, strict standards on maximum acceptable concentration of different mycotoxins in foods have been fixed at both national and international levels. Cameroon's access to international

---

Email: mounjouenpou@yahoo.fr_

food markets depends heavily on its ability to comply with these regulatory requirements.

This chapter gives an overview of mycotoxin contamination of Cameroon's food commodities, national regulations and initiatives taken by the government to protect the population, the legal and regulatory framework for food quality control, as well as factors limiting the effectiveness of the national health system.

2. Occurrence of Mycotoxins in Cameroon's Food Commodities

Many authors reported the occurrence of mycotoxins in several Cameroonian food commodities (Table 1), with often high levels of contamination by molds of the *Aspergillus* and *Penicillium* genera (Njobeh et al. 2009).

Maize is one of the most vulnerable food to mold contamination in Cameroon, with *Aspergillus* sp., *Fusarium* sp. and *Penicillium* sp. being the main fungal contaminants (Ngoko et al. 2001, 2008). These authors detected the presence of fumonisin B1 (FB1), Deoxynivalenol (DON) and Zearalenone (ZEN) at concentrations of 26000, 1300 and 1100 µg/kg, respectively. FB1 has the highest prevalence with increasing concentrations with storage time. The presence of these three mycotoxins in maize from Cameroon is also reported in concentrations which were within the previously mentioned range (Njobeh et al. 2010, Abia et al. 2013). Aflatoxins (AFs) were also detected by Njobeh et al. (2010) in 55% of their samples at concentrations between 0.1–15 µg/kg. This was not the case for Kana et al. (2013) who rather observed a prevalence of 9%, the maximal concentration obtained being 42 µg/kg. The content of this toxin appeared to depend on the agro-ecological zones. The average total aflatoxin level of positive maize samples from the Sahelian Zone was 2.4 µg/kg while positive samples from the Western High Plateau (characterized by its hot climate and high relative humidity) contained 11.9 µg/kg. The studies performed by Njumbe et al. (2014) with a more accurate and reliable method (LC-MS/MS) showed levels of contamination of maize which could go up to 5412 µg/kg for FBs, 645 µg/kg for aflatoxins B1 (AFB1) and 3842 µg/kg for DON. Several other studies assessed the presence of mycotoxins in maize derived products which are directly consumed by populations. In 2015, the analysis of *kutukutu*, a fermented maize-based dough, largely consumed in the northern part of the country, revealed an AFB1 content which in some cases exceeded the 2 µg/kg European Commission standard limit fixed for such products (Tchikoua et al. 2015). Nguegwouo et al. (2017) noticed that AFs and FBs contents of maize-based dishes (beer, porridge, fufu, etc.) were dependent on the production process as the sieving step seemed to lower mycotoxin concentrations. Similar results were obtained by Abia et al. (2017) who also detected low levels of mycotoxins (AFB1, the bacterial toxin cereulide, DON,

Table 1: An overview of mycotoxins occurrence in some Cameroonian food commodities

Food commodity	Method of analysis	Number of samples	Mycotoxin detected	Frequency of mycotoxin-presence on samples (%)	Range of concentration (μg/kg)	References
Maize	ELISA	18	Fumonisin B1	NS*	300–26000	Ngoko et al. (2001)
			Deoxynivalenol	NS	<100–1300	
			Zearalenone	NS	<50–110	-
	ELISA	18	Fumonisins	89	50–26000	Ngoko et al. (2008)
			Deoxynivalenol	NS	100–1300	-
			Zearalenone	NS	50–180	-
	TLC+ HPLC	40	Fumonisin B1	65	3684 (37–24225)	Njobeh et al. (2010)
			Zearalenone	78	69 (28–273)	-
			Deoxynivalenol	73	59 (18–273)	-
			Aflatoxins	55	1.5 (0.1–15)	-
Maize	LC-MS/MS	37	Fumonisin B1	100	508 (2–2313)	Abia et al. (2013)
	Aflatest immunoaffinity column+HPLC	77	Aflatoxins	9	1 (≤2–42)	Kana et al. (2013)
Maize kernels	LC-MS/MS	165	Fumonisins	74	(10–5412)	Njumbe et al. (2014)
			Aflatoxin B1	22	(6–645)	-

Food	n	Method	Mycotoxin	Incidence (%)	Concentration	References
Kutukutu (fermented maize-based dough)	29	ELISA	Zearalenone	14	(27–334)	-
			Deoxynivalenol	12	(27–3842)	-
Maize-based dishes	22	ELISA + LC-MS/MS	Aflatoxin B1	100	(≤2.8)	Tchikoua et al. (2015)
			Aflatoxins	100	(0.8–20)	Nguegwouo et al. (2017)
			Fumonisins	100	(10–5990)	
Maize-*fufu*	50	LC-MS/MS	Aflatoxin B1	24	0.9 (n.d–1.8)	Abia et al. (2017)
			Deoxynivalenol	100	23 (14–55)	-
			Fumonisin B1	100	151 (48–709)	-
			Nivalenol	100	268 (116–372)	-
			Patulin	30	105 (12–890)	-
			Zearalenone	100	49 (5–150)	-
Peanuts	16	TLC + HPLC	Fumonisin B1	19	517 (25–1498)	Njobeh et al. (2010)
			Zearalenone	63	70 (31–186)	-
			Deoxynivalenol	75	123 (17–270)	-
			Aflatoxins	75	6.5 (0.1–13)	-
	90	LC-MS/MS	Aflatoxin B1	29	(0.3–12)	Njumbe et al. (2014)
			Ochratoxin A	13	(6–125)	-
Peanuts meal	41	Aflatest immunoaffinity Column + HPLC	Aflatoxins	100	161 (39–950)	Kana et al. (2013)
Groundnuts	35	LC-MS/MS	Aflatoxin B1	97	47 (<LOQ–210)	Abia et al. (2013)

(Contd.)

Table 1: (*Contd.*)

Food commodity	Method of analysis	Number of samples	Mycotoxin detected	Frequency of mycotoxin-presence on samples (%)	Range of concentration (µg/kg)	References
Beans	TLC + HPLC	15	Fumonisin B1	20	727 (28–1351)	Njobeh et al. (2010)
			Zearalenone	33	48 (27–157)	-
			Deoxynivalenol	47	25 (13–35)	-
			Aflatoxins	33	2.4 (0.2–6.2)	-
Soybeans	TLC + HPLC	5	Fumonisin B1	40	195 (25–365)	Njobeh et al. (2010)
			Zearalenone	0	-	-
			Deoxynivalenol	40	110 (13–207)	-
			Aflatoxins	40	2.1 (0.2–3.9)	-
Cocoa beans	Immunoaffinity Column + HPLC	36	Ochratoxin A	NS	11.52 (5.3–21)	Mounjouenpou et al. (2012)
Arabica coffee	Immunoaffinity Column + HPLC	104	Ochratoxin A	NS	(0.12–124)	Nganou et al. (2014)
Robusta coffee	Immunoaffinity Column + HPLC	48	Ochratoxin A	75	0.6–18	Mounjouenpou et al. (2013)
Arabica coffee	Immunoaffinity Column + HPLC	51	Ochratoxin A	65	0.3–4.9	Mounjouenpou et al. (2013)
Green coffee beans	HPLC	7	Ochratoxin A	57	1.7 (1–2.5)	Romani et al. (2000)
Sorghum (Variety Damugari)	ELISA	NS	Aflatoxin B1	75	(0–230)	Djoulde (2013)
Sorghum (Variety Djigari)	ELISA	NS	Aflatoxin B1	45	(0–145)	Djoulde (2013)

Sample	Method	n	Mycotoxin	%	Value range	Reference
Sorghum beer (*Bil-bil*)	ELISA	70	Deoxynivalenol	100	450 (140–730)	Djoulde (2011)
			Fumonisin B1	79	150 (0–230)	-
Sorghum beer (*Kpata*)	ELISA	50	Deoxynivalenol	74	520 (0–680)	Djoulde (2011)
			Fumonisin B1	100	210 (0.5–340)	-
Stored Cassava chips	ELISA	72	Aflatoxins	33	5.2–15	Essono et al. (2009)
Cassava products (flakes + chips)	LC-MS/MS	165	Aflatoxin B1	25	6–194	Njumbe et al. (2014)
			Penicillic acid	6	25–184	-
Miscellaneous **	TLC + HPLC	6	Fumonisin B1	0	-	Njobeh et al. (2010)
			Zearalenone	17	67	-
			Deoxynivalenol	50	25 (13–35)	-
			Aflatoxins	17	0.3	-
Eggs	HPLC	62	Aflatoxins	45	0.82 ± 1.7	Tchana et al. (2010)
Cow milk	ELISA	63	Aflatoxin M1	16	0.006–0.53	-
Black pepper	ELISA	20	Ochratoxin A	10	1.5 (1.2–1.9)	Nguegwouo et al. (2018)
White pepper	ELISA	20		40	3.3 (1.8–4.9)	-
Breast milk	ELISA	42	Aflatoxin M1	38	7.4 (0.9–37)	Chuisseu et al. (2018)

* NS, Not specified; ELISA, Enzyme-Linked Immuno-Sorbent Assay; TLC, Thin Layer Chromatography; HPLC, High-Performance Liquid Chromatography; LC-MS/MS, Liquid Chromatography coupled with Magnetic Sector Mass Spectrometers

** Rice, pumpkin seeds "egusi", fermented cassava flakes "gari", fermented cassava flour "nkum nkum"

FB1, nivalenol, patulin, ZEN, etc.) in maize-*fufu* (also known as fufu-corn), a boiled maize-dough dish mostly consumed in the western highlands of Cameroon. This study equally revealed the presence of a mixture of cereulide, patulin and ZEN derivatives in a Cameroonian food.

Peanuts, beans and soybeans are often prone to fungal contamination. Non-mycotoxigenic as well as mycotoxigenic molds of *Aspergillus* and *Penicillium* genera have been isolated from samples from different parts of the country (Njobeh et al. 2009). An evaluation of the mycotoxin content in peanuts by Njobeh et al. (2010) showed average values of 517, 70, 123 and 6.5 µg/kg for FB1, ZEN, DON and AFs, respectively. A concentration of AFs ranging between 39–950 µg/kg was reported in poultry feed peanut meals (Kana et al. 2013). Njumbe et al. (2014) also detected OTA contamination in several peanut tested samples with concentrations ranging from 6 to 125 µg/ kg. Njobeh et al. (2010) also reported an average concentration of 727 and 195 µg/kg for FB1, 25 and 110 µg/kg for DON, 2.4 and 2.1 µg/kg for AFs, in beans and soybeans, respectively. These authors detected no ZEN in soybeans but an average content of 46.7 µg/kg in beans. Abia et al. (2013) detected FB1 in 18 of the 35 peanuts samples they collected and in all 10 soybeans samples, at a mean amount of 5 and 49 µg/kg, respectively.

Cocoa and coffee are food commodities with a high impact on the economy of many producing countries like Cameroon, where they are often contaminated by ochratoxin A (OTA). Due to their high level of consumption, strict standards have been defined for these foods, especially on the international markets. The presence of ochratoxigenic fungi like *Aspergillus niger*, and *Aspergillus carbonarius* has already been reported in cocoa samples from Cameroon (Mounjouenpou et al. 2008). Studying the effect of post-harvest treatment on the final OTA content in cocoa beans or their derived products (roasted cocoa, nibs, butter, cocoa powder, chocolate spread), Mounjouenpou et al. (2008, 2011a, b, 2012a) observed that pod damage, and late pod opening were aggravating factors for OTA contamination. Fermented dried cocoa from intact pods presented an OTA content below those from poor quality pods (intentionally or naturally damaged) which showed contents of up to 76 µg/kg (Mounjouenpou et al. 2011, 2012b). *A. carbonarius, A. niger* and *A. ochraceus* are fungal contaminant reported in coffee beans from Cameroon. Several studies were conducted on 104 coffee samples and concluded that just a few presented an average OTA content above the acceptable limit for roasted coffee which is 5 µg/kg (Mounjouenpou et al. 2013, Romani et al. 2000). They showed that their local Arabica coffee brand samples were all below the limit, and only few from the Robusta coffee were above the limit. Studies conducted by Romani et al. (2000) had already shown an OTA contamination of 1 - 2.5 µg/kg in green coffee beans from Cameroon.

The occurrence of mycotoxins in **sorghum and its derived products** has also been studied during the last decade. This cereal is a crop mainly produced and consumed in the northern part of Cameroon. No OTA, DON and FB1 were detected in grains collected in this region by Djoulde (2013). Only AFB1 was detected in sorghum cultivated in the rainy season (0–230 µg/kg). In some locally produced artisanal sorghum derived products (beer, flour, baby's beverage, and cake), mycotoxins were detected at levels ranging between 0–250 mg/kg for AFB1, 0–45 mg/kg for OTA, and 0–538 µg/kg for DON. In a previous work, Djoulde (2011) observed that local sorghum beers (*bil-bil* and *kpata*) were contaminated by both DON (0–730 µg/L) and FB1 (0–340 µg/L).

Apart from maize, peanuts, beans, soybeans, cocoa, coffee and sorghum, the mycotoxin content of other foods from Cameroon has also been studied. Essono et al. (2009) assessed the total aflatoxin content in **cassava chips**, a cassava derived product (obtained after fermentation and drying) which is widely consumed locally. They reported contamination of 33% of their samples at levels ranging between 5.2–15 µg/kg. A higher range of aflatoxin content (6–194 µg/kg) in cassava flakes and chips was noticed by Njumbe et al. (2014) who also reported the presence of penicillic acid inside these two products (25–184 µg/kg).

Tchana et al. (2010) observed AFs contamination in **egg samples** (45.2%) at a mean concentration of 0.82 µg/kg, contamination which varied from one agro-ecological zone to another. Indeed, the frequency of contamination was higher in the humid forest areas (25 to 53%) as compared to the littoral, savannah and steppe areas. These authors also noticed aflatoxin M1 (AFM1) contamination in 16% of their **cow milk samples** at levels varying from 0.006 to 0.525 µg/L.

More recently, the presence of OTA was detected by Nguegwouo et al. (2018) in 10% of black pepper samples (1.2–1.9 µg/kg) and 40% of white pepper samples (1.8–4.9 µg/kg) collected from Yaoundé markets.

3. Effect of Mycotoxins on Human Health and Associated International Regulations

Human contamination with mycotoxins can occur directly through the consumption of contaminated foods or indirectly through consumption of animals in which mycotoxins have been bio-accumulated. Mutagenic, hepatotoxic, dermonecrotic, nephrotoxic, haematotoxic, immunosuppressive, neurotoxic, carcinogenic and teratogenic effects have been reported on both humans and animals (Zinedine 2004) and vary from one mycotoxin to another. Besides these mycotoxins commonly found in foods and which are responsible for health hazards worldwide (Bennett and Klich 2003, Wild and Gong 2009), there are also new emerging mycotoxins (fusaroproliferin, gliotoxin, cyclopiazonic acid, penitrem A and verruculogen) whose toxicological effects are still unknown (Hajslova et al. 2011, Logrieco et al. 2009).

Due to their potential negative impact on health, norms on the level of mycotoxins in marketed and consumed foods have been established worldwide. Based on toxicological studies, a maximum acceptable content of each specific mycotoxin in different food matrices has been defined at both national and international levels.

Some African countries like South Africa, Morocco, Egypt and Zimbabwe have their regulations (CAC 2004, CAC 2005, FAO 2003, Van Egmond and Dekker 1997, National Mycotoxin Regulations 2016), but the most updated and worldwide used regulations are those of the European Union and USA (*Codex Alimentarius* Commission 2006 and 2008), detailed in Table 2, which serve as reference to many other countries from which they import foods. Indeed, they are the main importers of many crops and food products and are strict on the maximum acceptable mycotoxin limits.

The application of these measures helped in limiting the risk of mycotoxicoses. Nevertheless, this implies additional costs for quality control and eventual rejection at destination of contaminated foods. Unfortunately, many agricultural products destined for export are often put aside and destroyed due to their high mycotoxin contents thus affecting the economy of producing countries (Shane 1994, Wu 2007). Many of the countries exporting their food products are developing or underdeveloped countries, where monitoring of the entire production chain from farms to the final product is a challenge (Cardwell 2000, Ukwuru et al. 2017). Situations where exporting countries would retain higher risk commodities for their national consumption and therefore expose their populations also exist (Milićević et al. 2010). In fact, the risk of contaminated foods to be consumed locally after-market rejection or use for feeding animals is important leading to the possibility of exposure to high levels of contamination and/or appearance of residues in food from animal origin (milk).

4. National Regulations and Initiatives

The evolution of international food security and food safety policies show growing consensus on the duty of governments to guarantee food security by improving access to all sections of their populations to sufficient healthy foods of good nutritional quality.

Although endowed with an agricultural production potential that has guaranteed food self-sufficiency for several decades, ensuring food safety to Cameroon consumers remains a major issue as the country lacks a reliable food control system based on modern, internationally accepted concepts in the *Codex Alimentarius*.

The absence of a reliable food control system in Cameroon raises many difficulties with major economic implications. Cameroon is unable to consider the export of some of these products and therefore become a promising growth sector. This is the case for high-market value seafood, which are often rejected by importing European Union countries (Commission Regulation 146/2009 of 20/02/2009).

At present, Cameroon does not have national standards or regulatory limits for mycotoxins in food and feed. The current policy is to consider that

Table 2: Some international regulations on mycotoxins

Country/ Region	Mycotoxins	Foods	Maximum level (µg/kg)	References
European Union	Ochratoxin A	Cereals, dry fruits, wine, spices, oat, raisins, coffee, cocoa, soybeans, meat	0.5–15	European Commission (2006)
	Ochratoxin A	Wheat, barley and rye	5	*Codex Alimentarius* Commission (2015)
	Aflatoxins B1, G1, B2, G2	Maize, wheat, rice, spices, almonds, oil seeds, dried fruits, cheese	0.1–8 4–15	European Commission (2010)
	Aflatoxin M1	Milk, eggs, meat	0.5	*Codex Alimentarius* Commission (2015)
	Patulin	Apples, cherries, cereal grains, grapes, pears, bilberries	10–50	European Commission (2007)
		Apple juice	50	*Codex Alimentarius* Commission (2015)
	T-2 and HT-2	Cereals for direct human consumption	50	European Food Safety Authority (2011)
		Oats, barley (including malting barley) and maize	200	European Food Safety Authority (2011)
		Wheat, rye and other cereals	100	European Food Safety Authority (2011)
	Deoxynivalenol (DON)	Wheat, oats and maize	1250	European Commission (2007)
		Flour, meal, semolina and flakes derived from wheat, maize or barley	1000	*Codex Alimentarius* Commission (2015)
		Cereal grains (wheat, maize and barley) destined for further processing	2000	*Codex Alimentarius* Commission (2015)
	Fumonisins (FB1 and FB2)	Maize-based breakfast cereals and maize-based snacks	800	European Commission (2007)
		Raw maize grain	4000	*Codex Alimentarius* Commission (2015)
		Maize flour and maize meal	2000	*Codex Alimentarius* Commission (2015)

(Contd.)

Table 2: *(Contd.)*

Country/ Region	Mycotoxins	Foods	Maximum level (μg/kg)	References
		Maize intended for direct human consumption	1000	European Commission (2007)
	Zearalenone	Unprocessed cereals other than maize	100	European Commission (2007)
		Maize intended for direct human consumption, maize-based snacks and maize-based breakfast cereals	100	European Commission (2007)
		Cereal flour, bran and germ for direct human consumption	75	European Commission (2007)
USA	Total aflatoxins	Food for human consumption	20	FDA (Food and Drug Administration) 2011
		Corn, peanut products, cottonseed meal	20	USDA (United States Department of Agriculture) 2015
	Aflatoxin M1	Milk, milk products	0.5	FDA (Food and Drug Administration) 2011
	Aflatoxins B1, B2, G1, G2	Maize, wheat, rice, peanut, sorghum, pistachio, almond, ground nuts, tree nuts, figs, cottonseed, spices	20	USDA (United States Department of Agriculture) 2015
	Deoxynivalenol (DON)	Cereals, cereal products for human food	1000	USFDA (US Food and Drug Administration) 2010
	Total fumonisins (FB1, FB2 and FB3)	Cereals	2000 - 4000	Alshannaq and Yu. 2017
		Corn products and cleaned maize used for popcorn	2000 - 3000	Alshannaq and Yu. 2017
	Patulin	Apples, apple juice and concentrate	50	Alshannaq and Yu. 2017
	Ochratoxin A	Cereals, wheat, barley, and rye and derived products	5	*Codex Alimentarius* Commission 2008

any food contaminated with random substances will be unfit for human consumption. Hence, all food laws that prohibit the trade of falsified products consider any food product contaminated with random substances as a falsified product. In many countries, this general law has led to specific regulations that impose acceptable limits for certain mycotoxins (aflatoxins) in foods. In 1970, only 18 countries had laws and regulations on the limitation of aflatoxin levels in food products and this number increased to 50 in 1991.

However, in the absence of a national law on mycotoxins, several state initiatives exist.

4.1 National Food Safety Committee of Cameroon

The National Food Safety Committee of Cameroon (NFSC) was created by decree N° 0111/CAB/PM of March 2, 2004 as an ad hoc committee in charge of Food Safety (FS). Its role is to:

- Diagnose current food control systems in Cameroon;
- Reflect on new food quality control procedure techniques; and
- Propose measures to improve FS in Cameroon.

The main recommendation of this committee was the creation and operation of a national body responsible for the:

- Coordination of all FS activities;
- Development and harmonization of food control procedures;
- Organization of a surveillance system for the FS;
- Coordination and development of risk analysis activities;
- Development, updating and revision of food laws and regulations;
- Evaluation and monitoring of FS activities;
- Training and assistance to private companies;
- Training and retraining of state personnel;
- Approval of laboratories for food analyzes and carrying out of studies and surveys related to FS;
- Development and coordination of multi-sectorial FS projects;
- Preparation and participation of Cameroon in the work of the *Codex Alimentarius*; and
- Consumer information and education.

4.2 National Codex Alimentarius Committees

The National *Codex Alimentarius* Committees (**NCAC**) are national committees in charge of activities within the framework of the *Codex Alimentarius.* Their roles are an effective participation in the work of the Codex, improvement of the co-operation with the *Codex Alimentarius* Commission and its organs. They carry out studies for promotion of Codex activities, public awareness or FS (development of texts, harmonization of procedures, FS capacity building, awareness raising, information and education of various target audiences, consumers, economic operators, accreditation of laboratories).

4.3 National Committee of Codex Alimentarius and Food Safety in Cameroon

The National Committee of *Codex Alimentarius* and Food Safety in Cameroon (NCCAFS) was created by a Prime Ministerial Decree on February 14, 2008 and set up by MINIMIDT decree of August 19, 2008, in accordance with the recommendations of the NCCAFS *ad hoc* committee. Its competence included all issues relating to the *Codex Alimentarius* (studies, programming of activities and missions, review of technical files, consultation with stakeholders), as well as proposals for actions related to the development and amendment of Food Safety laws and regulations, harmonization of food control procedures and food safety oversight mechanisms, coordination and mobilization of human and financial resources for the implementation of activities related to NCCAFS.

As a result, two of the project's objectives were thus realized. However, that the grouping of closely related or interdependent functions within a single committee (such as participation in Codex work and coordination of the control services) raised other difficulties. Indeed, the preparation of Cameroon's participation in the work of the Codex involved experts from various backgrounds (public and private persons, scientists and consumers, etc.), while the coordination of control actions required greater discretion and more limited participation. These data may need to be taken into account in future when, having accumulated some practical experience, the Committee will also have to appreciate the limitations of its current organization. In any case, this creation could be considered as a first step and the expression of a political will to positively evolve these activities by a real coordination.

The implementation of recommendations of this committee required an in-depth technical diagnosis of the existing situation, as well as implementation of priority capacity building activities.

5. Legal and Regulatory Framework of Food Quality Control

In Cameroon, the food safety system is provided by several national institutions as well as decentralized local authorities, notably:

- The Ministry of Scientific Research and Innovation (MINRESI), through the Food and Nutrition Research Center of the Institute of Medical Research and Study of Medicinal Plants (CRAN / IMPM), the National Agency for Radio Protection (ANRP) and the Institute of Agricultural Research for Development (IRAD) ensure the quality control of foodstuffs.
- The Ministry of Agriculture and Rural Development (MINADER) that contributes to ensuring food security by facilitating the entry of good quality material and by ensuring the control of certain toxic residues (pesticide and mycotoxins) in food.

- The Minister of Public Health which is the main player in the quality control of medicines and other healthy products: food supplements, cosmetics, medical devices . . .
- The Ministry of Livestock, Fisheries and Animal Industries (MINEPIA), which through its National Veterinary Laboratory (LANAVET), ensures the safety of all imported and exported food and as such deal with issues relating to the search for chemical and biological residues in livestock, dairy and fish products; international trade in food products.
- Ministry of Mines, Industry and Technological Development (MINMITD) through the "Agency for Standards and Quality (ANOR)". ANOR is the standardization body that was created by Presidential Decree No. 2009/296 of 17 September 2009. It is an administrative public institution with legal personality and financial autonomy. Under the technical and financial supervision of MINIMIDT and MINFI (Ministry of Finance) respectively, its main missions are to ensure the development and certification of standards, certification and evaluation of conformity to standards, promotion of standards and quality approach to public, para-public and private sector organizations.
- The National Laboratory for Quality Control of Drugs and Expertise (NALAQACODE). Placed under the supervision of the Ministry of Public Health for the quality control of drugs, and MINIMIDT for the quality control of industrial products including food products, NALAQACODE provides physicochemical and microbiological control of agrifood and dietetic products, drinking water and hygienic beverages via the ANOR standards for food technology.
- The Ministry of Commerce (MINCOMMERCE). Cameroon is a member of the World Trade Organization (WTO), the credibility of controls on the flow of products in international trade is important. In collaboration with NALAQACODE, MINCOMMERCE promotes and defends a quality label for products intended for the local market and for export, monitoring the circuits for conservation and distribution of high consumption of food products, as well as the distribution of food products, development and application of standards for measuring and quality control instruments in liaison with the concerned ministerial departments. To this end, its main work force is the National Brigade of Controls and Fraud (BNCRF).
- Ministry of Finance (MINFI) and Cameroon Customs. They act in synergy for the development and application of customs legislations and regulations in the field of import and export of food products and the fight against illicit trafficking of products.

All these institutions are acting in synergy to fight against counterfeiting, and the sale of food products containing ingredients that are dangerous to human health or that do not respect the standards set for their activities, based on a well-defined legal framework.

6. Factors Limiting the Effectiveness of the National Food Security System

The report of 2004 on stakeholders in the food quality control sector identified a number of facts: the food safety system in Cameroon (including water), involves several institutions and decentralized local authorities, thereby creating other jurisdictional conflicts, which could negatively impact expected results.

6.1 Legal Component

The absence of any coherent legal basis on which the various sectorial texts could be based constitutes a major handicap. Each institutional operator (administration in charge of controls) sheltering behind one or more laws, sometimes very old but not explicitly repealed, conferring on it competences considered often as exclusive of any other intervention, which is translated by innumerable conflicts. Application texts were rarely promulgated. Laboratories applied indistinctly international standards or major economic partners (Codex, EU) when they existed, and industrial producers did the same. The rules of procedure to be imposed on the agents of control were varied and moreover unknown to the agents themselves. The texts relating to the rights of the persons visited were non-existent. Without this coherent legal basis, common to all administrations and known to operators, no serious action of reorganization implying coordination between the control services, could be envisaged.

6.2 Organizational Component

Due to the scattering and the legal stacking, little organized coordination or consultation exists between the services. The hegemonic will of certain administrative structures is sometimes clearly affirmed. This often results in leadership struggles and difficulties of institutional cooperation.

6.3 Control of Imported and Exported Products

The control of imported products is very inadequate, with the exception of phytosanitary controls, which seem to have been fairly well achieved at least in the Douala seaport. In addition, Cameroon has put in place administrative procedures that were justified, according to officials, by the desire to protect consumers. In addition to contravening the general principles of international trade, the modalities of application greatly contribute to doubts about their effectiveness. This also resulted in additional costs for importers. As for exports, the products are theoretically subject to official controls, the reliability of which is sometimes contested by the exporters themselves who saw them as additional administrative hassle (for example, the EU's position on seafood from Cameroon, currently not allowed on the EU territory).

Finally, whether it is imported or exported products, the crossing of land or sea borders does not always take place within a "regulatory" framework.

6.4 Controls on the National Territory

Very few statistics are available. The official control services are not equipped with sufficient material means (means of displacement, control) and the capacity of the staff is not regularly enhanced on new multi-disciplinary control techniques (legal, technical, technological, economic). The sometimes-implemented solution of imposing costly audits on operators by private companies is strongly criticized by the contractors with regards to the cost of such interventions and the fact that they create a mismatch in competition. In addition, these audits which are generally only carried out on premises, are not monitored, and target only a small proportion of the operators.

6.5 Control Laboratories

With the exception of the Pasteur Center of Cameroon in microbiology, very few public laboratories have equipment, infrastructure and personnel to properly fulfill their analytical control mission and as such only put forward the text of creation to justify a competence which was unfortunately only legal but nevertheless a source of conflicts.

7. Framework Law on Food Safety in Cameroon

To better regulate this sector and for better efficiency of sanitary control of food in Cameroon, the Head of State recently signed a framework law (Law N° 2018/020 of December 11, 2018) which fixes the principles and bases of food safety regulations.

This law aims to:

- Guarantee consumers that food is safe;
- Prevent and control foodborne illnesses;
- Make compulsory the declaration of foodborne illnesses;
- Promote national and international food trade by establishing an effective safety system based on scientific principles;
- Contribute to the establishment of specific standards for food consumed and marketed in Cameroon and ensure their application;
- Improve the quality of food produced on national territory, through implementation of good production, manufacturing and hygiene practices, a system of controlling health and phyto-sanitary risks;
- Facilitate integrated management of food safety, at the various stages of the food chain;
- Promote coordination mechanisms between the various competent food sector authorities and gradually prepare them for the integration of their activities in regional or international bodies, in particular Codex, the World Organization for Animal Health, the International Convention

for the Protection of Plants, the international Food Safety Authorities Network;

- Develop an effective official inspection and control system, based on standards, technical regulations and, where appropriate, scientific data;
- Support and stimulate the development of the food industry and encourage competitiveness on the national and international market;
- Promote the participation of consumers and the stakeholders involved in the food chain (consumption, processing and marketing), as well as in the application and execution of the national policy on food safety;
- Establish a framework which facilitates implementation of national and international requirements in the food sector;
- Strengthen the responsibility of any producer, processor or distributor in protecting consumer health, by implementing self-checking systems and authorizations for consumption;
- Ensure communication on risks, in connection with epidemiological surveillance networks; and health alert or watch systems; and
- Set up a national system of authorizations for the consumption of foodstuffs, animal feed, food additives and supplements.

This legal framework also describes the organization and functioning of the food safety system. In this case, it provides for a health security policy and strategy document, and defines the attributions and scope of intervention of each stakeholder. However, application texts and other specific texts that will facilitate the implementation of this framework law for better efficiency of the food safety system in Cameroon are still awaited.

8. Conclusion

The presence of mycotoxins in Cameroonian's food is important and may lead to severe health consequences for consumers. Even if the limited studies on the assessment of human exposure in Cameroon tend to show a limited health risk, it is important to note that some of the food samples analyzed presented mycotoxins contamination levels, which were above the regulatory standards.

Efforts should, therefore, be made by the government to educate producers on mitigation strategies (from farms to the final product), and populations on the risk associated with such contaminated products. Efforts should also focus on the establishment of national standards for better effectiveness of the national food security system and to better protect population. More studies on exposure need to be done and should also focus on young children and the elder people who may be the most sensitive groups of the population. The problem of multiple exposure/contamination of foods and subsequent synergistic/additive effects should also not be forgotten.

Acknowledgement

The author is thankful to the Minister of Scientific Research and Innovation of Cameroon and the General Manager of IRAD for their support, Dr Ehabe

E. Eugene and Dr Nguegwouo Evelyne for their precious contribution on this document.

References

Abia, W.A., Warth, B., Ezekiel, C.N., Sarkani, B., Turner, P.C., Marko, D. et al. 2017. Uncommon toxic microbial metabolite patterns in traditionally home-processed maize dish (fufu) consumed in rural Cameroon. Food Chemistry and Toxicology 107: 10–19. DOI: 10.1016/j.fct.2017.06.011

Abia, W.A., Warth, B., Sulyok, M., Krska, R., Tchaba, A.N., Njobeh, P.B. et al. 2013. Determination of multi-mycotoxin occurrence in cereals, nuts and their products in Cameroon by liquid chromatography tandem mass spectrometry (LC-MS/MS). Food Control 31: 438–453. DOI: 10.1016/j.foodcont.2012.10.006

Alshannaq, A. and Yu, J.H. 2017. Occurrence, toxicity, and analysis of major mycotoxins in food. International Journal of Environment Research Public Health 14.

Bennett, J.W. and Klich, M. 2003. Mycotoxins. Clinical Microbiology Revue 16: 497–516. DOI: 10.1128/CMR.16.3.497

C.A.C. (*Codex Alimentarius* Commission). 2004. Code of practice for the prevention and reduction of aflatoxin contamination in peanuts. CAC/RCP; 55-2004.

C.A.C. (*Codex Alimentarius* Commission). 2005. Code of practice for the prevention and reduction of aflatoxin contamination in tree nuts. CAC/RCP; 59-2005.

C.A.C. (*Codex Alimentarius* Commission). 2008. Codex Committee on Contaminants in Foods, Discussion Paper on Ochratoxin A in cocoa. Second Session (CX/CF08/2/15). The Hague (the Netherlands): CEC, 2008.

C.A.C. (*Codex Alimentarius* Commission). 2015. General standard for contaminants and toxins in food and feed (CODEX STAN 193-1995), Adopted in 1995, revised in 1997, 2006, 2008, 2009, Amended in 2010, 2012, 2013, 2014, 2015. 59 pp.

Cardwell, K.F. 2000. Mycotoxin contamination of foods in Africa: Antinutritional factors. Food Nutrition Bulletin 21: 488–492. DOI: 10.1177/156482650002100427

Cheli, F., Pinotti, L., Novacco, M., Ottoboni, M., Tretola, M. and Dell'Orto, V. 2017. Mycotoxins in Wheat and Mitigation Measures, Wheat Improvement, Management and Utilization, Ms. Ruth Wanyera (Ed.), InTech. DOI: https://doi.org/10.5772/67240

Chuisseu, D.D.P., Abia, W.A., Zibi, S.B., Simo, K.N., Ngantchouko, N.C.B., Tambo, E. et al. 2018. Safety of breast milk vis-a-vis common infant formula and complementary foods from western and centre regions of Cameroon from mycotoxin perspective. R. Adv. Food Sci. 1: 23–31.

Djoulde, D.R. 2011. Deoxynivalenol (DON) and fumonisins B1 (FB1) in artisanal sorghum opaque beer brewed in north Cameroon. African Journal of Microbiology Research 5: 1565–1567. DOI: 10.5897/AJMR10.709

Djoulde, D.R. 2013. Sustainability and effectiveness of artisanal approach to control mycotoxins associated with sorghum grains and sorghum-based food in Sahelian zone of Cameroon. pp. 137–151. *In:* Makun, H.A. (ed.). Mycotoxins and Food Safety in Developing Countries. Croatia: Intech. Available: http://dx.doi.org/10.5772/54789

E.C. (European Commission). 2006. Commission Recommendation 2006/576/EC of 17 August 2006 on the presence of deoxynivalenol, zearalenone, ochratoxin A, T-2 and HT-2 and fumonisins in products intended for animal feeding. Official Journal of European Union 229: 7–9.

E.C. (European Commission). 2007. Commission Regulation (EC) No. 1126/2007 on maximum levels for certain contaminants in foodstuffs as regards *Fusarium* toxins in maize and maize products. Official Journal of European Union 255: 14–17.

E.C. (European Commission). 2010. Commission Regulation (EU) No 178/2010 of 2 March 2010 amending Regulation (EC) No 401/2006 as regards groundnuts (peanuts), other oilseeds, tree nuts, apricot kernels, liquorice and vegetable oil. Official Journal of European Union 52: 32.

EFSA (European Food Safety Authority). 2011. Scientific opinion on the risks for animal and public health related to the presence of T-2 and HT-2 toxin in food and feed. EFSA Journal 9: 1–187.

Essono, G., Ayodele, M., Akoa, A., Foko, J., Filtenborg, S. and Olembo. 2009. Aflatoxin-producing *Aspergillus* spp. and aflatoxin levels in stored cassava chips as affected by processing practices. Food Control 20: 648–654. DOI: 10.1016/ j. food cont.2008.09.018

F.A.O. (Food and Agricultural Organization). 2003. Worldwide regulations for mycotoxins in food and feed in 2003. FAO Food and Nutrition Papers 81: 1–165. Available: http://doi.org/10.1017/CBO97811 07415324.004

F.D.A. (Food and Drug Administration). 2010. Guidance for Industry and FDA: Advisory Levels for Deoxynivalenol (DON) in Finished Wheat Products for Human Consumption and Grains and Grain By-Products Used for Animal Feed. US FDA: Silver Spring, MD, USA.

F.D.A. (Food and Drug Administration). 2011. U.S. Food and Drug Administration (FDA) mycotoxin regulatory guidance. *In:* U.S. Food and Drug Administration (FDA) and Feed Association 1-9.

Hajslova, J., Zachariasova, M. and Cajka, T., 2011. Analysis of multiple mycotoxins in food: Methods of molecular biology 747: 233–258. DOI: 10.1007/978-1-61779-136-9_10

Kana, J.R., Gnonlonfin, B.G.J., Harvey, J., Wainaina, J., Wanjuki, I., Skilton, R.A. et al. 2013. Assessment of aflatoxin contamination of maize, peanut meal and poultry feed mixtures from different agroecological zones in Cameroon. Toxins 5: 884–894. DOI: 10.3390/toxins5050884

Karlovsky, P., Suman, M., Berthiller, F. and De meester, J. 2016. Impact of food processing and detoxification treatments on mycotoxin contamination. Mycotoxin Research 32: 179–205.

Logrieco, A., Moretti, A. and Solfrizzo, M. 2009. Alternaria toxins and plant diseases: An overview of origin, occurrence and risks. World Mycotoxin Journal 2: 129–140. DOI: 10.3920/WMJ2009.1145

Milićević, D.R., Škrinjar, M. and Baltić, T. 2010. Real and perceived risks for mycotoxin contamination in foods and feeds: Challenges for food safety control. Toxins 2: 572–592.

Mounjouenpou, P., Gueule, D., Tondje, P.R., Fontana-Tachon, A. and Guiraud, J.P. 2008. Filamentous fungi producing achratoxin-A during cocoa processing in Cameroon. International Journal of Food Microbiology 121: 234–241. DOI: 10.1016/j.ijfoodmicro.2007.11.017

Mounjouenpou, P., Amang, A., Mbang, J., Ntoupka, M., Guyot, B., Fontana-Tachon, A. et al. 2011a. Biodiversity of potentially ochratoxin-A producing black Aspergilli related to the cocoa bean post-harvest processing. Journal of Natural Product and Plant Resource 3: 50–58.

Mounjouenpou, P., Gueule, D., Ntoupka, M., Durand, N., Fontana-Tachon, A., Guyot, B. et al. 2011b. Influence of post-harvest processing on ochratoxin-A content in cocoa and on consumer exposure in Cameroon. World Mycotoxin Journal 4: 141–146. DOI: 10.3920/WMJ2010.1255

Mounjouenpou, P., Amang, A., Mbang, J., Guyot, B. and Guiraud, J.P. 2012a. Traditional procedures of cocoa processing and occurrence of ochratoxin-A in the derived products. Journal of Chemical Pharmaceutical Research 4: 1332–1339.

Mounjouenpou, P., Gueule, D., Maboune, Tetmoun, S.A., Guyot, B., Fontana-Tachon, A. et al. 2012b. Incidence of pod integrity on the fungal microflora and ochratoxin-A production in cocoa. Journal of Biology and Life Science 3(1): 254–265.

Mounjouenpou, P., Durand, N., Guiraud, J.-P., Maboune Tetmoun, S.A., Gueule, D. and Guyot, B. 2013. Assessment of exposure to ochratoxin-A (OTA) through ground roasted coffee in two cameroonian cities: Yaounde and Douala. International Journal of Food Science Nutrition Engineering 3: 35–39. DOI: 10.5923/j.food.20130303.03

National Mycotoxin Regulations. 2016. South African Maize Crop Quality Report 2015/2016 Season Government Notice No. R. 1145, dated 8 October 2004. Published under Government Notice No. 987 of 05 September 2016. Available: http://www.sagl.co.za/Portals/0/Maize%20Crop%202015%20 2016/Page%20 73.pdf

Ngoko, Z., Daoudou Imele, H., Kamga, P.T., Mendi, S., Mwangi, M., Bandyopadhyay, R. et al. 2008. Fungi and mycotoxins associated with food commodities in Cameroon. Journal of Applied Bioscience 6: 164-168.

Ngoko, Z., Marasas, W.F.O., Rheeder, J.P., Shephard, G.S. and Wingfield, M.J. 2001. Fungal infection and mycotoxin contamination of maize in the humid forest and the western highlands of Cameroon. Phytoparasitica 29: 352–360. DOI: 10.1007/BF02981849

Nguegwouo, E., Etame, L., Tchuenchieu, A., Mouafo, H., Mounchigam, E., Njayou, N. et al. 2018. Ochratoxin A in black pepper, white pepper and clove sold in Yaoundé (Cameroon) markets: Contamination levels and consumers practices increasing health risk. International Journal of Food Contamination 5: 1–7.

Nguegwouo, E., Njumbe, E., Njobeh, P.B., Medoua, G.N., Ngoko, Z., Fotso, M. et al. 2017. Aflatoxin and fumonisin in corn production chain in bafia, centre Cameroon: Impact of processing techniques. Journal of Pharmacy and Pharmacology 5: 579–590.

Nguegwouo, E., Tchuenchieu, A., Mouafo, T.H., Fokou, E., Medoua Nama, G., De Saeger, S. et al. 2019. Mycotoxin contamination of food and associated health risk in Cameroon: A 25-years Review (1993-2018). EJNFS 9(1): 52–65.

Njobeh, P.B., Dutton, M.F., Koch, S.H., Chuturgoon, A., Stoev, S. and Seifert, K. 2009. Contamination with storage fungi of human food from Cameroon. International Journal of Food Microbiology 135: 193–198.

Njobeh, P.B., Dutton, M.F., Koch, S.H. and Chuturgoon, A. 2010. Simultaneous occurrence of mycotoxins in human food commodities from Cameroon. Mycotoxin Research 26: 47–57. DOI: 10.1007/s12550-009-0039-6

Njumbe, E., Hell, K. and De Saeger, S. 2014. A comprehensive study to explore differences in mycotoxin patterns from agro-ecological regions through maize, peanut, and cassava products: A case study. Cameroon. Journal of Agricultural Food Chemistry 62: 4789–4797. DOI: 10.1021/jf501710u

Romani, S., Sacchetti, G., Lopez, C.C., Pinnavaia, G.G. and Dalla Rosa, M. 2000. Screening on the occurrence of ochratoxin-A in green coffee beans of different origins and types. Journal of Agricultural Food Chemistry 48: 3616–3619.

Shane, S.H. 1994. Economic issues associated with aflatoxins. pp. 513–527. *In:* Eaton, D.L. and Groopman, J.D. (eds.). The Toxicology of Aflatoxins: Human Health, Veterinary, and Agricultural Significance. Academic Press: San Diego, CA, USA.

Tchana, A.N., Moundipa, P.F. and Tchouanguep, F.M. 2010. Aflatoxin contamination in food and body fluids in relation to malnutrition and cancer status in Cameroon. International Journal of Environmental Research Public Health 7: 178–188. DOI: 10.3390/ijerph7010178

Tchikoua, R., Tatsadjieu, N.L. and Mbofung, C.M.F. 2015. Effect of selected lactic acid bacteria on growth of aspergillus flavus and aflatoxin B1 production in kutukutu. Journal of Microbiology Research 5: 84–94. Available: http://doi:10.5923/j.microbiology. 20150503.02

Ukwuru, M.U., Ohaegbu, C.G. and Muritala, A. 2017. An overview of mycotoxin contamination of foods and feeds. Journal of Biochemistry Microbial Toxicology 1: 1-11.

USDA (United States Department of Agriculture). 2015. Mycotoxin Handbook. Grain Inspection, Packers and Stockyards Administration Federal Grain Inspection Service. September 17; 60.

USFDA (US Food and Drug Administration). 2010. Guidance for Industry and FDA: Advisory Levels for Deoxynivalenol (DON) in Finished Wheat Products for Human Consumption and Grains and Grain By-Products Used for Animal Feed. US FDA: Silver Spring, MD, USA.

Van Egmond, H.P. and Dekker, W.H. 1997. Worldwide regulations for mycotoxins in 1995 – A compendium. FAO Food and Nutrition paper, FAO, Rome, Italy; 64 pp.

Wild, C.P. and Gong, Y.Y. 2009. Mycotoxins and human disease: A largely ignored global health issue. Carcinogenesis 31: 71–82.

Wu, F. 2007. Measuring the economic impacts of Fusarium toxins in animal feeds. Animal. Feed Science Technology 137: 363–374.

Zain, M.E. 2011. Impact of mycotoxins on humans and animals. Journal of Saudi Chemistry Society 15: 129–144. DOI: 10.1016/j.jscs.2010.06.006

Zinedine, A. 2004. Occurrence and legislation of mycotoxins in food and feed from Morocco. Food Control 20: 334–344.

Mycotoxins during Consumer Food Storage

Philippe Dantigny[1*]**, Monika Coton**[1]**, Angélique Fontana**[2] **and Sabine Schorr-Galindo**[2]

[1] Univ Brest, Laboratoire Universitaire de Biodiversité et Écologie Microbienne, F-29280 Plouzané, France
[2] Qualisud Joint Research Unit, Univ Montpellier, CIRAD, Montpellier SupAgro, Univ d'Avignon, Univ de La Réunion, Montpellier, France

1. Introduction

In the farm to fork approach, the consumer is a key element at the end of the food chain. In this sense, food safety needs to be ensured and recommendations provided to consumers are always needed. Many recommendations exist such as choose foods processed for safety, cook foods thoroughly, eat cooked foods immediately, store cooked foods in cold storage, reheat cooked foods thoroughly, avoid contact between raw foods and cooked foods, wash hands repeatedly, keep all kitchen surfaces meticulously clean, protect foods from insects, rodents, and other animals, and use safe water (PAHO, 2019) but they are mainly directed to preserve foods from pathogenic bacteria, viruses and parasites. Some chemical hazards, in particular mycotoxins, are produced by biological agents, e.g. molds, mainly *Penicillium, Aspergillus* and *Fusarium* species. European regulations set maximum levels for certain mycotoxins, due to their known potential toxicities and chemical and thermal resistances, in some foods at different steps of the food chain between production and retail ((EC) No 1881/2006), but this cannot be done at the consumer level. All food categories can be prone to mold contamination including foods where bacterial growth may be limited like low water activity or low pH products that may or may not be treated thermally as well as low nutrient foods (Coton and Dantigny 2019). Mold-contaminated foods generally have a negative impact on product quality as consumers can easily detect visible fungal mycelium that develops on food surfaces. In most cases, foods are either discarded or trimmed. In fact, consumers often lack knowledge as to

*Corresponding author: philippe.dantigny@univ-brest.fr

how to handle moldy foods safely and they are unaware as to whether the contaminating mold is mycotoxigenic. If a mycotoxigenic mold is present, mycotoxins can potentially be produced and migrate into the product according to different biological, storage and product related factors (mold species, mycotoxin properties, water activity, temperature, food composition, . . .) (Coton and Dantigny 2019). Therefore, it is necessary to study mycotoxin migration in foods by quantifying their concentration and by correlating these concentrations to simple observations that can be made by consumers (i.e. visual aspect of fungal growth). Safe recommendations about moldy foods could be provided only after this step using scientific studies.

2. Existing Recommendations for Consumers

Some French web pages already provide recommendations to consumers regarding food products spoiled by molds. Among these, the website "Quoi dans mon assiette—What in my plate" (2016) translated the recommendations given by USDA (2013) "Molds on food: are they dangerous?" One of the objectives of the US website was to provide recommendations to consumer as to what to do if you encounter moldy foods in the household. The first page introduced consumers to mold related allergies and highlighted the risk of respiratory diseases due to spore inhalation. The following pages explained that some mold species are responsible for mycotoxin production but it was not underlined that mycotoxins cannot be destroyed by cooking practices. In contrast to the lack of information on mycotoxin heat resistance, the carcinogenicity of aflatoxins (a regulated mycotoxin in some foodstuffs) was largely described. Although this information is of paramount importance for food safety, aflatoxins are not the main consumer mycotoxins, i.e., mycotoxins that are produced during food storage at consumer's homes. Other chapters also provided information about what molds are, where can we find them, whether they are harmful or not, which are the most common, which molds are useful, whether they can develop in the fridge, and how mold growth can be controlled. It was recommended to buy foods in small quantities to avoid food waste. It was also mentioned that molds develop during cold storage (i.e. in the fridge) as well as in acidic conditions (i.e. low pH foods), but also that molds can be destroyed by heating food at 100°C. Finally, it was highlighted that molds can not only develop on the surface of foods but also that mycelium can penetrate deeply inside some foods such as bread. Linked to this, they eventually stated that very dangerous molds can produce mycotoxins that can migrate beyond the moldy zone of the food. Depending on the food category, recommendations to either discard or eat the product after trimming were provided and explained in Table 1.

The first food category was of highly perishable products (meats, ready to eat foods, some dairy products) that should be refrigerated and discarded in case of fungal spoilage. This is due to the possible development of pathogenic bacteria at the same time as molds. It was also recommended to discard bread and bakery products as fungal mycelium can penetrate, more or less deeply,

Table 1: Different food categories for which recommendations were given by the USDA (adapted from USDA, 2013)

Food category	Recommendation	Reason
Cold cuts, bacon, hot-dogs, poultry, ready-to-eat meals, cooked pasta, cooked grains, soft cheeses, yogurts, fresh creams/sour creams	Discard	High moisture foods. Contamination can occur below the surface. Bacterial contamination can also occur.
Breads and bakery products	Discard	Molds can develop within porous foods
Jams	Discard	Molds might produce mycotoxins, it is not recommended to eat jams after removing molds
Peanut butter, dried fruits	Discard	Foods without conservatives can be contaminated by molds
Soft fruits and vegetables (tomatoes, cucumbers, peaches,...)	Discard	High moisture fruits and vegetables can be contaminated by molds that penetrate into the food
Firm fruits and vegetables (cabbage, peppers, carrots,...)	Keep food but remove at least 1 inch around and below the moldy section*	Mold spots can be removed. Firm fruit have low humidity and molds cannot easily penetrate into the food
Hard pressed cheeses (produced without molds)	Keep food but remove at least 1 inch around and below the moldy section*	Molds generally cannot penetrate into the food
Mold ripened cheeses (Brie, Roquefort, Camembert,...)	Discard cheeses contaminated by molds other than those used for manufacture	Molds other than those used for manufacture might produce mycotoxins
Dry cured meats and hams	Keep food. Scrap the mold off the surface	Presence of fungi on the surface is normal at ambient temperature

*Avoid mold cross-contamination with the knife when trimming foods into porous products. Jams should also be discarded as mycotoxins can be produced by the typical contaminating mold species. Recommendations concerning fruits and vegetables differ depending on firmness of the products. Soft products should be discarded, whereas hard products could be trimmed as molds will not penetrate deeply. In contrast to jam, the potential production and mycotoxin migration in firm products was omitted. Hard and semi-hard cheeses could be trimmed one inch around and below the molded zone. However, in European countries where the metric system is enforced, is it really possible to directly substitute one inch for one centimeter? In fact, the fraction of the product that should be trimmed will depend on the mold lesion size and it can be expected that more mycotoxins will be produced in the case of a larger lesion. This means that there is a higher potential for mycotoxin migration to occur and to go deeper into the food matrix.

3. Mycotoxin Migration in Foods

Studies on mycotoxin migration in foods are listed in Table 2. According to the USDA (2013), most of these moldy foods, i.e., bread, jams, crème fraiche, tomato purees, should be discarded. Concerning aflatoxins produced by *Aspergillus flavus*, they remained concentrated and mainly close to the mycelium that penetrated within the bread. Mycelial penetration can be invisible to consumers, especially for white bread, due to the uncolored mycelium. Other studies that concerned the majority of food categories, i.e., jams, creams, tomato purees and cheeses, quantified mycotoxin migration in the molded foods for two weeks incubation at 20°C (Olsen et al. 2017). In these conditions, the lesion size depends on the food products but also the mold species. In fact, the majority of these studies quantified mycotoxins within and outside the lesions as a function of time, which is a widely used microbiological approach. However, this approach is not suitable for consumers, as they usually do not know for how long the foods were stored. The only available information for a given consumer is the lesion size and aspect. Studies on mycotoxin migration should therefore be based on lesion size and aspect that can also be linked to storage time. Many of the described studies have also been carried out at 20°C, which is a correct value for room temperature food storage, but lower temperatures should also be chosen when appropriate, as many foods, such as hard and semi-hard cheeses, may be stored in the fridge, or fruits and vegetables in the cellar. In this sense, a semi-hard French cheese was recently studied using a worst case scenario approach (Coton et al. 2019) as described in the last section of this chapter. This was done to improve the experimental conditions used for mycotoxin migration experiments and to mimic as closely as possible consumer habits when it comes to food storage. The overall goal is to be able to provide simple consumer recommendations to prevent mycotoxin exposure.

About 2/3 of the studies listed were linked to *Penicillium* species while the others dealt with *Aspergillus* and *Fusarium* species. The latter species are more often encountered in the case of mold-spoiled vegetables. In most cases,

Table 2: List of studies on mycotoxin migration in various foods

Food	Fungal species	Mycotoxin(s) studied	References
Whole wheat bread	*Aspergillus parasiticus*	Aflatoxins	Reiss 1981
	Aspergillus ochraceus	OTA	
	Penicillium chrysogenum	CIT	Frank 1968
	Aspergillus flavus	Aflatoxins	Rychlik and Schieberle 2001
Bread	*Penicillium expansum*	PAT	Frank 1968
	Aspergillus flavus	Aflatoxins	
Apple jam	*Penicillium crustosum*	Penitrem A, ROQC	Olsen et al. 2019
	Penicillium roqueforti	ROQC	
	Penicillium expansum	PAT, ROQC	Olsen et al. 2017
Blueberry jam	*Penicillium expansum*	PAT, ROQC	
	Penicillium crustosum	Penitrem A, ROQC	
Reduced fat crème fraiche (15% fat)	*Penicillium expansum*	CIT, PAT	Olsen et al. 2019
	Penicillium roqueforti	ROQC	Olsen et al. 2017
	Aspergillus versicolor	CPA	
Crème fraiche (34% fat)	*Penicillium roqueforti*	ROQC	
	Penicillium crustosum	Penitrem A, ROQC	
Gouda cheese	*Penicillium commune*	CPA, Rugulovasin A	
	Penicillium crustosum	Penitrem A, ROQC	
	Penicillium roqueforti	ROQC	
	Aspergillus versicolor	CPA	

(Contd.)

Table 2: *(Contd.)*

Food	Fungal species	Mycotoxin(s) studied	References
Comté cheese	*Penicillium verrucosum*	OTA, CIT	Coton et al. 2019
Tilsit cheese	*Aspergillus flavus*	Aflatoxins	Frank 1968
Peppers	*Fusarium proliferatum*	Fumonisins, Beauverin	Monbaliu et al. 2010
Tomatoes	*Penicillium expansum*	PAT	Rychlik and Schieberle 2001
Tomato purée	*Penicillium crustosum*	Penitrem A, ROQC	Olsen et al. 2019
Apples	*Penicillium expansum*	PAT	Rychlik and Schieberle 2001, Marín et al. 2006, Bandoh et al. 2009, Coton et al. 2020
Dry cured ham	*Aspergillus ochraceus* *Penicillium* spp	OTA OTA, CPA,…	Escher et al. 1973 Peromingo et al. 2019
Dry cured sausages	*Penicillium* spp	OTA, CPA,…	Peromingo et al. 2019
Italian-type salami	*Aspergillus westerdijkiae*	OTA	Parussolo et al. 2019

regulated mycotoxins for which Europe set maximum levels in foodstuffs (EC 1881/2006)) were targeted such as OTA, aflatoxins, patulin. However, a few other mycotoxins for which no regulation exists were also included in some cases (e.g. citrinin, penitrem A, roquefortine C, and cyclopiazonic acid). Multiple studies have dealt with patulin in apples (Rychlik and Schieberle 2001, Marín et al. 2006, Bandoh et al. 2009) but the main objectives were to assess the impact of sorting rotten fruits, using various chemical processes (Bandoh et al. 2009, Baert et al. 2012), to limit high patulin concentrations in apple juice. More recently, patulin was quantified in *Penicillium expansum* molded Golden delicious apples for different lesion sizes and after room temperature and cold storage conditions to mimic consumer habits (Coton et al. 2020); this was done using the worst case scenario approach described in the last section of this chapter. In this study, patulin was quantified at highest levels within lesions but still diffused around the moldy zones and according to storage temperature.

4. Food Products to be Considered

Obviously, food should not be contaminated by pathogenic bacteria during manufacture, transportation or retail, and spoilage should only occur at consumer's homes if conditions are favorable. Any highly perishable food product that should be kept refrigerated are excluded from these studies. They include high moisture foods that may be prone to mold growth, but also bacterial development including pathogens which cannot be detected by the consumer. These products are also characterized by a "best before date" on packaging, e.g. cold cuts, bacon, hot-dogs, poultry, ready-to-eat meals, cooked pasta, cooked grains, soft-cheeses, yogurts, fresh creams/sour creams.

4.1 Fruits and Vegetables

USDA (2013) segregated between firm fruits and vegetables with low humidity such as cabbage, peppers and carrots, as mold spots can easily be removed since they cannot penetrate into the foods, and soft fruits and vegetables such as tomatoes, cucumbers and peaches that should be directly discarded as molds will penetrate into these high moisture products. However, this difference does not seem precise enough as for example: is water melon a soft or firm fruit? Also, it was not clear either whether recommendations were based on firmness or moisture content. Recommendations also suggested that fruits and vegetables are only contaminated by molds on their surface. A more logical approach would be to classify fruits and vegetables based on their probability of being spoiled by mycotoxigenic molds. For example, *P. expansum* is known to be the causative agent of blue rot in pome fruits and it actually causes between 70 and 80% of decay in stored fruits (Viñas et al. 1993). *P. expansum* is also considered as the major patulin producer (Morales et al. 2007) and all strains produce this mycotoxin (Andersen et al. 2004) although at variable levels. In contrast to *P. expansum*, *Penicillium italicum* and

Penicillium digitatum are common contaminants of citrus fruits but have not been reported to produce mycotoxins. Accordingly, it can be recommended to remove mold spots from this kind of fruit and to eat the unrotten portion if the organoleptic properties stay acceptable. Fruits that are quite acidic may be studied because bacteria are more sensitive to low pH than molds. In contrast to fruits, more care should be taken when providing recommendations to consumers regarding moldy vegetables as they can also be contaminated by bacteria.

4.2　Low Moisture Foods

Low moisture foods, e.g., bread, bakery products, jams, hard and semi-hard cheeses have not been reported to be contaminated by pathogenic bacteria. They are characterized by a "best before date" written on the packaging and are usually stored at ambient temperatures, except for cheeses and according to countries. USDA recommends to discard bread because of the porous properties of this type of product. It could still be of interest to study mycotoxin migration in breads according to the moldy lesion size and visible aspect of the molded bread.

5.　Molds to be Considered

In temperate climates, *Penicillium* species are the main producers of mycotoxins in processed foods. The major mycotoxins produced by *Penicillium* species in cheese are listed in Table 3. In the case of Camembert and Roquefort cheeses, fungal ripening cultures are intentionally added to the cheeses during production and are well identified and have a very long history of safe use. For other cheeses, mycotoxin contamination can occur during cheese processing if a mycotoxigenic mold is present or even come from contaminated milk (i.e. aflatoxin M_1 (AFM1) contamination in cow milk). OTA and AFM1 are considered as the most toxic mycotoxins that can be detected in cheese. It should be noted that, contrary to OTA, AFM1 was not reported in Table 3 because its presence in cheese did not result from mold contaminated cheese but from aflatoxin B_1 contaminated cow feed. This mycotoxin is known to be further metabolized by cows into AFM1 that is secreted in their milk. According to different authors (Lund et al. 1995, Garnier et al. 2017), the major fungal contaminants of cheese are *Penicillium*, i.e., *P. commune, P. palitans, P. nalgiovense,* and *P. verrucosum. P. commune* and *P. roqueforti* are the most frequent contaminants of cheddar, although other species such as *P. chrysogenum, P. expansum, P. solitum, P. viridicatum* and *P. brevicompactum* can be found (Hocking 1994).

In hard cheeses from Norway, some *Penicillium* species such as *P. roqueforti, P. commune, P. palitans* and *P. solitum* can be found (Kure and Skaar 2000). *P. commune, P. roqueforti* and *P. verrucosum* are the most frequent in Turkish cheeses (Hayaloglu and Kirbag 2007). In Spain, cheeses are mostly contaminated by *Penicillium* species, i.e., *P. brevicompactum, P. granulatum* (=

P. glandicola), and *P. verrucosum* (Barrios et al. 1998). Finally, *P. commune, P. roqueforti, P. brevicompactum* and *P. verrucosum* are the major sources of the contamination of goat and sheep cheeses (Montagna et al. 2004).

P. verrucosum, which is the only OTA producer among the *Penicillium* species cited above and in Table 3, is probably responsible for OTA production in blue-veined and soft cheeses. It should be highlighted that some of the other *Penicillium* species described in cheeses are not necessarily mycotoxin producers in this food. For example, patulin production by *P. expansum* in apples is very well documented while no relationship exists to link this species to patulin in cheese. It should be pointed out that the different mycotoxins detected in cheeses were either produced during processing or were originally present in the milk. The objective of one of the case studies reported in the next section by Coton et al. (2019) was to quantify mycotoxins produced in a mold contaminated cheese during storage by the consumer, either by mimicking refrigerated conditions or room temperature storage.

Table 3: List of major mycotoxins produced by *Penicillium* species in cheese

Penicillium	*Mycotoxins*	*References*
P. brevicompactum	Mycophenolic acid	Frisvad and Filtenborg 1989, Frisvad and Samson 2004
P. camemberti	Cyclopiazonic acid	Hymery et al. 2014
P. chrysogenum	Roquefortine C, PR toxin	Frisvad and Samson 2004
P. commune	Cyclopiazonic acid	El-Banna et al. 1987, Polonelli et al. 1987, Frisvad and Filtenborg 1989
P. expansum	Patulin, citrinin	Harwig et al. 1973, Ciegler et al. 1977
P. glandicola	Penitrem A, patulin, roquefortine C	Frisvad and Samson 2004
P. palitans	Viridicatin	Ciegler and Hou 1970
P. roqueforti	PR toxin, Roquefortine C, mycophenolic acid, isofumigaclavine A, patulin, penicillic acid	Frisvad and Samson 2004, Nielsen et al. 2006, Hymery et al. 2014
P. verrucosum	OTA	Pitt and Hocking 2009
P. viridicatum	Naphtoquinones	Pitt and Hocking 2009

Penicillium species that can spoil food products other than cheeses reported in paragraph 2 and are listed in Table 4. Of course, many species can contaminate cheeses and other food products. In the case of dried cured meats and hams, the USDA (2013) recommended to scrap off the surface of these products as different fungal species can be present. However, in many cases, *P. nalgiovense* dominates and this species is deliberately inoculated onto the surface of dried cured sausages during ripening for technological interests. This species is not known to produce any mycotoxins and has a long history of

Table 4: List of major *Penicillium* species contaminating food products other than cheeses (from Pitt and Hocking 2009)

	Food products	Penicillium species
Fruits	Citrus fruits	*P. italicum, P. digitatum, P. ulaiense*
	Pomaceae	*P. expansum, P. solitum, P. purpurogenum*
	Stone fruits	*P. expansum*
	Solanaceae	Not reported
	Cucurbitaceae	*Penicillium* spp
	Berries	Not reported
	Figs	*Penicillium* spp
	Tropical fruits	*Penicillium* sp, *P. funiculosum, P. purpurogenum*
Vegetables	Peas, Beans	Not reported
	Onions	*P. purpurogenum, P. allii*
	Potatoes, Roots, Greens	Not reported
Intermediate moisture foods	Fresh pastas	*P. crustosum, P. solitum*
	Bakery products	*P. roqueforti, P. brevicompactum, P. chrysogenum, P. crustosum, P. glabrum, P. commune*
	Dry cured meats	*P. nordicum, P. chrysogenum, P. expansum, P. roqueforti, P. rugulosum, P. variabile, P. viridicatum, P. commune, P. solitum, P. olsonii, P. nalgiovense, P. brevicompactum, P. aurantiogriseum, P. chrysogenum*
Low moisture foods	Jams	*P. coryphilum*
	Dried fruits	*P. citrinum, Penicillium* spp

safe use. However, this species should not be confused with other *Penicillium* species such as *Penicillium nordicum* that is commonly identified on meat products and produces OTA. At present, no report exists on the contamination of dried cured meats and hams by *P. nordicum* at the consumer level and what levels of mycotoxins may be produced and potentially migrate into the foods.

In contrast to *Penicillium*, only a few links between food products and *Aspergillus* species were reported although some exceptions exist such as peanuts (*A. flavus*), grapes (*A. carbonarius*) and coffee (*A. ochraceus/A. westerdijkiae*). These species are also very common on many other substrates and both spoilage and mycotoxin production can occur (Hocking 2006). However, *Aspergillus* develop more rapidly, at higher temperatures and at lower water activities than *Penicillium*. Therefore, *Aspergillus* appear more adapted to contaminate foodstuffs in tropical countries. In this sense, it seems more relevant to study mycotoxin migration for mycotoxins produced by *Penicillium* rather than those produced by *Aspergillus* for food matrices stored by European consumers. Despite this fact, *Aspergillus parasiticus, A. flavus* and *A. ochraceus* have all been described as aflatoxin and OTA producers in bread (Table 2). *A. ochraceus* can also develop on hams and potentially produce OTA (Escher et al. 1973). *Aspergillus versicolor* and *A. flavus* have already been isolated from dairy products, but less frequently than *Penicillium* species (Frank 1968, Olsen et al. 2017). These authors indicated that the mycotoxin concentrations in these products were low. Recent studies (Olsen et al. 2019) concerning jams have also shown mycotoxin production by *Penicillium*, but also referred to aflatoxin production in jams inoculated with *A. parasiticus* (Pensalla et al. 1978)

Mycotoxins produced by *Fusarium* spp. are rather considered as field mycotoxins as this species is a frequent contaminant of cereal and grain crops. However, some species can produce mycotoxins in moldy vegetables during consumer food storage. Finally, *Alternaria alternata* can produce many mycotoxins such as tenuazonic acid, alternariol, alternariol monomethyl ether, altenuene, and altertoxins. These mycotoxins have been frequently studied due to their mutagenic activity, particularly that of altertoxin III, as its mutagenicity is approximately 10 times lower than that of aflatoxin B_1 (Bottalico and Logrieco 1998). *A. alternata* is a major pathogen of fresh tomatoes, but no studies have focused on the production and migration of mycotoxins during storage.

6. Mycotoxins to be Considered

According to Driehuis (2015), field and storage mycotoxins produced in cereals can be distinguished. The major toxinogenic molds capable of producing field-derived mycotoxins are different *Fusarium* species, *A. flavus*, *A. parasiticus* and *Claviceps* species. The most frequently occurring *Fusarium* mycotoxins are deoxynivalenol (DON), nivalenol, T2-toxin and HT2-toxin (all belonging to the trichothecene group of mycotoxins), zearalenone, fumonisins, enniatins and beauvericin. Though *Aspergillus* is often classified as a mold associated with mycotoxin production during storage of food commodities, it can infect crops in the field under favorable conditions, especially in subtropical and warm temperate climates. *A. flavus* and *A. parasiticus* are associated with aflatoxin production in many crops, including maize. *P. expansum* can also produce patulin in apples prior harvesting, and as such should not be considered as only a storage mycotoxin. Molds associated with spoilage and mycotoxin formation during grain storage belong to *Aspergillus* and *Penicillium* genera. The most relevant species are *A. flavus*, associated with aflatoxins, *A. ochraceus* and *P. verrucosum* associated with OTA, and *P. roqueforti* and *P. paneum*, associated with roquefortin C, mycophenolic acid, as well as a number of other mycotoxins.

By analogy to field and storage mycotoxins, the concept of mycotoxins at consumer stage has been explained in this chapter. It is clear that field mycotoxins cannot be considered as mycotoxins at consumer stage, but these mycotoxins are the same as storage mycotoxins as the same fungal species can be encountered. However, storage mycotoxins are usually produced in raw materials such as cereals, apples, peanuts or grapes while mycotoxins at consumer stage are produced in processed foods in consumer's homes.

7. Worst Case Scenario

In the case studies reported in the next paragraph, the worst case scenario was privileged to ensure food safety. First, it was assumed that consumers have poor knowledge about molds. For example, consumers cannot differentiate a non-toxigenic mold, such as *Mucor* or *Rhizopus*, from a toxigenic mold, such as some *Penicillium* species. It should also be underlined that many consumers believe *Penicillium* to be a harmless genus because of the well-known *Penicillium notatum* species that is an antibiotic producer (i.e. penicillin). Generally speaking, they are not aware of mycotoxins, their potential toxicity or their ability to diffuse into a food matrix. It was also assumed that the food product was contaminated by a toxigenic species even if its occurrence is not frequent in an everyday situation. Both ambient and cool storage (8°C) conditions were chosen to mimic consumer habits as well as to simulate a poorly regulated fridge. Different species and isolates were systematically tested for their ability to produce mycotoxin(s) *in situ* in the selected food products as strong matrix effects have already been described when it comes to mycotoxin production profiles. The studied mycotoxins were also selected according to their long-term toxicity which was evaluated by IARC (International Agency for Research on Cancer) based on carcinogenicity, i.e., group 1, carcinogenic to humans; group 2A, probably carcinogenic to humans; group 2B, possibly carcinogenic to humans; group 3, not classifiable as to its carcinogenicity to humans. Accordingly, case studies examined the migration of aflatoxins, OTA, citrinin and patulin in selected food products.

When studying mycotoxin migration, a standardized and robust method is necessary and should be based on a worst case scenario to help ensure food safety at the consumer level. In this sense, the choice of mycotoxigenic mold species is key and the highest mycotoxin or multi-mycotoxin producing strain should be selected and the most toxic mycotoxins studied. Food contamination should also mimic as closely as possible how a given species commonly contaminates a given food (via a lesion, surface contamination…) and the visual aspect of mold growth or lesion size should be monitored during storage. This data can then be linked to mycotoxin migration data, if relevant, and also be used to provide simple consumer recommendations based on robust scientific data. Many factors are known to affect mycotoxin production and migration into a food matrix and include food related factors (composition, humidity, pH,…), biological factors (fungal species and strain, growth, lesion size, mycotoxin properties…) and consumer level factors (storage conditions) (Coton and Dantigny 2019). In this context, consumer food storage habits should be privileged (room temperature or cold storage rather than optimal fungal growth conditions) and relative humidity kept constant to prevent drying or excessive changes in the physico-chemical properties of the food. Mycotoxin migration should be monitored according to the mold lesion size rather than only time, when possible, and mycotoxin content should be determined in increasingly wider or deeper fractions within, around and below the surface mycelium. This approach is based on a 3D mycotoxin migration hypothesis as described in the two case studies below. Finally, mold growth can also be monitored within the food product to determine whether mycotoxin migration and/or mycelium penetration into the food is occurring as mycotoxins can either remain near the fungal mycelium or migrate much further into the food product if conditions are favorable.

In this sense, a semi-hard French cheese was recently studied using this worst case scenario approach (Coton et al. 2019). To do so, different *Penicillium* species known to commonly contaminate cheeses were selected and their mycotoxin profiles were determined directly on cheese slices. A *P. verrucosum* strain produced the highest levels of two regulated mycotoxins, OTA and citrinin, and was therefore used for the migration study. After contaminating the cheese cubes surfaces with *P. verrucosum* spores (a suspension of 10^6 spores/mL were spread onto the surface of each cheese cube to mimic a typical contamination under the cheese packaging), they were incubated at ambient and in cold storage conditions to mimic consumer habits as stated above. Mycotoxins were then quantified in 2 mm depth-wise cheese slices, up to 2 cm in depth, and the visual aspect of fungal growth was monitored (white mycelium versus blue mold appearance). By the end of the study (28 days for room temperature or 42 days for cold storage), mycotoxins could be detected up to 1.6 cm in depth and the highest levels accumulated when a blue mold aspect was visible on the cheese surface. Based on these findings, trimming was considered to be acceptable provided that only a white mycelium appeared while in the case of a blue mold appearance, it was recommended to discard the cheese to prevent mycotoxin exposure (Coton et al. 2019).

More recently, the same approach was used to study patulin migration in *P. expansum* contaminated Golden delicious apples for different mold lesion sizes and after room temperature and cold storage (Coton et al. 2020). The highest patulin producing *P. expansum* strain was selected for the experiments and fungal induced lesions were monitored on and within apples up to 3 cm in size. To contaminate apples, a 10^6 spores/mL suspension was inoculated onto the surface of the apple just under the peel to induce the moldy lesion. Fungal induced lesions were measured daily until they reached 1, 2 and 3 cm in diameter. For each lesion size, 5 cm diameter cylinder plugs, with lesions centered, were removed and then the plug was cut into 1 cm sized depths up to 5 cm. For each fraction, a 1, 2, 3, 4 and 5 cm sized disk was cut to obtain increasingly sized rings in order to quantify mycotoxins based on a 3D migration hypothesis. Once the lesion size reached 3 cm, we considered that consumers would no longer eat the remaining portion of the apple as other fungal metabolites, such as geosmin, would be produced leaving a very unpleasant moldy taste. Patulin was systematically quantified at the highest levels within lesions but it also diffused 1 cm around these zones both in width and in depth for both storage temperatures. Mold induced lesions also appeared much faster at 20°C versus cold storage. Simple consumer recommendations were therefore proposed and at least a 1 cm zone around the necrosis area should be removed to avoid patulin exposure and cold storage used to delay fungal growth.

Experiments were also conducted with the same approach of worst case scenario on bread and jam.

In the case of bread, inoculations were carried out both on the inside sections and crust using 10^6 spores/mL suspension of isolates originating from a naturally molded sandwich bread or of *P. verrucosum* NRRL 5571 reference strain known to produce OTA and citrinin. The production of mycotoxins was observed for the different strains during contamination and growth was possible at temperatures over 20 °C but not less than 8 °C. The OTA levels found were higher than the maximum levels authorized for products derived from cereals (3 μg/kg) with colony sizes above 1 cm. In addition, a phenomenon of co-occurrence of citrinin and ochratoxin A was observed on inoculated bread. The production of mycotoxins was greater when molds developed on the crust of the sandwich bread. Mycotoxins were also found in the inner sections of the inoculated slice and on adjacent slices without macroscopic observation of the mold. However, mold presence was confirmed microscopically. These data therefore suggest that there was migration of the fungus rather than toxin migration as already observed in a previous study (Reiss 1981). This migration and the production of toxins was confirmed for up to 3 cm horizontally for the inner portion and 4 cm vertically on the crust, the adjacent sections were also affected by this migration. These initial results tend to confirm the fact that slices of sandwich bread should not be eaten if they become slightly moldy and that it is better to avoid the adjacent slices.

For jam, fig jam was chosen because of how sensitive this fruit is to contamination by *Aspergillus* and *Penicillium* species and their possible contamination by AFB1, OTA or citrinin (Buchanan et al. 1975, Hagagg et al. 2018). Inoculations were carried out with different isolates (*Penicillium* spp.) obtained from naturally molded fig jams during their simulated conservation (opening and sampling with a spoon during breakfasts) and reference strains of *A. flavus* (*A. flavus* NRRL 3518, AFB1 producer) and *P. verrucosum* (*P. verrucosum* NRRL 5571, producer of OTA and citrinin) with a 10^6 spores/mL spore suspension. Preliminary results confirmed that the toxinogenic molds that developed on jams during their storage were capable of producing mycotoxins. Refrigerated temperature (8 °C) inhibited the growth of *A. flavus* and AFB1 production but not OTA by *Penicillium* strains. A temperature above 20 °C, a_w above 0.95, reduced sugar concentration (by 30% compared to conventional concentrations) and incubation for more than 10 days would be favorable conditions for *Aspergillus* and *Penicillium* development and the subsequent production of AFB1 and OTA at levels risky for human health.

Even if there are no regulations for jams at present, the maximum values authorized on certain fruit matrices were reached or even exceeded. Mycotoxin production was observed up to 2 cm around and under the fungal colonies, but more in-depth analyses did not reveal any migration of the toxins which is likely due to too short incubation times. Recommendations would be to pay more attention to jams with reduced sugar content in unrefrigerated storage for mold colonies over 2 cm in diameter, which would indicate a development period of more than 10 days which is often associated with quantifiable mycotoxin levels and diffusion into the matrix.

8. Conclusions

Mycotoxins produced during storage of foods at consumer's homes were not studied until recently. However, this step should be included into the farm to fork approach. Much information on consumer behavior is still lacking when it comes to moldy foods. How are food products stored, how do consumers handle moldy food products? In addition, no report exists on which products are spoiled by molds at consumer's homes, which molds cause spoilage, and which mycotoxins are produced. Therefore, it is very difficult to provide consumers with simple and accurate recommendations. On the one hand, these recommendations should be based on very simple and limited observations such as the extent and color of the moldy section of the food. It should also be underlined that storage time is not a relevant observation, because consumers do not know *a priori* for how long the food product has been stored. On the other hand, mycotoxin production and accumulation in foods are very complex phenomena. The type and concentration of mycotoxins may depend on the food, mold species, but also the isolate. At present, recommendations based on the worst case scenario have been provided to consumers for some foods such as semi-hard cheese or apples, but more studies should be conducted on other foods and categories.

References

Andersen, B., Smedsgaard, J. and Frisvad, J.C. 2004. *Penicillium expansum*: Consistent production of patulin, chaetoglobosins, and other secondary metabolites in culture and their natural occurrence in fruit products. Journal of Agricultural and Food Chemistry 52: 2421–2428. doi.org/10.1021/jf035406k

Baert, K., Devlieghere, F., Amiri, A. and De Meulenaer, B. 2012. Evaluation of strategies for reducing patulin contamination of apples juice using a farm to fork risk assessment model. International Journal of Food Microbiology 154: 119–129. doi.org/10.1016/j.ijfoodmicro.2011.12.015

Bandoh, S., Takeuchi, M., Ohsawa, K., Higashihara, K., Kawamoto, Y. and Gotoa, T. 2009. Patulin distribution in decayed apple and its reduction. International

Biodeterioration and Biodegradation 63: 379–382. doi.org/10.1016/j. ibiod.2008.10.010

Barrios, M., Medina, L. and Lopez, M. 1998. Fungal biota isolated from Spanish cheeses. Journal of Food Safety 18: 151–157. doi.org/10.1111/j.1745-4565.1998. tb00210.x

Bottalico, A. and Logrieco, A. 1998. Toxigenic *Alternaria* species of economic importance. pp. 65–108. *In:* Sinha, K.K. and Bhatnagar, D. (eds.). Mycotoxins in Agriculture. Marcel Dekker, New York.

Buchanan, J.R., Sommer, N.F. and Fortlage, R.J. 1975. *Aspergillus flavus* infection and aflatoxin production in fig fruits. Applied Microbiology 30(2): 238–241.

Ciegler, A. and Hou, C.T. 1970. Isolation of viridicatin from *Penicillium palitans*. Archives für Mikrobiology. 75: 261–267.

Ciegler, A., Vesonder, R.F. and Jackson, L.K. 1977. Production and biological activity of patulin and citrinin from *Penicillium expansum*. Applied and Environmental Microbiology 33: 1004–1006.

Coton, M. and Dantigny, P. 2019. Mycotoxin migration in moldy foods. Current Opinion in Food Science 29: 88–93. doi.org/10.1016/j.cofs.2019.08.007

Coton, M., Auffret, A., Poirier, E., Debaets, S., Coton, E. and Dantigny, P. 2019. Production and migration of ochratoxin-A and citrinin in Comté cheese by an isolate of *Penicillium verrucosum* selected among *Penicillium* spp. mycotoxin producers in YES medium. Food Microbiology 82: 551–559. doi.org/10.1016/j. fm.2019.03.026

Coton, M., Bregier, T., Poirier, E., Debaets, S., Arnich, N., Coton, E. et al. 2020. Production and migration of patulin in *Penicillium expansum* molded apples during cold and ambient storage. International Journal of Food Microbiology 313: 108377. doi.org/ 10.1016/j.ijfoodmicro.2019.108377

Driehuis, F. 2015. Mycotoxins in high moisture grain silages and ensiled grain by-products. 17th International Silage Conference. Piracicaba, Sao Paulo, Brazil, 1-3 July. https://www.researchgate.net/publication/307907385_Mycotoxins_ in_high_moisture_grain_silages_and_ensiled_grain_by-products. Accessed 03/12/2019.

El-Banna, A.A., Pitt, J.I. and Leistner, L. 1987. Production of mycotoxins by *Penicillium* species. Systematic and Applied Microbiology 10: 42–46.

Escher, F.E., Koehler, P.E. and Ayres, J.C. 1973. Production of ochratoxins-A and B on country cured ham. Applied Microbiology 26: 27–30.

European Commission Regulation (EC) No 1881/2006 of 19 December 2006 setting maximum levels for certain contaminants in foodstuffs.

Frank, H.K. 1968. Diffusion of aflatoxins in foodstuffs. Journal of Food Science 33: 98–100.

Frisvad, J.C. and Filtenborg, O. 1989. Terverticillate *Penicillia*: Chemotaxonomy and mycotoxin production. Mycologia 81: 837–861. doi.org/10.2307/3760103

Frisvad, J.C. and Samson, R.A. 2004. Polyphasic taxonomy of *Penicillium* subgenus *Penicillium*. A guide to identification of food and air-borne terverticillate *Penicillia* and their mycotoxins. Studies in Mycology 49: 1–173.

Garnier, L., Valence, F., Pawtowski, A., Auhustsinava-Galerne, L., Frotté, N., Baroncelli, R. et al. 2017. Diversity of spoilage fungi associated with various French dairy products. International Journal of Food Microbiology 241: 191–197. doi.org/10.1016/j.ijfoodmicro.2016.10.026

Hagagg, L.F., Abdel, E.H.A.H.M. and Awany, N.M. 2018. Identification of fungi and detection of mycotoxins associated with infected fig fruits. Journal of Applied Plant Protection 7(1): 1–10. doi.org/10.3390/toxins11060322

Harwig, J., Chen, Y.K., Kennedy, B.P.C. and Scott, P.M. 1973. Occurrence of patulin and patulin-producing strains of *Penicillium expansum* in natural rots of apple in Canada. Canadian Institute of Food Science and Technology 6: 22–25.

Hayaloglu, A.A. and Kirbag, S. 2007. Microbial quality and presence of moulds in Kuflu cheese. International Journal of Food Microbiology 115: 376–380. doi. org/10.1016/j.ijfoodmicro.2006.12.002

Hocking, A.D. 1994. Fungal spoilage of high-fat foods. Food Australia 46: 30–33.

Hocking, A.D. 2006. *Aspergillus* and related teleomorphs, 736 pp. *In:* Clive de Blackburn (eds.). Food Spoilage Microorganisms. Woodhead Publishing.

Hymery, N., Vasseur, V., Coton, M., Mounier, J., Jany, J.-L., Barbier, G. et al. 2014. Filamentous fungi and mycotoxins in cheese: A review. Comprehensive Reviews in Food Science and Food Safety 13: 437–456. doi.org/10.1111/1541–4337.12069

Kure, C.F. and Skaar, I. 2000. Mould growth on the Norwegian semi-hard cheeses Norvegia and Jarlsberg. International Journal of Food Microbiology 62: 133–137. doi.org/10.1016/s0168–1605(00)00384–6

Lund, F., Filtenborg, O. and Frisvad, J.C. 1995. Associated mycoflora of cheese. Food Microbiology 12: 173–180.

Marín, S., Morales, H., Hassan, A., Ramos, A.J. and Sanchis, V. 2006. Patulin distribution in the tissue of *Penicillium expansum*-contaminated Fuji and Golden apples. Food Additives and Contaminants 23: 1316–1322. doi. org/10.1080/02652030600887610

Monbaliu, S., Van Poucke, K., Heungens, K., Van Peteghem, C. and de Saeger, S. 2010. Production and migration of mycotoxins in sweet pepper analyzed by multimycotoxin LC-MS/MS. Journal of Agricultural and Food Chemistry 58: 10475–10479. doi.org/10.1021/jf102722k

Morales, H., Sanchis, V., Rovira, A., Ramos, A.J. and Marín, S. 2007. Patulin accumulation in apples during postharvest: Effect of controlled atmospheres and fungicide treatments. Food Control 18: 1443–1448. doi.org/10.1016/j. foodcont.2006.10.008

Montagna, M.T., Santacroce, M.P., Spilotros, G., Napoli, C., Minervini, F., Papa, A. et al. 2004. Investigation of fungal contamination in sheep and goat cheeses in southern Italy. Mycopathologia 158: 245–249. doi.org/10.1023/b:myco.0000041897.17673.2c

Nielsen, K.F., Sumarah, M.W., Frisvad, J.C. and Miller, J.D. 2006. Production of metabolites from the *Penicillium roqueforti* complex. Journal of Agriculture and Food Chemistry 54: 3756–3763. doi.org/10.1021/jf060114f

Olsen, M., Gidlund, A. and Sulyok, M. 2017. Experimental mould growth and mycotoxin diffusion in different food items. World Mycotoxin Journal 10: 153–161. doi.org/10.3920/WMJ2016.2163

Olsen, M., Lindqvist, R., Bakeeva, A., Leong, S.L. and Sulyok, M. 2019. Distribution of mycotoxins produced by *Penicillium* spp. inoculated in apple jam and crème fraiche during chilled storage. International Journal of Food Microbiology 292: 13–20. doi.org/10.1016/j.ijfoodmicro.2018.12.003

PAHO, 2019. WHO "Golden Rules" for Safe Food Preparation. https://www.paho.org/disasters/index.php?option=com_content&view=article&id=552:who-golden-rules-for-safe-food-preparation&Itemid=0&lang=en. Accessed 12/11/2019.

Parussolo, G., Oliveira, M.S., Garcia, M.V., Bernardi, A.O., Lemos, J.G., Stefanello, A. et al. 2019. Ochratoxin-A production by *Aspergillus westerdijkiae* in Italian-type salami. Food Microbiology 83: 134–140. doi.org/10.1016/j.fm.2019.05.007

Pensala, O., Niskanen, A. and Lindroth, S. 1978. Aflatoxin production in black currant, blueberry and strawberry jams. Journal of Food Protection 41: 344–347.

Peromingo, B., Sulyok, M., Lemmens, M., Rodriguez, A. and Rodriguez, M. 2019. Diffusion of mycotoxins and secondary metabolites in dry-cured meat products. Food Contaminants 101: 144–150.

Pitt, J.I. and Hocking, A.D. 2009. Fungi and Food Spoilage 3rd ed. Springer, Dordrecht, Germany, 519 pp.

Polonelli, L., Morace, G., Rossi, R., Castagnola, M. and Frisvad, J.C. 1987. Antigenic characterization of *Penicillium camemberti* and related common cheese contaminants. Applied Environmental and Microbiology 53: 872–878.

Quoi dans mon assiette (2016). Peut-on manger des aliments avec des moisissures sans risque? Lesquels jeter? https://quoidansmonassiette.fr/peut-on-manger-des-aliments-moisis-moisissures-pourris-sans-risque/ Accessed 06/20/2018.

Reiss, J. 1981. Studies on the ability of mycotoxins to diffuse in bread. XV. Mycotoxins in foodstuffs. European Journal of Microbiology and Biotechnology 12: 239–241.

Rychlik, M. and Schieberle, P. 2001. Model studies on the diffusion behavior of the mycotoxin patulin in apples, tomatoes, and wheat bread. European Food Research and Technology 212: 274–278. doi.org/10.1007/s002170000255

USDA, 2013. Molds on food: Are they dangerous? https://www.fsis.usda.gov/wps/wcm/connect/a87cdc2c-6ddd-49f0-bd1f-393086742e68/Molds_on_Food.pdf?MOD=AJPERES. Accessed 06/20/2018.

Viñas, I., Vela, E. and Sanchis, V. 1993. Capacidad productora de patulina de cepas Penicillium expansum procedentes de centrales hortofructicolas de Lleida. Revista Iberoamericana de Micologia 10: 30–32.

Biodiversity of Aflatoxigenic *Aspergillus* section *Flavi* Species According to Food Matrices and Geographic Areas

Carvajal-Campos Amaranta[1], Manizan Ama Lethicia[2], Didier Montet[3,4], Sophie Lorber[5], Olivier Puel[5] and Catherine Brabet[3,4*]

[1] Université de Montpellier, Faculté des Sciences, Campus Triolet, Place Eugène Bataillon, CC 023, 34095 Montpellier Cedex 5, France
[2] Laboratoire de Biotechnologie et Microbiologie des Aliments, UFR des Sciences et Technologie des Aliments, Université Nangui Abrogoua, 02 BP 801 Abidjan 02, Côte d'Ivoire
[3] CIRAD, UMR QualiSud, 73 rue Jean-François Breton, 34398 Montpellier Cedex 5, France
[4] Qualisud, Université Montpellier, CIRAD, Montpellier SupAgro, Université d'Avignon, Université de La Réunion, Montpellier, France
[5] Toxalim (Research Centre in Food Toxicology), Université de Toulouse, INRAE, ENVT, INP-Purpan, UPS, Toulouse, France

1. Introduction

Aflatoxins (AFs) are polyketide-derived metabolites produced by fungi on a wide range of crops (cereals, oilseeds, tree nuts, spices, dried fruits, etc.), both in the field and the post-harvest. As chemical stable molecules, resistant to conventional thermal or technological processes, they will pass through the whole food or feed supply chain to the final processed products (Kumar et al. 2017, Pankaj et al. 2018). There are more than 20 known AFs and derivatives, yet the most hazardous AFs include the four naturally-occurring AFB1, AFB2, AFG1 and AFG2, and the hydroxylated metabolites of AFB1 and AFB2, AFM1 and AFM2, which are produced through biotransformation in the liver and excreted in the milk of humans and mammals (Kumar et al. 2017). AFs are known for their high acute and chronic toxicity to both humans and animals, and are considered among the most dangerous mycotoxins with carcinogenic,

*Corresponding author: catherine.brabet@cirad.fr

hepatotoxic, immunotoxic, teratogenic and mutagenic effects (Kowalska et al. 2017). AFB1 is the most prevalent and toxic aflatoxin, and is classified as a Group 1 human carcinogens (IARC 2012). The intake of high amounts of AFs can cause acute intoxication (aflatoxicosis) associated with jaundice, vomiting, hemorrhages, abdominal pain, acute liver failure, problems with absorption of nutrients, and can be lethal (IARC 2015). Reported outbreaks in western India (1974) and in Kenya (2004) caused the death of 106 and 125 people, respectively (Lewis et al. 2005, Azziz-Baumgartner et al. 2005). Chronic exposure to low levels of AFs is associated with high risk of hepatocellular carcinoma (HCC), immunosuppression, teratogenic and mutagenic effects, reduction of nutrient absorption, child stunting, detrimental effects in the endocrinal system, and liver failure (Stack and Carlson 2003, Turner et al. 2005, Bbosa et al. 2013). More than five billion people worldwide are at risk of chronic exposure to AFs through contaminated foods (Wu and Glucu 2012). Humans' exposure to AFs mainly occurs by direct intake of contaminated foods of vegetal or animal origin (Bhat et al. 2010, IARC 2015). Hence, the presence of AFs in foodstuff and feedstuff is a public health issue associated with detrimental effects in economy. In addition, commodities are generally contaminated by several mycotoxins, and this co-occurrence may result in a greater toxicity to humans caused by the possible additive or synergistic effects of these compounds (Grenier and Oswald 2011). The main aflatoxin-producing fungi belong to *Aspergillus* section *Flavi* of the *Circumdati* subgenus, and only five AF producers do not belong to this section, *A. ochraceoroseus*, *A. rambellii* (*A.* section *Ochraceorosei*), *A. astellatus*, *A. olivicola*, and *A. venezuelensis* (*A.* section *Nidulantes*) (Varga et al. 2015). The aim of the present chapter is to describe the characteristics of the aflatoxigenic *Aspergillus* section *Flavi* species isolated from food or feed, and to address the biodiversity according to food matrices and geographic areas.

2. Characteristics of the Aflatoxigenic *Aspergillus* section *Flavi* Species

The major producers of AFs worldwide are *Aspergillus flavus* and *A. parasiticus* and for a while they were recognized as the only AF producers, and later, diversity surveys pointed *A. nomius* as the third main AF producer for its frequency in spoiled food (Kurtzman et al. 1987, Perrone et al. 2014). However, the implementation of polyphasic species identification showed a more complex story. Over the last two decades, this approach demonstrated a high biodiversity and plasticity within the section *Flavi* (Varga et al. 2011, Soares et al. 2012, Taniwaki et al. 2012, Frisvad et al. 2019). Currently this section encloses 34 species, from which 19 are aflatoxigenic (Frisvad et al. 2019).

2.1 Morphological Characteristics

Macro- and micro-morphological characters of the 19 aflatoxigenic *Aspergillus* section *Flavi* species on Malt Extract Agar (MEA) and Czapek Yeast Agar

(CYA) after seven days of incubation, in the dark at 25°C are summarized in Table 1. These two media are recommended as standard for *Aspergillus* and commonly used in taxonomic studies on this genus (Samson et al. 2014). Most aflatoxigenic *Aspergillus* species grow well on MEA and CYA at 25°C attaining colony diameter of more than 50 mm after seven days; *A. austwickii* is the slowest growing species (colony diameter MEA: 45–47 mm, CYA: 4648 mm). Colony surface is mostly deep, floccose, rarely plane, velvety or powdery. Most species sporulate and produce sclerotia with the exception of *A. pipericola* (no sporulation) and *A. arachidicola, A. luteovirescens, A. novoparasiticus, A. pseudocaelatus* and *A. pseudonomius* (no sclerotia). Conidia are mainly in shades of (dark) yellow-green, but also in shades of brown. Conidial heads are uniseriate with smooth or rough conidia, biseriate with rough conidia, or uniseriate to biseriate with smooth or slightly rough, smooth to rough or rough conidia. Sclerotia size varies between large (400–2000 µm), intermediate (*A. sergii*: 513-551 µm) and small (< 400 µm). Sclerotia morphology can be globose to ellipsoidal, and they become dark brown or black with time (Varga et al. 2011, Frisvad et al. 2019). Morphological characters are useful to differentiate some aflatoxigenic species within section *Flavi,* but for closely related species a proper description based only on these characters is a challenge. For this reason, a polyphasic approach that includes the morphological, chemical (mycotoxins and other extrolite production) and molecular characteristics is necessary to identify and characterize the *Aspergillus* section *Flavi* species (Varga et al. 2011).

2.2 Mycotoxins and Other Extrolites

Aflatoxigenic species from section *Flavi* produce an ample spectrum of secondary metabolites, besides of AFs some important and emergent mycotoxins include cyclopiazonic acid (CPA), tenuazonic acid and aflatrems (Table 2). From the nineteen aflatoxigenic species, three produce only AFB (AFB1 and AFB2: *A. flavus, A. pseudotamarii*; AFB1: *A. togoensis*), the other species produce both AFB (B1, B2) and AFG (G1, G2). *A. flavus* is the main AFB1 producer across the world and therefore the best known species from *Flavi* section, its AF production potential varies from non-aflatoxigenic to aflatoxigenic strains, with a high incidence of non-aflatoxigenic strains (60–70%) (Varga et al. 2011). AF production potential is not associated with virulence or competitive ability during crop infection. This plasticity and the lack of production of AFG are driven by several genetic differences, including single polymorphisms and large deletions in the AF biosynthesis gene cluster (Ehrlich 2004, Chang and Ehrlich 2010). *A. flavus* L-morphotype generally presents a 0.8 kb deletion in the CypA/NorB (*AflU/AflF*) region of the aflatoxin biosynthetic gene cluster (BGC), whereas *A. flavus* S-morphotype present of 1.5 kb (Ehrlich and Yu 2010). The CYPA protein encoded by *AflU* is required for AFG production (Ehrlich et al. 2004). A third group, gathering the strains responsible of the lethal outbreak in Kenya, was highlighted with a 2.2 kb deletion (Probst et al. 2012). AF production by *A. flavus* L-morphotype isolates is variable, while *A. flavus* S-morphotype isolates produce high quantities

Table 1: Morphological characteristics of aflatoxigenic *Aspergillus* section *Flavi* species on MEA and CYA after seven days of incubation in the dark at 25°C

Aflatoxigenic Aspergillus species	Medium	Colony diameter (mm)	Colony surface and conidial color	Sclerotia (μm)	Conidial head morphology and conidial surface	References
A. aflatoxiformans*	MEA	47–50	Colonies moderately deep, mycelium floccose and white. Conidia yellow-green, moderately dense on MEA and sparse on CYA	100–250	Uniseriate Smooth	Frisvad et al. 2019
	CYA	50–51				
A. arachidicola	MEA	60–65	Colonies velvety. Conidia olive to olive brown, abundant	Not observed	Uniseriate or biseriate Rough	Pildain et al. 2008
	CYA					
A. austwickii	MEA	45–47	Colonies moderately deep, mycelium floccose and white. Conidia yellow-green, moderately dense on CYA	100–300	Uniseriate Smooth	Frisvad et al. 2019
	CYA	46–48				
A. flavus	MEA	50–70	Colonies similar to those on CYA although usually less dense	<400 (S-type) 400–800 (L-type) or not observed	Typically biseriate, rare uniseriate Finely rough, rarely smooth	Pitt and Hocking 2009, Frisvad et al. 2019
	CYA	60–70	Colonies plane, sparse to moderately dense, velutinous in marginal areas at least, often floccose centrally, rare deep; mycelium white. Conidia greyish green, yellow green or olive yellow, sometimes yellow			
A. korhogoensis*	MEA	37–60	Colonies deeply floccose with a dominant white aerial mycelium. Conidia dull yellowish green	<400	Uniseriate or biseriate Smooth or slightly rough	Carvajal-Campos et al. 2017
	CYA	57–80				

Species	Medium	Temp	Colony characteristics	Sclerotia	Seriation	Reference
A. luteovirescens*	MEA	65	Colonies loose and deep, yellow-green becoming darker at maturity	Generally not observed, when present large	Mostly biseriate	Peterson et al. 2001, Frisvad et al. 2019
A. minisclerotigenes	MEA	60–70	Colonies floccose, white mycelium Conidia light-greyish green, sparse	150–300	Normally biseriate, rare uniseriate Smooth to rough	Pildain et al. 2008
	CYA		Colonies velvety Conidia light-greyish green			
A. mottae	MEA	>70	Colonies plane, mycelium white Conidia yellow-green, scare	249–371	Normally biseriate, rare uniseriate Smooth to finely rough	Soares et al. 2012
	CYA					
A. nomius	MEA	40–70	Colonies mostly floccose. Conidia green	< 500	Mostly biseriate Rough	Kurtzman et al. 1987, Doster et al. 2009
	CYA	52–60	Colonies greyish green			
A. novoparasiticus	MEA	56 – 60	Colonies powdery Greenish-yellow to olive	Not observed	Usually uniseriate, rarely biseriate	Gonçalves et al. 2012
	CYA	58 – 63				
A. parasiticus	MEA	50–65	Colonies similar to those on CYA but usually less dense	400–800	Mostly uniseriate Rough	Pitt and Hocking 2009
	CYA	50–70	Colonies plane, low, dense and velutinous, mycelium white Conidia dark yellowish green			
A. pipericola	MEA	61–72	Colonies moderately deep, mycelium floccose and white	75–250	Biseriate Rough	Frisvad et al. 2019
	CYA	58–72	Sporulation absent			

(Contd.)

Table 1: *(Contd.)*

Aflatoxigenic Aspergillus species	Medium	Colony diameter (mm)	Colony surface and conidial color	Sclerotia (μm)	Conidial head morphology and conidial surface	References
A. pseudocaelatus	MEA	60–65	Colonies velvety. Conidia olive to olive brown, abundant	Not observed	Uniseriate or biseriate Rough	Varga et al. 2011
	CYA					
A. pseudonomius	MEA	60–65	Colonies floccose with dominant aerial mycelium. Poor sporulation	Not observed	Uniseriate Rough	Varga et al. 2011
	CYA					
A. pseudotamarii	MEA	60–70	Colonies mostly floccose. Conidia olive green	1000–2000	Biseriate Rough	Ito et al. 2001
A. sergii	MEA	55	Colonies plane, velvety and dense. Conidia in a uniform, dense layer but sparse in the areas of sclerotium production, light green	513–551	Uniseriate Rough	Soares et al. 2012
	CYA					
A. texensis	CYA	71	Colonies velvety. mycelium white. Conidia yellow-green, sparse	130–300	-	Singh et al. 2018
A. togoensis	MEA	>50	Conidia yellow-brown to orange-brown	Large sclerotia	Biseriate	Samson and Seifert 1985, Frisvad et al. 2019
	CYA	>50				
A. transmontanensis	MEA	55–57	Colonies similar to growth on CYA with conidial heads more dense and floccose	458–609	Mostly uniseriate Rough	Soares et al. 2012
	CYA		Colonies dense and velutinous. Conidia in a uniform, dense layer but sparse in the areas of sclerotium production, dark yellow-green			

* A. parvisclerotigenus, A. bombycis and A. korhogoensis synonimyzed as A. aflatoxiformans, A. luteovirescens and A. cerealis, respectively, by Frisvad et al. 2019.

Table 2: Production of mycotoxins and other secondary metabolites by aflatoxigenic *Aspergillus* section *Flavi* species

Aflatoxigenic *Aspergillus* species	Aflatoxins	Other mycotoxins	Other secondary metabolites	References
A. aflatoxiformans	AFB1 AFB2 AFG1 AFG2	Aflatrems, cyclopiazonic acid, versicolorins	Aspergillic acid, aflavarins, aflavinines, aspirochlorin, kojic acid, paspaline, paspalinine, metabolite gfn	Frisvad et al. 2019
A. arachidicola		Versicolorins	Aspergillic acid, aspirochlorin, chrysogine, ditryptophenaline kojic acid, miyakimides, parasiticolide, "NO2" metabolite, parasitocolides	Pildain et al. 2008, Varga et al. 2011, Frisvad et al. 2019, Iamanaka et al. 2019
A. austwickii		Aflatrems, cyclopiazonic acid, versicolorins	Aflavarins, kojic acid, paspaline, paspalinine, metabolite gfn	Frisvad et al. 2019
A. korhogoensis		Aflatrems, cyclopiazonic acid, 3-O-methylsterigmatocystin, sterigmatocystin, versicolorins	Aspergillic acid, aflavarins, asparasones, asparasone A aflavinines, kojic acid, leporin B, norsolorinic acid, paspaline, paspalinine	Carvajal-Campos et al. 2017
A. luteovirescens		Tenuazonic acid (some strains)	Aspergillic acid, kojic acid For some strains: an altersolanol, chrysogine	Pildain et al. 2008, Varga et al. 2011, Frisvad et al. 2019
A. minisclerotigenes		Aflatrems, cyclopiazonic acid	Aspergillic acid, aflavarins, aflavinines, kojic acid, paspalinine, parasiticolides. For some strains: aflavazole	Pildain et al. 2008, Frisvad et al. 2019
A. mottae		Cyclopiazonic acid, 3-O-methylsterigmatocystin, versicolorins	Aspergillic acid, an aflavinin, kojic acid, parasiticol, paspalinine	Soares et al. 2012, Frisvad et al. 2019

Table 2: (*Contd.*)

Aflatoxigenic Aspergillus species	Aflatoxins	Other mycotoxins	Other secondary metabolites	References
A. nomius		3-O-methylsterigmatocystin, tenuazonic acid, versicolorins	Aspergillic acid, anominine, aspernomine, kojic acid, a miyakamide, pseurotin, parasiticol, paspaline, paspalinine, pseurotin A	Kurtzman et al. 1987, Frisvad et al. 2019
A. novoparasiticus		Tenuazonic acid	Aspergillic acid, aspirochlorin, ditryptophenaline, kojic acid, miyakamides, parasiticolide, crysogine, a tetracyclic compound, ustilagoidin	Goncalvez et al. 2012, Frisvad et al. 2019, Iamanaka et al. 2019
A. parasiticus			Aspergillic acid, kojic acid, parasperone, parasiticol, parasiticolide A and B	Frisvad et al. 2019
A. pipericola		Aflatrem, cyclopiazonic acid	Aflavinins, aflavarins, paspaline, paspalinine	Frisvad et al. 2019
A. pseudocaelatus		Cyclopiazonic acid, tenuazonic acid	Aspirochlorin, dirtyptophenaline, kojic acid	Varga et al. 2011, Frisvad et al. 2019
A. pseudonomius		Tenuazonic acid	Aspergillic acid, chrysogine, kojic acid, a miyakamide	Varga et al. 2011, Frisvad et al. 2019
A. sergii		Aflatrem, cyclopiazonic acid, 3-O-methylsterigmatocystin, sterigmatocystin, versicolorins,	Aspergillic acid, aflavazole, an aflavarin, aflavinines, asperfuran, kojic acid, leporin B, paspalinine	Soares et al. 2012, Frisvad et al. 2019
A. texensis		Cyclopiazonic acid	Aspergillic acid	Singh et al. 2018

Species			References
A. transmontanesis		Aspirochlorin, kojic acid, a miyakamide	Soares et al. 2012, Frisvad et al. 2019
A. flavus	AFB1 AFB2	Aflatrems (only in sclerotium producers), cyclopiazonic acid	Aspergillic acid, asperfuran, aspirochlorin, citreoisocoumarin, ditryptophenaline, flavimin, kojic acid, miyakamides, ustilaginoidin C, ustiloxin B. Only in sclerotium producers: aflavarins, aflavinines, paspaline and paspalinine	Amaike and Keller 2011, Umemura et al. 2013, Arroyo-Manzanares et al. 2015, Cary et al. 2015a,b, Frisvad et al. 2019
A. pseudotamarii		Cyclopiazonic acid, tenuazonic acid	Aflavinines, kojic acid, For some strains: an altersolanol, aspirochlorin, paspaline and paspalinine	Ito et al. 1999, Varga et al. 2011, 2015, Frisvad et al. 2019
A. togoensis	AFB1	Sterigmatocystin	A bisiderin, paspaline, paspalinine. For some strains: paxilline	Frisvad et al. 2019

of AFs (Ehrlich 2004). Recently *A. pseudotamarii* genome was sequenced and BGC analysis showed it lacks the first 600 base pairs in the *AflU* gene encoding for CYPA protein (Kjærbølling et al. 2020). *A. parasiticus*, the second main AF producer, also shows AF production plasticity, 3 to 6% of strains are considered non-aflatoxigenic (Chang et al. 2007). For the other AFBG species, the production seems to occur in all isolates.

Besides of AFs, some of their biosynthetic intermediates, like versicolorins, 3-O-methylsterigmatocystin and sterigmatocystin, are considered to be potentially toxic and reported in several aflatoxigenic species (Table 2). Though, these species are weak producers, as the intermediates are mainly transformed into AFs. Sterigmatocystin is considered an emerging mycotoxin and classified as 2B carcinogen for its potentially carcinogenic, mutagenic and teratogenic effects (IARC 2012, Bertuzzi et al. 2017). It is reported to be produced by *A. korhogoensis, A. sergii* and *A. togoensis*. The latter species is the only species from *Flavi* section capable to storage sterigmatocystin (Varga et al. 2015). Versicolorin A was shown as cytotoxic and genotoxic for lungs (Jakšić et al. 2012), renal and hepatic cells (Theumer et al. 2018). Recently, versicolorin A was reported as more cytotoxic than AFB1 for human intestinal cells. The toxic effects of 3-O-methylsterigmatocystin are less acute than AFs, sterigmatocystin and versicolorin A (Jakšić et al. 2012).

Cyclopiazonic acid is an important mycotoxin from section *Flavi* that has a synergetic effect with AFs. It might play a role in niche adaptation, providing an advantage in fungal fitness under specific environmental conditions (Georgianna et al. 2010). It is produced by *A. flavus* and closely related species, as well as *A. pseudocaelatus* and *A. pseudotamarii* (Table 2). Some species from section *Flavi* unable to produce CPA are believed to have deletions of the biosynthetic pathway silencing its production, such as *A. parasiticus, A. nomius* and their closely related species (Moore et al. 2016). Tenuazonic acid is a non-mutagenic mycotoxin that causes tremors, diarrhea, vomiting and hemorrhages. In section *Flavi*, it is reported in the clades *A. nomius* and *A. tamarii* (Table 3) and some strains of *A. novoparasiticus*. Aflatrem is another emergent mycotoxin, classified as a potent tremorgenic compound that causes neurological disorders. It seems that it might interfere with the release of neurotransmitters by receptors in the central and peripheral nervous systems (Zhang et al. 2004).

Two common extrolites produced by several aflatoxigenic species in section *Flavi* are kojic acid and aspergillic acid; the first organic acid is used in various industrial applications, especially in cosmetic and health care industries (Mohamad et al. 2010). Kojic acid is only not produced by *A. togoensis* and *A. pipericola*, whereas aspergillic acid is not produced by *A. austwickii, A. pipericola, A. pseudocaelatus, A. transmontanesis, A. pseudotamarii* and *A. togoensis* (Table 2). The lack of aspergillic acid production is characteristic of *A. tamarii* clade (Varga et al. 2011, Frisvad et al. 2019). Leporins are also found in several *Flavi* section species and are proven to have anti-insectan activity (Cary et al. 2015 a, b). Sclerotia extrolites include some mycelium extrolites and some unique extrolites; among them some reported in aflatoxigenic species are AFs and aflatrems (already mentioned above), aflavazole, aflavinines, aflavarins, anominine, aspernomine, and paspalines (Table 2) (Cary et al. 2015b, Frisvad et al. 2019). Aflavinines and aflavarins are frequently reported, especially in species closely related with *A. flavus*. They have anti-insectan activity (Gloer et al. 1988), and are suggested to have key ecological roles in species survival (TePaske and Gloer 1992, Cary et al. 2015b).

2.3 Molecular Markers and Phylogenetic Analyses

The inclusion of molecular markers for species identification in *Aspergillus* section *Flavi* is proven to be advantageous, especially for cryptic species identification. It helped not only to unmask diversity but also to understand the relationships within *Aspergillus* section *Flavi*. Genes tested with fruitful results include *benA*, *cmdA* and *RPB2*. These markers are advantageous because they have conserved and variable regions and are widely described in literature (primers and sequences) (Varga et al. 2011, Frisvad et al. 2019). Phylogenetic studies showed that most aflatoxigenic species are derived species. A recent study performed by Frisvad et al. (2019) showed that aflatoxigenic species are grouped in three main clusters: *A. flavus*-clade, *A. tamarii*-clade and *A. nomius*-clade (Table 3). *A. togoensis* is a basal species and is not considered as a food contaminant.

Species from *A. flavus*-clade can be sub-divided in two main clusters, one that includes species with *A. flavus* overlapping traits and the other that includes species with *A. parasiticus* overlapping traits, and *A. mottae* as its basal species (Table 3). *A. flavus* sub-clade species number has increased dramatically over the last decade, and as aforementioned, it is composed of cryptic species (overlapping morphological, genetic and chemotype traits) (Frisvad et al. 2005, 2019, Varga et al. 2011). Within this sub-group four clusters are formed: (1) *A. flavus*, (2) *A. minisclerotigenes* and *A. texensis*, (3) *A. pipericola* and (4) a cluster including *A. aflatoxiformans*, *A. korhogoensis* and *A. austwickii* (Varga et al. 2011, Carvajal-Campos et al. 2017, Singh et al. 2018, Frisvad et al. 2019). *A. oryzae*, considered as a domesticated variant of *A. flavus* for a long time, forms a subgroup based on the three above-mentioned genes. The origin of *A. oryzae* is controversial. A recent study based on 200 monocore genes supports an earlier hypothesis that it is closer to *A. aflatoxiformans* and *A. minisclerotigenes* (Kjaerbolling et al. 2020), thus, another study based on 82 genomes of *A. oryzae* strains used in the manufacture of Asian fermented foods confirmed that it formed a monophyletic cluster nested in *A. flavus* clade (Watarai et al. 2019). AFBG strains with small sclerotia are commonly misclassified as *A. flavus* S_{BG} morphotype; however phylogenetic surveys revealed that this classification includes several cryptic species related to *A. flavus*: *A. minisclerotigenes*, *A. texensis*, *A. aflatoxiformans*, *A. korhogoensis*, *A. pipericola* and *A. austwickii* (Pildain et al. 2008, Frisvad et al. 2005, 2019, Varga et al. 2011, Soares et al. 2012, Perrone et al. 2014, Carvajal-Campos et al. 2017, Singh et al. 2018, Singh and Cotty 2019). *A. parasiticus* sub-clade species are AFBG producers. Their phylogenetic relationships are more complicated to explain as they present slight differences depending on the molecular markers used. Anyhow, *A.*

Table 3: Main clades of aflatoxigenic *Aspergillus* section *Flavi*

Clade	Sub-clade	Species
A. flavus-clade	Group of species closely related to *A. flavus sensu stricto*	*A. flavus, A. aflatoxiformans, A. austwickii, A. korhogoensis, A. minisclerotigenes, A. pipericola, A. texensis*
	Group of species closely related to *A. parasiticus*	*A. parasiticus, A. arachidicola, A. novoparasiticus, A. sergii, A. transmontanensis*
	Basal species to the *A. flavus* and *A. parasiticus* groups	*A. mottae*
A. tamarii-clade	-	*A. tamarii, A. caelatus, A. pseudocaelatus, A. pseudotamarii*
A. nomius-clade	-	*A. nomius, A. luteovirensces, A. pseudonomius*

parasiticus and *A. sojae* (domesticated species) cluster together. Both species are closely related to *A. arachidicola* and *A. novoparasiticus*, the latter being morphologically similar to *A. parasiticus*. *A. sergii* and *A. transmontanensis* are the basal species of this sub-clade. Finally, *A. mottae* sets as the basal species of the *A. flavus* and *A. parasiticus* sub-clades, though it resembles the species from the *A. parasiticus* sub-clade, sharing morphological, genetic and chemical traits (Varga et al. 2011, Soares et al. 2012, Carvajal-Campos et al. 2017, Frisvad et al. 2019).

A. tamarii-clade species produce AFBG with the exception of *A. tamarii* and *A. caelatus*. Mature colonies of *A. pseudotamarii* resemble morphologically to the *A. tamarii* and *A. pseudocaelatus* overlap traits with its sister taxon, *A. caelatus* (Varga et al. 2011).

A. nomius-clade is composed only by AFBG species. *A. pseudonomius* (Varga et al. 2011) is the sibling species of *A. nomius*, and both overlap traits, risking misidentification (Peterson et al. 2001, Moore et al. 2016). *A. luteovirescens* also share traits with species from its cluster (Peterson et al. 2001). Although *A. luteovirescens* produces AFBG, it is not considered as a pathogen for humans or animals as it occurs less frequently in food and feedstuff (Varga et al. 2015, Moore et al. 2016). *A. nomius, A. pseudonomius,* and *A. luteovirescens* lack production of CPA, which has been suggested as a fixed trait in the cluster (Varga et al. 2011, Frisvad et al. 2015).

3. Aflatoxigenic *Aspergillus* section *Flavi* Biodiversity According to Food Matrices, Production Chain Stage and Geographical Areas

The presence of aflatoxigenic *Aspergillus* section *Flavi* species in various foodstuff (cereals, dried fruits, legumes, oilseeds, roots and tubers, spices, tree nuts, etc.) and on all continents (Africa, Asia, Oceania, Europe and America) are highlighted by several studies (Table 4). Species distribution is not homogenous, a higher diversity is found in tropical and subtropical regions, as environmental conditions enhance their development (Schatzmayr and Streit 2013), *Aspergillus* section *Flavi* species physiological requirements of moisture are around 0.85 to 0.99 a_w, while optimal temperatures are between 28 and 42°C (Medina et al. 2015, Yogendrarajah et al. 2016). Aflatoxigenic species of section *Flavi* are mainly saprophytic and inhabit numerous ecological niches, like soil, decaying vegetation, hay and seeds, agricultural fields, and stored products. Fungal contamination of staples occurs during pre-harvest steps (crops and recollect) and/or in post-harvest steps (mainly storage and manufacturing production processes) (Rodrigues et al. 2012).

3.1 Biodiversity According to Food Matrices and Geographical Areas

The aflatoxigenic *Aspergillus* section *Flavi* species reported worldwide in the main food matrices are listed in Table 4. Species in this section were also reported in other food crops like *A. flavus* from acha in Nigeria, canola in Serbia, mustard seeds and chilgoza pine nuts in India (Tomar-Balhara et al. 2006, Sharma et al. 2013, Škrinjar et al. 2013, Ezekiel et al. 2014). In sugarcane, besides of *A. flavus*, *A. novoparasiticus* was reported in Egypt, Brazil, Japan and USA while *A. parasiticus* and *A. arachidicola* only in Asian countries and Brazil, respectively (Takahashi et al. 2004, Kumeda et al. 2003, Ahmed et al. 2010, Garber and Cotty 2014, Iamanaka et al. 2019, Abdallah et al. 2020). *A. flavus*, *A. parasiticus* and *A. nomius* were isolated from cocoa beans in Brazil (Copetti et al. 2011) and *A. aflatoxiformans* from edible mushrooms in Nigeria (Ezekiel et al. 2013a).

3.2 Cereals

3.2.1 Maize

Worldwide, *A. flavus* is reported as the most common species from section *Flavi* through the maize production chain (field, storage, derived products) (Table 4), including aflatoxigenic and non-aflatoxigenic isolates (Probst et al. 2014, Kachapulula et al. 2017). Maize plants are sensitive to drought and temperature, by consequence prone to fungal infection under plant stress conditions. *A. flavus* is considered to have a commensal relationship with this crop (Kornher 2018, Taniwaki et al. 2018); nevertheless, it is not the only species from section *Flavi* contaminating maize, *A. parasiticus* is the second most important contaminant, and occasionally some other species are reported (Table 4). In Africa, the presence of aflatoxigenic species on maize are mainly reported at post-harvest steps, with *A. flavus*, *A. parasiticus* and S_{BG} isolates as the most important contaminates. *A. flavus* L-morphotype strains are more common, including aflatoxigenic and non-aflatoxigenic strains (Probst et al. 2014, Kachapulula et al. 2017). S_{BG} isolates in the region are more probably *A. aflatoxiformans* and *A. minisclerotigenes* (in this order). *A. minisclerotigenes*

Table 4: Aflatoxigenic *Aspergillus* section *Flavi* species according to food matrices, production chain stage and geographical areas

Crops	Production chain step	Geographic zone		Aflatoxigenic *Aspergillus species*	References
Cereals					
Barley	**Post-harvest:** Grains**	Asia	Pakistan	*A. flavus*	Fakhrunnisa et al. 2006
		Europe	Spain	*A. flavus* *A. parasiticus*	Mateo et al. 2011
Maize	**Post-harvest:** kernels from storage or market	Africa	Nigeria, South Africa	*A. flavus*	Atehnkeng et al. 2008, Chilaka et al. 2012, Ekwomadu et al. 2018
			Egypt, Kenya	*A. flavus**	El-Shanshoury et al. 2014, Gachara et al. 2018, Abbas et al. 2020
			Cameroon	*A. flavus* *A. parasiticus*	Njobeh et al. 2009
			Kenya	*A. flavus* *A. minisclerotigenes* *A. parasiticus*	Okoth et al. 2018
			Ghana	*A. flavus* *A. parasiticus* S_BG isolates	Agbetiameh et al. 2018
			Nigeria	*A. korhogoensis*	Frisvad et al. 2019
	Non-specified step: at harvest and kernels from market		Cameroon, Ghana, Mali, Rwanda, Sierra Leona, Tanzania, Uganda	*A. flavus*	Probst et al. 2014

Step	Region	Country	Species	References
		Malawi, Zambia	*A. flavus* *A. parasiticus*	Probst et al. 2014, Kachapulula et al. 2017
		Burkina Faso, Ethiopia, Senegal	*A. flavus* S_{BG} isolates	Diedhiou et al. 2011, Probst et al. 2014
		Congo, Kenya, Nigeria	*A. flavus* *A. minisclerotigenes* S_{BG} isolates	Probst et al. 2014
		Mozambique, Zambia, Zimbabwe	*A. flavus* *A. parasiticus* S_{BG} isolates	Probst et al. 2014
		Somalia	*A. minisclerotigenes*	Probst et al. 2014
Pre-harvest: at harvest	Asia	China, Pakistan	*A. flavus*	Saleem et al. 2012, Mamo et al. 2018
Post-harvest: fresh kernels, corn meal		South Korea	*A. flavus*	Frisvad et al. 2019
		China, Vietnam	*A. flavus**	Gao et al. 2007, Trung et al. 2008, Tran-Dinh et al. 2009
		Pakistan, Philippines	*A. flavus* *A. parasiticus*	Ibrahim et al. 2016, Balendres et al. 2019
Non-specified step		China	*A. flavus* *A. arachidicola* *A. novoparasiticus* *A. pseudonomius*	Rasheed et al. 2019
Pre-harvest: at harvest	Central and South America	Mexico, Venezuela	*A. flavus*	Mazzani et al. 2008, Ortega-Beltran et al. 2015

(Contd.)

Table 4: (Contd.)

Crops	Production chain step	Geographic zone		Aflatoxigenic Aspergillus species	References
			Argentina	A. flavus, A. parasiticus	Etcheverry et al. 1999, Nesci and Etcheverry 2002
	Post-harvest: fresh kernels, corn meal		Colombia	A. flavus	Acuña et al. 2005
			Argentina, Ecuador	A. flavus, A. parasiticus	Etcheverry et al. 1999, Pacin et al. 2003
			Argentina	A. flavus, A. parasiticus, S_{BG} isolates	Camiletti et al. 2017
	Pre-harvest: at harvest	Europe	France	A. flavus, A. parasiticus	Bailly et al. 2018
	Post-harvest: kernels from storage and market		Hungary, Italy, Serbia	A. flavus	Giorni et al. 2007, Dobolyi et al. 2013, Mauro et al. 2013, Baranyi et al. 2015a
			France, Hungary	A. flavus, A. parasiticus	Sebők et al. 2014, Bailly et al. 2018
			Portugal	A. flavus, A. mottae	Soares et al. 2012
			Serbia	A. pseudonomius	Jakić-Dimić et al. 2009
	Pre-harvest: at harvest	North America	USA	A. flavus	Probst et al. 2014
				A. flavus, A. parasiticus	Jaime-Garcia and Cotty 2004, 2010

Substrate	Processing step	Continent	Country	Species	Reference
	Post-harvest: kernels**			A. flavus	Frisvad et al. 2019
	Non-specified step: kernels			A. texensis	Singh et al. 2018
	Post-harvest: kernels**	Oceania	Australia	A. flavus	Egmond and Jonker 2004
Millet	Post-harvest: grains from storage or markets	Africa	Nigeria	A. flavus	Osamwonyi and Wakil 2012, Ezekiel et al. 2014
Rice	Post-harvest: grains from market	Africa	Egypt	A. flavus*	El-Shanshoury et al. 2014
		Asia	Nigeria	A. aflatoxiformans A. austwickii A. korhogoensis	Frisvad et al. 2019
	Pre-harvest: field		China	A. flavus	Mamo et al. 2018
	Post-harvest: grains from market		Cambodia, India, Indonesia, Malaysia, Saudi Arabia, Thailand	A. flavus	Jayaraman and Kalyanasundaram 1990, Reddy et al. 2010, Al-Husnan et al. 2019
			South Korea, Pakistan	A. flavus A. parasiticus	Park et al. 2005, Ibrahim et al. 2016
	Non-specified step		Thailand	A. flavus	Frisvad et al. 2019
			Philippines	A. flavus A. parasiticus	Sales and Takumi 2005
	Pre-harvest: fields	South America	Brazil	A. flavus A. arachidicola A. novoparasiticus A. pseudocaelatus	Katsurayama et al. 2018
	Post-harvest: grains from market			A. flavus	Katsurayama et al. 2018

(Contd.)

Table 4: (*Contd.*)

Crops	Production chain step	Geographic zone		Aflatoxigenic Aspergillus species	References
Sorghum	**Post-harvest:** stored grains	Africa	Ethiopia	*A. flavus* *A. parasiticus*	Weledesemayat et al. 2016
	Non-specified step: seeds	Asia	Turkey, India, Pakistan,	*A. flavus*	Fakhrunnisa et al. 2006, Turgay and Ünal 2009, Divakara et al. 2014
	Post-harvest: stored grains	Europe	Serbia	*A. flavus*	Jakić-Dimić et al. 2009
Wheat	**Pre-harvest:** kernels**	Africa	Algeria	*A. flavus*	Riba et al. 2008, 2010
	Post-harvest: stored grains, grains and flour from market		Algeria, Egypt, Morocco	*A. flavus*	Riba et al. 2008, 2010, Dahab et al. 2016, Ennouari et al. 2018
			Egypt	*A. flavus**	El-Shanshoury et al. 2014
	Pre-harvest: grains**	Asia	Pakistan	*A. flavus*	Fakhrunnisa et al. 2006, Hussain et al. 2013
	Post-harvest: stored grains, grains and flour from market		India, Iran, Saudi Arabia, Turkey	*A. flavus*	Sinha and Sinha 1990, Saberi-Riseh et al. 2004, Doolotkeldieva 2010, Al-Wadai et al. 2013, Joshaghani et al. 2013
			Iran, Pakistan	*A. flavus* *A. parasiticus*	Kachuei et al. 2009, Ibrahim et al. 2016
	Pre-harvest: grains**	South America	Argentina, Brazil	*A. flavus*	González et al. 2008, Savi et al. 2014

	Post-harvest: stored grains	Europe	Serbia	*A. flavus*	Jakić-Dimić et al. 2009
			Slovakia	*A. parasiticus*	Dovičičová et al. 2012
		North America	Canada	*A. flavus*	Wallace and Sinha 1962
Dried Fruits					
Apricot	Post-harvest: dried fruit from market	Africa	Egypt	*A. flavus*	Zohri and Abdel-Gawad 1993
Figs		Asia	Iraq	*A. flavus* *A. parasiticus*	Saadullah and Abdullah 2015
		Africa	Egypt	*A. flavus*	Zohri and Abdel-Gawad 1993
			Algeria	*A. flavus, A. parasiticus*	Ait Mimoune et al. 2018
		Asia	Iran, Turkey	*A. flavus*	Şenyuva et al. 2008, Javanmard 2010
			Iraq	*A. flavus* *A. parasiticus*	Saadullah and Abdullah 2015
Plum		Africa	Egypt	*A. flavus*	Zohri and Abdel-Gawad 1993
		Asia	Iraq	*A. flavus* *A. parasiticus*	Saadullah and Abdullah 2015
Raisins		Africa	Egypt	*A. flavus*	Zohri and Abdel-Gawad 1993
		Asia	Iraq	*A. flavus* *A. parasiticus*	Saadullah and Abdullah 2015

(Contd.)

Table 4: *(Contd.)*

Crops	Production chain step	Geographic zone		Aflatoxigenic Aspergillus species	References
Legumes					
Beans	**Post-harvest:** dried grains from market	Africa	Cameroon	*A. flavus* *A. parasiticus*	Njobeh et al. 2009
		South America	Brazil	*A. flavus* *A. parasiticus*	Costa and Scussel 2002
Soybeans		Africa	Cameroon	*A. flavus* *A. parasiticus*	Njobeh et al. 2009
Oilseeds					
Cotton	**Pre-harvest:** fields	North America	USA	*A. flavus*	Cotty 1997
	Post-harvest: cotton balls				Klich et al. 1986
	Post-harvest: cotton seeds, cake and meal	Asia	India	*A. flavus*	Mazen et al. 1990, Tomar et al. 2012
Peanuts	**Pre-harvest:** at harvest	Africa	Egypt	*A. flavus*	Embaby and Abdel-Galel 2014
			Ethiopia, Nigeria	*A. flavus* *A. parasiticus*	Mohamed and Chala 2014, Guchi 2015, Wartu et al. 2015
	Post-harvest: dried peanuts, peanuts from storage; peanuts, butter or cake from markets		Côte d'Ivoire	*A. aflatoxiformans* *A. korhogoensis*	Carvajal-Campos et al. 2017, Manizan 2019

Stage	Continent	Country/Region	Species	References
		Algeria, Benin, Cameroon, Congo, Côte d'Ivoire, Egypt, Ethiopia, Nigeria, South Africa	*A. flavus* *A. parasiticus*	Njobeh et al. 2009, Sultan and Magan 2010, Adjou et al. 2012, Boli et al. 2014, Kamika et al. 2014, Mohamed and Chala 2014, Wartu et al. 2015, Ait Mimoune et al. 2018
		Botswana, Ghana, Nigeria	*A. flavus* *A. parasiticus* S_{BG} isolates	Mphande et al. 2004, Ezekiel et al. 2013b, Agbetiameh et al. 2018,
		Ethiopia	*A. flavus* *A. parasiticus* *A. nomius*	Guchi 2015
		Nigeria	*A. aflatoxiformans*	Frisvad et al. 2019
Pre-harvest: fields	Asia	China, Indonesia, Vietnam	*A. flavus*	Tran-Dinh et al. 2009, Mamo et al. 2018, Frisvad et al. 2019
Post-harvest: stored peanuts and peanut-based products		Malaysia	*A. flavus* *A. nomius*	Norlia et al. 2018
Pre-harvest: plants and grains from field	South America	Argentina, Brazil	*A. flavus* *A. arachidicola* *A. minsclerotigenes* *A. parasiticus*	Vaamonde et al. 2003, Barros et al. 2003, Pildain et al. 2008, Martins et al. 2017
Post-harvest: drying processes and storage		Argentina	*A. flavus*	Vaamonde et al. 2003, Barros et al. 2003, Pildain et al. 2008
Pre-harvest: fields	North America	USA	*A. flavus* *A. parasiticus*	Jaime-Garcia and Cotty 2004, 2010

(Contd.)

Table 4: (*Contd.*)

Crops	Production chain step	Geographic zone		Aflatoxigenic Aspergillus species	References
	Post-harvest: drying processes and storage				Horn 2007
	Pre-harvest: fields	Oceania	Australia	*A. flavus* *A. minisclerotigenes* *A. parasiticus*	Pitt and Hocking 2006, Frisvad et al. 2019
	Post-harvest: drying and storage				Pitt and Hocking 2006, Frisvad et al. 2019
Sesame	**Post-harvest:** grains from storage and market	Africa	Nigeria	*A. flavus* *A. aflatoxiformans*	Ezekiel et al. 2014
			Nigeria	*A. aflatoxiformans* *A. austwickii*	Frisvad et al. 2019
			Senegal	*A. flavus* S_{BG} isolates	Diedhiou et al. 2011
	Post-harvest: Grains**	Central America	Mexico	*A. aflatoxiformans*	Frisvad et al. 2019
Roots and Tubers					
Cassava	**Post-harvest:** chips	Africa	Benin	*A. flavus*	Gnonlonfin et al. 2008
			Benin	*A. flavus* *A. aflatoxiformans* *A. novoparasiticus*	Adjovi et al. 2014
			Cameroon	*A. flavus* *A. parasiticus* *A. nomius*	Onana et al. 2013

Yam		Africa	Benin, Nigeria	*A. flavus*	Bankole and Mabekoje 2004, Gnonlonfin et al. 2008
Spices					
Chili peppers	**Post-harvest:** market	Africa	Benin, Mali, Nigeria, Togo	*A. flavus*	Hell et al. 2009, Ezekiel et al. 2019
			Morocco	*A. flavus* *A. minisclerotigenes*	El Mahgubi et al. 2013
			Nigeria	*A. flavus* *A. aflatoxiformans* *A. minisclerotigenes A. parasiticus*	Singh and Cotty 2019
		Asia	Pakistan	*A. flavus* *A. parasiticus*	Ibrahim et al. 2016
			India, Lebanon	*A. flavus*	Sharfun-Nahar et al. 2004, Makhlouf et al. 2019
Cumin	**Post-harvest:** market	Africa	Morocco	*A. flavus* *A. minisclerotigenes*	El Mahgubi et al. 2013
Curry	**Post-harvest:** market	Asia	Lebanon	*A. flavus*	Makhlouf et al. 2019
Pepper	**Post-harvest:** market	Africa	Morocco	*A. flavus* *A. minisclerotigenes*	El Mahgubi et al. 2013
	Post-harvest: market	Asia	Lebanon, Turkey	*A. flavus*	Probst et al. 2014, Makhlouf et al. 2019
	Post-harvest: market	South America	Brazil	*A. flavus*	Freire et al. 2000

(Contd.)

Table 4: (*Contd.*)

Crops	Production chain step	Geographic zone		Aflatoxigenic Aspergillus species	References
Tree Nuts					
Almonds	**Post-harvest:** storage and market	Africa	Algeria	*A. flavus* *A. parasiticus*	Ait Mimoune et al. 2018
	Pre-harvest: orchards and at harvest	Europe	Portugal	*A. flavus* *A. minisclerotigenes A. parasitcus* *A. sergii*	Rodrigues et al. 2009, 2012, Soares et al. 2012
	Post-harvest: drying, storage			*A. flavus* *A. minisclerotigenes A. parasitcus* *A. sergii* *A. transmontanensis*	Rodrigues et al. 2009, 2012, Soares et al. 2012
	Pre-harvest: orchards and at harvest	North America	USA	*A. flavus* *A. parasiticus*	Purcell et al. 1980, Bayman et al. 2002, Doster et al. 2014
	Post-harvest: drying, storage				Bayman et al. 2002, Doster et al. 2014
Brazil nuts	**Pre-harvest:** at harvest	South America	Peru	*A. flavus*	Taniwaki et al. 2017
			Brazil	*A. flavus* *A. nomius*	Taniwaki et al. 2017

Commodity	Step	Continent	Country	Species	References
	Post-harvest: drying, storage		Brazil	*A. flavus* *A. arachidicola* *A. luteovirescens* *A. nomius* *A. parasiticus* *A. pseudocaelatus* *A. pseudonomius* *A. pseudotamarii*	Freire et al. 2000, Pacheco et al. 2010, Freitas-Silva et al. 2011, Reis et al. 2012, Andersson 2012, Baquião et al. 2013, Calderari et al. 2013, Massi et al. 2014, Taniwaki et al. 2017
Hazelnuts	**Pre-harvest:** orchards and at harvest	Europe	Italy	*A. flavus* *A. parasiticus*	Rodrigues et al. 2012, Prencipe et al. 2018
	Post-harvest: drying process, storage				
Nuts	**Post-harvest**	Europe	Croatia, Hungary	*A. flavus*	Baranyi et al. 2015a
Pistachios	**Pre-harvest:** orchards	Asia	Turkey	*A. flavus*	Heperkan et al. 1994, Kabirian et al. 2011
			Iran	*A. flavus** *A. parasiticus*	Rahimi et al. 2007
	Post-harvest: drying, storage		Turkey	*A. flavus*	Heperkan et al. 1994, Kabirian et al. 2011
	Pre-harvest: orchards	North America	USA	*A. flavus*	Bayman et al. 2002
	Post-harvest: market				Bayman et al. 2002

* The survey results may include other species closely related to *A. flavus*

S_{BG} isolates correspond to *A. flavus* closely related species that produce AFBG and have small sclerotia (*A misclerotigenes, A. aflatoxiformans, A. korhogoensis* or an unknown non-characterized species) and were generally identified as *A. flavus*.

** Step not specified

was properly identified in Kenya, Congo, Nigeria and Somalia, whereas *A. korhogoensis* was reported in Nigeria (Table 4).

In Asia, *A. flavus* is the main contaminant but not all strains are AF producers; other species reported less frequently include *A. parasiticus, A. arachidicola, A. novoparasiticus* and *A. pseudonomius* (Table 4). Surveys in China and Vietnam may unmask diversity, as *A. flavus* isolates have been reported with atypical chemotypes and morphological traits (Gao et al. 2007, Trung et al. 2008, Tran-Dinh et al. 2009). Rasheed et al. (2019) showed that diversity in the region is sub-estimated by reporting *A. arachidicola, A. novoparasiticus* and *A. pseudonomius*; whereas in Philippines, Balendres et al. (2019) reported *A. flavus* and *A. parasiticus* from maize and soil field samples. Similarly, in Pakistan, surveys from maize samples reported *A. flavus* and *A. parasiticus* (Table 4), while in Iran samples from maize soil fields showed *A. flavus* as the most frequent species, followed by *A. parasiticus* and *A. nomius* (Razzaghi-Abyaneh et al. 2006). In Central and South America, *A. flavus* and *A. parasiticus* are pointed as the major aflatoxigenic species infecting maize (Table 4). Ortega-Beltran et al. (2015) reported *A. flavus* S- and L-morphotypes and a S_{BG} strain from maize soil in Sonora (Mexico). In Argentinean crops, *A. flavus* L-morphotype was frequent in pre-harvest steps, whereas *A. parasiticus* aflatoxigenic strains in pre-planted crops (soil with plant debris). Camiletti et al. (2017) suggested that infection by aflatoxigenic *Aspergillus* section *Flavi* species increases while the flowering period is hot in some Argentinean regions. It is highly possible that the S_{BG} isolate reported in Argentina corresponds to *A. minisclerotigenes* because it has been isolated from the same region in other staples. In Europe, over the last decade, maize contamination by aflatoxigenic species from section *Flavi* has become an increasing issue. Climate change in Europe has a seasonal impact, as seasons are becoming hotter and drier, especially in the Southern, Central and Eastern regions, increasing the chances of *A. Flavi* species colonization (EFSA 2013, Battilani et al. 2016, Moretti et al. 2019). Despite the fact that *Fusarium* spp. such as *F. verticillioides* and *F. graminearum* are still the main contaminants of maize in Europe, the reports of AFs are rising (Giorni et al. 2007, Covarelli et al. 2011, Gallo et al. 2012, Mauro et al. 2013, Dobolyi et al. 2013, Baranyi et al. 2015a, Bailly et al. 2018, Kos et al. 2018). In European maize, *A. flavus* and *A. parasiticus* are the most frequent species (Table 4), yet *A. parasiticus* infection may be favored as cooler temperature suits its physiological requirements better (Horn 2007, Bailly et al. 2018). Also, *Aspergillus* section *Flavi* diversity seems to rise in storage conditions, at present *A. flavus* (non-aflatoxigenic and aflatoxigenic strains) and *A. parasiticus* (mainly aflatoxigenic strains) have been reported in fields and at harvesting steps while *A. flavus* (non-aflatoxigenic and aflatoxigenic strains), *A. parasiticus, A. pseudonomius* and *A. mottae* in silos and storage samples (Table 4). In the USA, *A. flavus* is the main species recovered from maize fields (soil, debris and ears) and stored maize kernels, followed by *A. parasiticus* and less frequent *A. texensis*. *A. flavus* diversity in the region includes S- and L-morphotypes, and aflatoxigenic and non-aflatoxigenic strains (Bayman et al. 2002, Jaime-Garcia and Cotty 2004,

2010), being the S-morphotype more frequent and generally aflatoxigenic. In Australia, *Fusarium* spp. are prone to contaminate maize as environmental conditions favors its development, yet *A. flavus* has been reported to occur in soil and kernels (Egmond and Jonker 2004).

3.2.2   Rice

Globally, rice is the second most produced and consumed cereal after maize, and *A. flavus* the most frequent contaminant species from the section *Flavi*. In Asia, *A. flavus* is the most common species reported, followed by *A. parasiticus* (Table 4), yet its identification has been based principally in morphological characteristics. In Africa, Frisvad et al. (2019) reported the presence of *A. aflatoxiformans* and two rare species from Nigerian markets, *A. korhogoensis* and *A. austwickii*; while in South America, Katsurayama et al. (2018) reported *A. novoparasiticus*, *A. arachidicola*, *A. pseudocaelatus* and *A. flavus* from rice and rice plantation soil in Brazil (Table 4). The proportion of aflatoxigenic *A. flavus* strains reported was low (1.5% of the isolated *A. flavus*), yet most of the strains produced CPA (Katsurayama et al. 2018). In Europe, rice is produced in some southern regions, like Spain and France, and even though surveys targeting *Aspergillus* section *Flavi* diversity in this staple do not exist to our knowledge, a study by Suárez-Bonnet et al. (2013) isolated the four AFs B and G from Spanish rice samples and traces of AFGs from French rice samples, suggesting a contamination caused by other species than *A. flavus.*

3.2.3   Wheat and Barley

Wheat and barley are the third and fourth cereals more produced and consumed around the world. Both cereals are more prone to be infected by *Fusarium* spp. hence most studies target these species; despite some reports based on morphology include *A. flavus* and *A. parasiticus* (Table 4).

3.2.4   Millet and Sorghum

Millet and sorghum are better adapted to harsh environments than other cereals and are common crops in Asia and Africa (FAO 1995). The main aflatoxigenic species from section *Flavi* reported to occur in these staples are *A. flavus* and *A. parasiticus* was identified only in stored sorghum grains from Ethiopia (Table 4), but reports of AFBG occurrence in grains and byproducts suggest species misidentification.

3.3   Dried Fruits and Legumes

Contamination of these food matrices by aflatoxigenic species from *Flavi* section was reported from market samples. Based on morphological description, *A. flavus* is the most common identified species in dried apricot, figs, raisins and plums collected in North Africa and Asia, followed by *A. parasiticus*. Both *A. flavus* and *A. parasiticus* were reported in all dried beans and soybeans samples from Cameroon and Brazil (Table 4).

3.4 Oilseeds

3.4.1 Peanuts

A. parasiticus and *A. flavus* are commonly reported in peanuts through the production chain (Table 4), both reported as commensal species of peanut plant (Pitt et al. 2012). The presence of fungal propagules in soil increases infection risk, notably if peanut crop suffers drought stress or related factors, raising the risk of high levels of AF production before harvest. Depending on the region, other species were also isolated in high frequencies (Table 4). In addition, the methods used for identification might fall to underestimate cryptic species and interesting populations, like S_{BG} species (Table 4). In Africa, many studies indicated the presence of *A. flavus* and *A. parasiticus* in peanuts during pre- and post-harvest stages including manufactured products such as cake and butter. Other species reported to infect peanuts during post-harvest include *A. aflatoxiformans*, *A. korhogoensis* and *A. nomius* (Table 4). Slow drying and poor storage conditions are a serious issue in the humid tropics, as moisture absorption might favor aflatoxigenic fungal infection after harvest (Pitt et al. 2012, Taniwaki et al. 2018). In Asia, *A. flavus* was found to be the most common species isolated from peanuts along the production chain and the most common mycotoxigenic component from the mycobiota (Table 4). In Malaysia, *A. nomius* and an atypical *A. flavus* L-morphotype were isolated in post-harvest steps (Table 4). In South America, studies are scarce though peanuts are widely consumed. In Argentina, the principal infectious species during pre-harvest from *Aspergillus* section *Flavi* were *A. flavus* (aflatoxigenic and non-aflatoxigenic strains) and *A. parasiticus*. Yet, *A. arachidicola* and *A. minisclerotigenes* were also present, whereas only *A. flavus* was identified during post-harvest (Table 4). In Brazil, throughout the peanut production chain, *A. flavus* was the most frequent species from section *Flavi*, followed by *A. parasiticus* (Martins et al. 2017). Nevertheless, species prevalence varied among the production chain stages. *A. flavus* prevalence was higher at sorting stage (78.3%), followed by drying (63.2%), threshing (54.5%), blanching (47.6%), field (40.7%) and ready-to-eat products (31.3%), while *A. parasiticus* was reported at drying (12.5%) and in ready-to-eat (1.3%) samples. This study also tested the aflatoxigenic potential of *A. flavus* strains, around 50% of samples were aflatoxigenic, and were isolated more frequently from fields than from processing plants (83% and 31%, respectively) (Martins et al. 2017). In northern USA, *A. flavus* and *A. parasiticus* were reported as the main peanut contaminants, both occurring in fields and after harvest (Table 4). *A. flavus* follows the same trend observed for other crops in USA, being S-morphotype more common. In Australia, studies on peanuts have reported the presence of *A. flavus*, *A. parasiticus* and *A. minisclerotigenes* (the latter especially in soil samples), during pre- and post-harvest steps (Pitt and Hocking 2006).

3.4.2 Cotton and Sesame

Reports from these staples are poor, probably the best know from them is cotton from USA. Cotton is contaminated by *A. flavus* (Table 4), mainly by

A. flavus S-morphotype isolates (Cotty 1997). Sesame contamination during post-harvest steps in Africa are caused by *A. flavus* and S_{BG} isolates (*A. aflatoxiformans* and the newly described species *A. austwickii*) (Table 4).

3.5   Roots and Tubers

Cassava and yam are important staples in Africa. Cassava is ranked as the third most important food crop in tropical regions after rice and maize. The conditions of production and storage of cassava, yam, and their traditional derivatives favor contamination and development of fungi (Adjovi et al. 2014). *A. flavus* is the most frequent species isolated, though *A. aflatoxiformans, A. novaparasiticus, A. nomius* and *A. parasiticus* were also reported from cassava chips (Table 4).

3.6   Spices

Post-harvest practices and environmental conditions during growth make spices susceptible to *Aspergillus* section *Flavi* fungal infection, principally by *A. flavus* and S_{BG} species (Table 4). In Asia and South America (based on morphological identification), *A. flavus* is the only spice contaminant (pepper and curry); in Central America, *A. pipericola* was isolated from a Mexican pepper sample (Frisvad et al. 2019) whereas in Africa, besides this species, *A. minisclerotigenes* has been identified (pepper and cumin) (Table 4). Chili peppers (paprika and chili) are frequently reported to present AFs, among the aflatoxigenic species *A. flavus* is most reported, followed by *A. minisclerotigenes, A. parasiticus* and *A. aflatoxiformans* (Table 4). Singh and Cotty (2019) reported that the majority of aflatoxigenic fungi isolated from chili peppers in Nigeria were *A. flavus* L-morphotype (76.7%), followed by *A. aflatoxiformans* (8.3%), *A. minisclerotigenes* (8.0%), and a non-identified S_{BG} lineage (2.8%).

3.7   Tree Nuts

3.7.1   Brazil Nuts

Brazil nuts are produced in the Amazon rainforest in Brazil, Bolivia and Peru and are vulnerable to the contamination of several aflatoxigenic, *i.e. A. flavus, A. nomius, A. pseudonomius, A. parasiticus, A. luteovirescens, A. arachidicola, A. pseudocaelatus* and *A. pseudotamarii* (Table 4). Incidence of Brazil nut contamination by these species can reach 100% (Calderari et al. 2013, Taniwaki et al. 2017). *A. flavus* and *A. nomius* are suggested to contaminate Brazil nuts before they are collected in dry seasons, while other *Flavi* species contaminate them during rainy season, drying processes, storage and markets (Taniwaki et al. 2017).

3.7.2   Pistachios, Almonds, Hazelnuts and Nuts

The most common aflatoxigenic species identified in pistachios during pre- and post-harvest is *A. flavus*, both in Asia and USA. Only in the survey from Iran, besides *A. flavus, A. parasiticus*, and an unknown AFBG strain were isolated (Table 4). *Aspergillus* section *Flavi* surveys on almonds showed higher diversity. In Europe (Portugal), the species reported are *A. flavus, A. minisclerotigenes, A. parasiticus, A. sergii* and *A. transmontanensis*; from these, *A. transmontanesis* was isolated only from plant processor environments (Table 4), while in USA and Algeria *A. flavus* and *A. parasiticus* have been identified in this crop (orchards and post-harvest steps). Hazelnuts in Europe (Italy) were contaminated by *A. flavus*, mostly non-aflatoxigenic strains, followed by *A. parasiticus* whereas *A. flavus* is the only section *Flavi* species reported from nuts in Europe (Table 4). Surveys showed differences between orchards and raw products in market outlets, the latter presented higher ratios of contamination, while considering the matrices pistachios are more sensitive, followed by walnuts and almonds (Bayman et al. 2002, Doster et al. 2014).

3.8 Biodiversity According to Food Production Chain Stages

In crops and pre-harvest stages, fungal infection is favored under humid and hot environmental conditions. In addition, cropping system may increase the incidence of fungal propagules in soil, if plant debris is left or incorporated into soil it becomes a source of infection and favors spore inoculation into the next planting cycle. The increase of fungal propagule number from year to year is reported for section *Flavi* species (Nesci and Etcheverry 2002, Jaime-Garcia and Cotty 2004, 2010). Insects also facilitate fungal infection by dispersing spores and wounding kernels and plants (Rychlik et al. 2014, Aiko and Mehta 2015). The main aflatoxigenic *Aspergillus* section *Flavi* species reported in the pre-harvest stages of the food production chain are *A. flavus* and *A. parasiticus*, which seem to have commensal relationships with certain crops (maize and peanuts) (Taniwaki et al. 2018). Some other species identified in these stages were *A. arachidicola, A. minisclerotigenes, A. nomius, A. novoparasiticus, A. pseudocaelatus* and *A. sergii* (Table 4).

In post-harvest a key step is the drying process, if the raw material is not properly dried, mold can easily develop. Moisture enhances *Aspergillus* section *Flavi* development, so its control is a delicate step in the food production chain to prevent AFs presence in final products. This control is easiest to be achieved in developed countries than in developing countries, especially for small producers (Marin et al. 2012, Gowda et al. 2013, Baranyi 2015b). The diversity of *Flavi* species is higher in post-harvest and first storage steps: *A. flavus, A. parasiticus, A. minisclerotigenes, A. aflatoxiformans, A. korhogoensis, A. austwickii, A. transmontanensis, A. sergii, A. luteovirescens, A. mottae, A. nomius, A. pipericola, A. pseudonomius, A. arachidicola, A. pseudocaelatus,* and *A. pseudotamarii* (Table 4).

A. flavus and *A. parasiticus* are the most frequent species found in commodities along the production chain, *i.e.* in pre- and post-harvest, and in a less extent *A. nomius* in Brazil nuts, *A. minisclerotigenes* in peanuts and almonds and *A. sergii* in almonds (Table 4).

4. Biodiversity of Aflatoxigenic *Aspergillus* section *Flavi* Species and Climate Change

Food security has become a very important issue across the globe and the potential effects of climate change on production and quality of food crops, including mycotoxins, is under scope, especially from a risk assessment perspective (Magan et al. 2011). Crop growth and its interaction with beneficiary and pathogenic and/or toxigenic microorganisms vary from year to year, mainly depending on local weather, making the agricultural sector particularly exposed to climate change (Moore and Lobell 2015). Climate change is a driver for distribution shifts in *Aspergillus* section *Flavi*, as it modifies the environmental conditions, resulting in new suitable niches favoring species development and also facilitating their colonization into temperate regions (Perrone et al. 2014); hence, climate change is an emerging issue worldwide for food and feed safety (Battilani et al. 2016). Projections of climate change suggest that in Africa and Oceania the suitable areas for agriculture will decrease, whereas some areas in Asia and Latin America tropical forest will become savanna (Paterson and Lima 2011). The possible change in patterns of AFs occurrence in crops due to climate change is a matter of concern that may require anticipatory actions, as these phenomena can lead to more health risk in affected areas (Battilani et al. 2016). However, it is difficult to create accurate projections as the knowledge of *Aspergillus* section *Flavi* diversity is poor (distribution, frequency and poor information of the life history and ecology). Although some species are found rarely, their total AF production is high and they produce other mycotoxins, like CPA that has additive effects. For instance, in Africa under climatic changing conditions there is a possibility that *A. minisclerotigenes*, *A. aflatoxiformans* and *A. korhogoensis* could expand their home ranges, as the scenarios of climate change suggest drier and warmer conditions, which may favor their frequency. The case of Africa is interesting, over the last three decades several studies performed in the region have increased the comprehension of the effects of AFs in human populations, unmasked species diversity of *Aspergillus* section *Flavi*, and helped to identify sensitive steps of production chain prone to mold contamination. These studies have shown a rapid diversification of species closely related to *A. flavus* (*A. minisclerotigenes*, *A. aflatoxiformans*, *A. korhogoensis* and *A. austwickii*). *A. flavus*, *A. minisclerotigenes* and *A. aflatoxiformans* contaminate staples in field, during harvesting and storage; whereas, *A. korhogoensis* and *A. austwickii* at storage steps (Cotty and Cardwell 1999, Carvajal-Campos et al. 2017, Frisvad et al. 2019). Rapid radiation in Africa may be a unique phenomenon linked with environmental conditions and evolutive pressures; or it can be just an artifact caused by the diversity underestimation in other world regions, originated by the reduced number of diversity surveys in Asia, Oceania (Pacific Islands) and Central and South America. Lack of studies is caused mainly by two factors: (1) diversity surveys are costly, time consuming and need special skills; (2) several studies are performed with the *a priori* that the only aflatoxigenic species are *A. flavus* and *A. parasiticus*. Likewise, several species identification surveys are based only on morphology and AF production, so misidentification of species occurs; as *A. nomius* in Brazil, which was considered to be the main contaminant of Brazil nuts, and later corrected by *A. pseudonomius* (Baquião et al. 2014, Massi et al. 2014).

References

Abbas, A., Hussien, T. and Yli-Mattila, T. 2020. A polyphasic approach to compare the genomic profiles of aflatoxigenic and non-aflatoxigenic isolates of *Aspergillus* Section *Flavi*. Toxins 12(1): 56. DOI: 10.3390/toxins12010056.

Abdallah, M.F., Audenaert, K., Lust, L., Landschoot, S., Bekaert, B., Haesaert, G. et al. 2020. Risk characterization and quantification of mycotoxins and their producing fungi in sugarcane juice: A neglected problem in a widely-consumed traditional beverage. Food Control 108: 106811. DOI: 10.1016/j.foodcont.2019.106811.

Acuña, C.A., Díaz, G.J. and Espitia, M.E. 2005. Aflatoxinas en maíz: Reporte de caso en la costa atlántica colombiana. Revista de la Facultad de Medicina Veterinaria y Zootecnia 52(2): 156–162. DOI: 10.15446/rfmvz.

Adjovi, Y.C.S., Bailly, S., Gnonlonfin, B.J.G., Tadrist, S., Querin, A., Sanni, A. et al. 2014. Analysis of the contrast between natural occurrence of toxigenic Aspergilli of the *Flavi* section and aflatoxin B1 in cassava. Food Microbiology 38: 151–159. DOI: 10.1016/j.fm.2013.08.005.

Adjou, E.S., Yehouenou, B., Sossou, C.M., Soumanou, M.M. and de Souza, C.A. 2012. Occurrence of mycotoxins and associated mycoflora in peanut cake product (kulikuli) marketed in Benin. African Journal of Biotechnology 11(78): 14354–14360. DOI: 10.5897/AJB12.324.

Agbetiameh, D., Ortega-Beltran, A., Awuah, R.T., Atehnkeng, J., Cotty, P.J. and Bandyopadhyay, R. 2018. Prevalence of aflatoxin contamination in maize and groundnut in Ghana: Population structure, distribution, and toxigenicity of the causal agents. Plant Disease 102(4): 764–772. DOI: 10.1094/PDIS-05-17-0749-RE.

Aiko, V. and Mehta, A. 2015. Occurrence, detection and detoxification of mycotoxins. Journal of Biosciences 40(5): 943–954. DOI: 10.1007/s12038-015-9569-6.

Ait Mimoune, N., Arroyo-Manzanares, N., Gámiz-Gracia, L., García-Campaña, A.M., Bouti, K., Sabaou, N. et al. 2018. *Aspergillus* section *Flavi* and aflatoxins in dried figs and nuts in Algeria. Food Additives & Contaminants: Part B 11(2): 119–125. DOI: 10.1080/19393210.2018.1438524.

Ahmed, A., Dawar, S. and Tariq, M. 2010. Mycoflora associated with sugar cane juice in Karachi city. Pakistan Journal of Botany 42(4): 2955–2962.

Al-Husnan, L., Al-Kahtani, M. and Farag, R.M. 2019. Bioinformatics analysis of aflatoxins produced by *Aspergillus* sp. in basic consumer grain (corn and rice) in Saudi Arabia. Potravinarstvo Slovak Journal of Food Sciences 13(1): 65–75. DOI: 10.5219/1020.

Al-Wadai, A.S., Al-Othman, M.R., Mahmoud, M.A. and Abd El-Aziz, A.R.M. 2013. Molecular characterization of *Aspergillus flavus* and aflatoxin contamination of wheat grains from Saudi Arabia. Genetics and Molecular Research 12(3): 3335–3352. DOI: 10.4238/2013.September.3.10.

Amaike, S. and Keller, N.P. 2011. *Aspergillus flavus.* Annual Review of Phytopathology 49: 107–133. DOI: 10.1146/annurev-phyto-072910-095221.

Andersson, E. 2012. A small growth study of *Aspergillus* section *Flavi*, and potentially aflatoxigenic fungi and aflatoxin occurrence in Brazil nuts from local markets in Manaus, Brazil. M.S. Thesis, Swedish University of Agricultural Sciences, Upssala.

Arroyo-Manzanares, N., Di Mavungu, J.D., Uka, V., Malysheva, S.V., Cary, J.W., Ehrlich, K.C. et al. 2015. Use of UHPLC high-resolution Orbitrap mass spectrometry to investigate the genes involved in the production of secondary metabolites in *Aspergillus flavus.* Food Additives & Contaminants: Part A 32: 1656–1673. DOI: 10.1080/19440049.2015.1071499.

Atehnkeng, J., Ojiambo, P.S., Donner, M., Ikotun, T., Sikora, R.A., Cotty, P.J. et al. 2008. Distribution and toxigenicity of *Aspergillus* species isolated from maize kernels from three agro-ecological zones in Nigeria. International Journal of Food Microbiology 122(1): 74–84. DOI: 10.1016/j.ijfoodmicro.2007.11.062.

Azziz-Baumgartner, E., Lindblade, K., Gieseker, K., Rogers, H.S., Kieszak, S., Njapau, H. et al. 2005. Aflatoxin Investigative Group. Case-control study of an acute aflatoxicosis outbreak, Kenya, 2004. Environmental Health Perspectives 113: 1779–1783. DOI: 10.1289/ehp.8384.

Bailly, S., Mahgubi, A.E., Carvajal-Campos, A., Lorber, S., Puel, O., Oswald, I.P. et al. 2018. Occurrence and identification of *Aspergillus* Section *Flavi* in the context of the emergence of aflatoxins in French maize. Toxins 10(12): 525. DOI: 10.3390/toxins10120525.

Balendres, M.A.O., Karlovsky, P. and Cumagun, C.J.R. 2019. Mycotoxigenic fungi and mycotoxins in agricultural crop commodities in the Philippines: A Review. Foods 8(7): 249. DOI: 10.3390/foods8070249.

Bankole, S.A. and Mabekoje, O.O. 2004. Mycoflora and occurrence of aflatoxin B1 in dried yam chips from markets in Ogun and Oyo States, Nigeria. Mycopathologia 157(1): 111–115. DOI: 10.1023/B:MYCO.0000012211.31618.18.

Baquião, A.C., De Oliveira, M.M., Reis, T.A., Zorzete, P., Danielle D.A. and Correa, B. 2013. Monitoring and determination of fungi and mycotoxins in stored Brazil nuts. Journal of Food Protection 76(8): 1414–1420. DOI: 10.4315/0362-028X.JFP-13-005.

Baranyi, N., Jaksić, D.D., Palágyi, A., Kiss, N., Kocsubé, S., Szekeres, A. et al. J. 2015a. Identification of *Aspergillus* species in Central Europe able to produce G-type aflatoxins. Acta Biologica Hungarica 66: 339–347. DOI: 10.1556/018.66.2015.3.9.

Baranyi, N., Kocsubé, S. and Varga, J. 2015b. Aflatoxins: Climate change and biodegradation. Current Opinion in Food Science 5: 60–66. DOI: 10.1016/j.cofs.2015.09.002.

Barros, G., Torres, A., Palacio, G. and Chulze, S. 2003. *Aspergillus* species from section *Flavi* isolated from soil at planting and harvest time in peanut-growing regions of Argentina. Journal of the Science of Food and Agriculture 83(13): 1303–1307. DOI: 10.1002/jsfa.1539.

Battilani, P., Toscano, P., Van der Fels-Klerx, H.J., Moretti, A., Leggieri, M.C., Brera, C. et al. 2016. Aflatoxin B1 contamination in maize in Europe increases due to climate change. Scientific Reports 6: 24328. DOI: 10.1038/srep24328.

Bayman, P., Baker, J.L. and Mahoney, N.E. 2002. *Aspergillus* on tree nuts: Incidence and associations. Mycopathologia 155(3): 161–169. DOI: 10.1023/A: 1020419226146.

Bbosa, G., Kitya, D., Lubega, A., Ogwal-Okeng, J., Anokbonggo, W.W. and Kyegombe, D.B. 2013. Chapter 12: Review of the biological and health effects of aflatoxins on body organs and systems. pp. 240–265. *In:* M. Razzaghi-Abyanedh (ed.). Aflatoxins – Recent Advances and Future Prospects. IntechOpen, Rijeka.

Bertuzzi, T., Romani, M., Rastelli, S., Mulazzi, A. and Pietri, A. 2017. Sterigmatocystin occurrence in paddy and processed rice produced in Italy in the years 2014–2015 and distribution in milled rice fractions. Toxins 9(3): 86. DOI: 10.3390/toxins9030086.

Bhat, R., Rai, R.V. and Karim, A.A. 2010. Mycotoxins in food and feed: Present status and future concerns. Comprehensive Reviews in Food Science and Food Safety 9(1): 57–81. DOI: 10.1111/j.1541-4337.2009.00094.x.

Boli, Z.A., Zoue, L.T., Alloue-Boraud, W.M., Kakou, C.A. and Koffi-Nevry, R. 2014. Proximate composition and mycological characterization of peanut butter sold in retail markets of Abidjan (Côte d'Ivoire). Journal of Applied Biosciences 72(1): 5822–5829. DOI: 10.4314/jab.v72i1.99667.

Calderari, T.O., Iamanaka, B.T., Frisvad, J.C., Pitt, J.I., Sartori, D., Pereira, J.L. et al. 2013. The biodiversity of *Aspergillus* section *Flavi* in Brazil nuts: From rainforest to consumer. International Journal of Food Microbiology 160(3): 267–272. DOI: 10.1016/j.ijfoodmicro.2012.10.018.

Camiletti, B.X., Torrico, A.K., Maurino, M.F., Cristos, D., Magnoli, C., Lucini, E.I. et al. 2017. Fungal screening and aflatoxin production by *Aspergillus* section *Flavi* isolated from pre-harvest maize ears grown in two Argentine regions. Crop Protection 92: 41–48. DOI: 10.1016/j.cropro.2016.10.012.

Carvajal-Campos, A., Manizan, A.L., Tadrist, S., Akaki, D.K., Koffi-Nevry, R., Moore, G.G. et al. 2017. *Aspergillus korhogoensis*: A novel aflatoxin producing species from the Côte d'Ivoire. Toxins 9(11): 353. DOI: 10.3390/toxins9110353.

Cary, J.W., Uka, V., Han, Z., Buyst, D., Harris-Coward, P.Y., Ehrlich, K.C. et al. 2015a. A secondary metabolite gene cluster containing a hybrid PKS–NRPS is necessary for synthesis of the 2-pyridones, leporins. Fungal Genetics and Biology 81: 88–97. DOI: 10.1016/j.fgb.2015.05.010.

Cary, J.W., Han, Z., Yin, Y., Lohmar, J.M., Shantappa, S., Harris-Coward, P.Y. et al. 2015b. Transcriptome analysis of *Aspergillus flavus* reveals veA-dependent regulation of secondary metabolite gene clusters, including the novel aflavarin cluster. Eukaryotic Cell 14: 983–997. DOI: 10.1128/EC.00092-15.

Chang, P.K., Matsushima, K., Takahashi, T., Yu, J., Abe, K., Bhatnagar, D. et al. 2007. Understanding nonaflatoxigenicity of *Aspergillus sojae*: A windfall of aflatoxin biosynthesis research. Applied Microbiology and Biotechnology 76(5): 977–984. DOI: 10.1007/s00253-007-1116-4.

Chang, P.K. and Ehrlich, K.C. 2010. What does genetic diversity of *Aspergillus flavus* tell us about *Aspergillus oryzae*? International Journal of Food Microbiology 138(3): 189–199. DOI: 10.1016/j.ijfoodmicro.2010.01.033.

Chilaka, C.A., De Kock, S., Phoku, J.Z., Mulunda, M., Egbuta, M.A. and Dutton, M.F. 2012. Fungal and mycotoxin contamination of South African commercial maize. Journal of Food, Agriculture & Environment 10(2): 296–303.

Copetti, M.V., Iamanaka, B.T., Pereira, J.L., Fungaro, M.H. and Taniwaki, M.H. 2011. Aflatoxigenic fungi and aflatoxin in cocoa. International Journal of Food Microbiology 148(2): 141–144. DOI: 10.1016/j.ijfoodmicro.2011.05.020

Costa, L.L.F. and Scussel, V.M. 2002. Toxigenic fungi in beans (*Phaseolus vulgaris* L.) classes black and color cultivated in the state of Santa Catarina, Brazil. Brazilian Journal of Microbiology 33(2): 138–144. DOI: 10.1590/S1517-83822002000200008.

Cotty, P.J. 1997. Aflatoxin-producing potential of communities of *Aspergillus* section *Flavi* from cotton producing areas in the United States. Mycological Research 101(6): 698–704. DOI: 10.1017/S0953756296003139.

Cotty, P.J. and Cardwell, K.F. 1999. Divergence of West African and North American Communities of *Aspergillus* Section *Flavi*. Applied and Environmental Microbiology 65: 2264–2265.

Covarelli, L., Beccari, G. and Salvi, S. 2011. Infection by mycotoxigenic fungal species and mycotoxin contamination of maize grain in Umbria, central Italy. Food and Chemical Toxicology 49(9): 2365–2369. DOI: 10.1016/j.fct.2011.06.047.

Dahab, N.F.A., Abdel-Hadi, A.M., Abdul-Raouf, U.M., El-Shanawany, R.A. and Hassane, A.M.A. 2016. Qualitative detection of aflatoxins and aflatoxigenic fungi in wheat flour from different regions of Egypt. IOSR Journal of Environmental Science, Toxicology and Food Technology 10(7): 20–26.

Diedhiou, P.M., Bandyopadhyay, R., Atehnkeng, J. and Ojiambo, P.S. 2011. *Aspergillus* colonization and aflatoxin contamination of maize and sesame kernels in two agro-ecological zones in Senegal. Journal Phytopathology 159: 268–275. DOI: 10.1111/j.1439-0434.2010.01761.x.

Divakara, S.T., Aiyaz, M., Puttaswamy, H., Chandra, N.S. and Ramachandrappa, N.S. 2014. *Aspergillus flavus* infection and contamination in sorghum seeds and their biological management. Archives of Phytopathology and Plant Protection 47(17): 2141–2156. DOI: 10.1080/03235408.2013.869892

Dobolyi, C.S., Sebők, F., Varga, J., Kocsubé, S., Szigeti, G., Baranyi, N. et al. 2013. Occurrence of aflatoxin producing *Aspergillus flavus* isolates in maize kernel in Hungary. Acta Alimentaria 42(3): 451–459. DOI: 10.1556/AAlim.42.2013.3.18.

Doolotkeldieva, T.D. 2010. Microbiological control of flour-manufacture: Dissemination of mycotoxins producing fungi in cereal products. Microbiology Insights 3: 1–15. DOI: 10.4137/MBI.S3822.

Doster, M.A., Cotty, P.J. and Michailides, T.J. 2009. Description of a distinctive aflatoxin-producing strain of *Aspergillus nomius* that produces submerged sclerotia. Mycopathologia 168: 193–201. DOI: 10.1007/s11046-009-9214-8.

Doster, M.A., Cotty, P.J. and Michailides, T.J. 2014. Evaluation of the atoxigenic *Aspergillus flavus* strain AF36 in pistachio orchards. Plant Disease 98(7): 948–956. DOI: 10.1094/PDIS-10-13-1053-RE.

Dovičičová, M., Tančinová, D., Labuda, R. and Sulyok, M. 2012. *Aspergillus parasiticus* from wheat grain of Slovak origin and its toxigenic potency. Czech Journal of Food Sciences 30: 483–487.

EFSA Panel on Contaminants in the Food Chain (CONTAM). 2013. Scientific opinion on the risk for public and animal health related to the presence of sterigmatocystin in food and feed. EFSA Journal 11: 3254. DOI: 10.2903/j.efsa.2013.3254..

Egmond, H.P. and Jonker, M.A. 2004. Current situation on regulations for mycotoxins. JSM Mycotoxins 2003(Suppl 3): 1–15.

Ehrlich, K.C., Chang, P.K., Yu, J. and Cotty, P.J. 2004. Aflatoxin biosynthesis cluster gene cypA is required for G aflatoxin formation. Applied and Environmental Microbiology 70: 6518–6524. DOI: 10.1128/AEM.70.11.6518-6524.2004.

Ehrlich, K.C. and Yu, J. 2010. Aflatoxin-like gene clusters and how they evolved. pp. 65–75. *In:* M. Rai and A. Varma (eds.). Mycotoxins in the Food, Feed and Bioweapons. Berlin, Springer Verlag.

Ekwomadu, T.I., Gopane, R.E. and Mwanza, M. 2018. Occurrence of filamentous fungi in maize destined for human consumption in South Africa. Food Science & Nutrition 6(4): 884–890. DOI: 10.1002/fsn3.561.

El Mahgubi, A.E., Puel, O., Bailly, S., Tadrist, S., Querin, A., Ouadia, A. et al. 2013. Distribution and toxigenicity of *Aspergillus* section *Flavi* in spices marketed in Morocco. Food Control 32: 143–148. DOI: 10.1016/j.foodcont.2012.11.013.

El-Shanshoury, A.R., El-Sabbagh, S.M., Emara, H.A. and Saba, H.E. 2014. Occurrence of moulds, toxicogenic capability of *Aspergillus flavus* and levels of aflatoxins in maize, wheat, rice and peanut from markets in central delta provinces, Egypt. International Journal of Current Microbiology and Applied Sciences 3(3): 852–865.

Embaby, E.M. and Abdel-Galel, M.M. 2014. Detection of fungi and aflatoxins contaminated peanut samples (*Arachis hypogaea* L.). Journal of Agricultural Technology 10(2): 423–437.

Ennouari, A., Sanchis, V., Rahouti, M. and Zinedine, A. 2018. Isolation and molecular identification of mycotoxin producing fungi in durum wheat from Morocco. Journal of Materials and Environmental Sciences 9: 1470–1479. DOI: 10.26872/jmes.2018.9.5.161.

Etcheverry, M., Nesci, A., Barros, G., Torres, A. and Chulze, S. 1999. Occurrence of *Aspergillus* section *Flavi* and aflatoxin B1 in corn genotypes and corn meal in Argentina. Mycopathologia 147(1): 37–41. DOI: 10.1023/A: 1007040123181.

Ezekiel, C.N., Sulyok, M., Frisvad, J.C., Somorin, Y.M., Warth, B., Houbraken, J. et al. 2013a. Fungal and mycotoxin assessment of dried edible mushroom in Nigeria. International Journal of Food Microbiology 162(3): 231–236. DOI: 10.1016/j.ijfoodmicro.2013.01.025.

Ezekiel, C.N., Sulyok, M., Babalola, D.A., Warth, B., Ezekiel, V.C. and Krska, R. 2013b. Incidence and consumer awareness of toxigenic *Aspergillus* section *Flavi* and aflatoxin B1 in peanut cake from Nigeria. Food Control 30(2): 596–601. DOI: 10.1016/j.foodcont.2012.07.048.

Ezekiel, C.N., Udom, I.E., Frisvad, J.C., Adetunji, M.C., Houbraken, J., Fapohunda, S.O. et al. 2014. Assessment of aflatoxigenic *Aspergillus* and other fungi in millet and sesame from Plateau State, Nigeria. Mycology 5(1): 16–22. DOI: 10.1080/21501203.2014.889769.

Ezekiel, C.N., Ortega-Beltran, A., Oyedeji, E.O., Atehnkeng, J., Kössler, P., Tairu, F. et al. 2019. Aflatoxin in chili peppers in Nigeria: Extent of contamination and control using atoxigenic *Aspergillus flavus* genotypes as biocontrol agents. Toxins 11(7): 429. DOI: 10.3390/toxins11070429.

Fakhrunnisa, M.H. and Ghaffar, A. 2006. Seed-borne mycoflora of wheat, sorghum and barley. Pakistan Journal of Botany 38(1): 185–192.

FAO. 2003. Worldwide Regulations for Mycotoxins in Food and Feed in 2003. FAO Nutr. Pap. Rome, pp. 1–165.

Freire, F.C., Kozakiewicz, Z. and Paterson, R.R. 2000. Mycoflora and mycotoxins in Brazilian black pepper, white pepper and Brazil nuts. Mycopathologia 149(1): 13–19. DOI: 10.1023/a: 1007241827937.

Freitas-Silva, O., Teixeira, A., da Cunha, F.Q., de Oliveira Godoy, R.L. and Venâncio, A. 2011. Predominant mycobiota and aflatoxin content in Brazil nuts. Journal für Verbraucherschutz und Lebensmittelsicherheit 6(4): 465–472. DOI: 10.1007/s00003-011-0703-6.

Frisvad, J.C., Skouboe, P. and Samson, R.A. 2005. Taxonomic comparison of three different groups of aflatoxin producers and a new efficient producer of aflatoxin B1, sterigmatocystin and 3-O-methylsterigmatocystin, *Aspergillus rambellii* sp. *nov.* Systematic and Applied Microbiology 28(5): 442–453. DOI: 10.1016/j.syapm.2005.02.012.

Frisvad, J.C., Hubka, V., Ezekiel, C.N., Hong, S.B., Nováková, A., Chen, A.J. et al. 2019. Taxonomy of *Aspergillus* section *Flavi* and their production of aflatoxins, ochratoxins and other mycotoxins. Studies in Mycology 93: 1–63. DOI: 10.1016/j.simyco.2018.06.001.

Gachara, G.W., Nyamache, A.K., Harvey, J., Gnonlonfin, G.J.B. and Wainaina, J. 2018. Genetic diversity of *Aspergillus flavus* and occurrence of aflatoxin contamination in stored maize across three agro-ecological zones in Kenya. Agriculture & Food Security 7(1): 52. DOI: 10.1186/s40066-018-0202-4.

Gallo, A., Stea, G., Battilani, P., Logrieco, A.F. and Perrone, G. 2012. Molecular characterization of an *Aspergillus flavus* population isolated from maize during the first outbreak of aflatoxin contamination in Italy. Phytopathologia Mediterranea 51(1): 198–206.

Gao, J., Liu, Z. and Yu, J. 2007. Identification of *Aspergillus* section *Flavi* in maize in northeastern China. Mycopathologia 164(2): 91–95. DOI: 10.1007/s11046-007-9029-4.

Garber, N.P. and Cotty, P.J. 2014. *Aspergillus parasiticus* communities associated with sugarcane in the Rio Grande Valley of Texas: Implications of global transport and host association within *Aspergillus* Section *Flavi.* Phytopathology 104: 462–471. DOI: 10.1094/PHYTO-04-13-0108-R

Georgianna, D.R., Fedorova, N.D., Burroughs, J.L., Dolezal, A.L., Bok, J.W., Horowitz-Brown, S. et al. 2010. Beyond aflatoxin: Four distinct expression patterns and functional roles associated with *Aspergillus flavus* secondary metabolism gene clusters. Molecular Plant Pathology 11(2): 213–226. DOI: 10.1111/j.1364-3703.2009.00594.x.

Giorni, P., Magan, N., Pietri, A., Bertuzzi, T. and Battilani, P. 2007. Studies on *Aspergillus* section *Flavi* isolated from maize in northern Italy. International Journal of Food Microbiology 113(3): 330–338. DOI: 10.1016/j.ijfoodmicro.2006.09.007

Gloer, J.B., TePaske, M.R., Sima, J.S., Wicklow, D.T. and Dowd, P.F. 1988. Antiinsectan aflavinine derivative from sclerotia of *Aspergillus flavus*. Journal of Organic Chemistry 53: 5457–5460. DOI: 10.1021/jo00258a011.

Gnonlonfin, G.J.B., Hell, K., Fandohan, P. and Siame, A.B. 2008. Mycoflora and natural occurrence of aflatoxins and fumonisin B1 in cassava and yam chips from Benin, West Africa. International Journal of Food Microbiology 122(1–2): 140–147. DOI: 10.1016/j.ijfoodmicro.2007.11.047.

Gonçalves, J.S., Ferracin, L.M., Vieira, M.L.C., Iamanaka, B.T., Taniwaki, M.H. and Fungaro, M.H.P. 2012. Molecular analysis of *Aspergillus* section *Flavi* isolated from Brazil nuts. Journal of Microbiology and Biotechnology 28(4): 1817–1825. DOI: 10.1007/s11274-011-0956-3.

Gowda, N.K.S., Swamy, H.V.L.N. and Mahajan, P. 2013. Recent advances for control, counteraction and amelioration of potential aflatoxins in animal feeds. pp. 129–140. *In:* M. Razzaghi-Abyanedh (ed.). Aflatoxins – Recent Advances and Future Prospects. IntechOpen, Rijeka.

González, H.H.L., Moltó, G.A., Pacin, A., Resnik, S.L., Zelaya, M.J., Masana, M. et al. 2008. Trichothecenes and mycoflora in wheat harvested in nine locations in Buenos Aires province, Argentina. Mycopathologia 165(2): 105–114. DOI: 10.1007/s11046-007-9084-x.

Grenier, B. and Oswald, I. 2011. Mycotoxin co-contamination of food and feed: Meta-analysis of publications describing toxicological interactions. World Mycotoxin Journal 4(3): 285–313. DOI: 10.3920/WMJ2011.1281.

Guchi, E. 2015. Effect of storage time on occurrence of *Aspergillus* species in groundnut (*Arachis hypogaea* L.) in Eastern Ethiopia. Journal of Applied & Environmental Microbiology 3(1): 1–5. DOI: 10.12691/jaem-3-1-1

Hell, K., Gnonlonfin, B.G.J., Kodjogbe, G., Lamboni, Y. and Abdourhamane, I.K. 2009. Mycoflora and occurrence of aflatoxin in dried vegetables in Benin, Mali and Togo, West Africa. International Journal of Food Microbiology 135(2): 99–104. DOI: 10.1016/j.ijfoodmicro.2009.07.039.

Heperkan, D., Aran, N. and Ayfer, M. 1994. Mycoflora and aflatoxin contamination in shelled pistachio nuts. Journal of the Science of Food and Agriculture 66 (3): 273–278. DOI: 10.1002/jsfa.2740660302.

Horn, B.W. 2007 Biodiversity of *Aspergillus* section *Flavi* in the United States: A review. Food Additives and Contaminants 24(10): 1088–1101. DOI: 10.1080/02652030701510012.

Hussain, M., Ghazanfar, M.U., Hamid, M.I. and Raza, M. 2013. Seed borne mycoflora of some commercial wheat (*Triticum aestivum* L.) cultivars in Punjab, Pakistan. International Journal of Phytopathology 2(2): 97–101. DOI: 10.33687/phytopath.002.02.0198.

Iamanaka, B.T., de Souza Lopes, A., Martins, L.M., Frisvad, J.C., Medina, A., Magan, N. et al. 2019. *Aspergillus* section *Flavi* diversity and the role of *A. novoparasiticus* in aflatoxin contamination in the sugarcane production chain. International Journal of Food Microbiology 293: 17–23. DOI: 10.1016/j.ijfoodmicro.2018.12.024

IARC. 2012. Monographs on the evaluation of carcinogenic risks to humans: Chemical agents and related occupations. A review of human carcinogens. Lyon, France. 2012. International Agency for Research on Cancer 100F: 224–248.

IARC. 2015. Mycotoxin control in low- and middle-income countries. *In:* Wild, C.P., Miller, J.D. and Groopman, J.D. (Eds.). International Agency for Research on Cancer, Working Group Report Volume 9, pp. 56.

Ibrahim, F., Jalal, H., Khan, A.B., Asghar, M.A., Iqbal, J., Ahmed, A. et al. 2016. Prevalence of Aflatoxigenic *Aspergillus* in Food and Feed Samples from Karachi, Pakistan. Journal of Infection and Molecular Biology 4(1): 1–8. DOI: 10.14737/journal.jimb/2016/4.1.1.8.

Ito, Y., Peterson, S.W., Wicklow, D.T. and Goto, T. 2001. *Aspergillus pseudotamarii*, a new aflatoxin producing species in *Aspergillus* section *Flavi*. Mycological Research 105: 233–239. DOI: 10.1017/S0953756200003385.

Jaime-Garcia, R. and Cotty, P.J. 2004. *Aspergillus flavus* in soils and corncobs in south Texas: Implications for management of aflatoxins in corn-cotton rotations. Plant Disease 88(12): 1366–1371. DOI: 10.1094/PDIS.2004.88.12.1366

Jaime-Garcia, R. and Cotty, P.J. 2010. Crop rotation and soil temperature influence the community structure of *Aspergillus flavus* in soil. Soil Biology and Biochemistry 42(10): 1842–1847. DOI: 10.1016/j.soilbio.2010.06.025.

Jakić-Dimić, D., Nešić, K. and Petrović, M. 2009. Contamination of cereals with aflatoxins, metabolites of fungi *Aspergillus flavus*. Biotechnol. Biotechnology in Animal Husbandry 25(5-6-2): 1203–1208.

Jakšić, D., Puel, O., Canlet, C., Kopjar, N., Kosalec, I. and Klarić M.Š. 2012. Cytotoxicity and genotoxicity of versicolorins and 5-methoxysterigmatocystin in A549 cells. Archives of Toxicology 86: 1583–1591. DOI: 10.1007/s00204-012-0871-x.

Javanmard, M. 2010. Occurrence of mould counts and *Aspergillus* species in Iranian dried figs at different stages of production and processing. Journal of Agricultural Science and Technology 12: 331–338.

Jayaraman, P. and Kalyanasundaram, I. 1990. Natural occurrence of toxigenic fungi and mycotoxins in rice bran. Mycopathologia 110: 81–85. DOI: 10.1007/BF00446995.

Joshaghani, H., Namjoo, M., Rostami, M., Kohsar, F. and Niknejad, F. 2013. Mycoflora of fungal contamination in wheat storage (Silos) in Golestan Province, north of Iran. Jundishapur Journal of Microbiology 6(4): e6334. DOI: 10.5812/jjm.6334.

Kabirian, H.R., Afshari, H., Moghadam, M.M. and Hokmabadi, H. 2011. Evaluation of pistachio contamination to *Aspergillus flavus* in Semnan Province. Journal of Nuts 2(3): 01–06. DOI: 10.22034/jon.2011.515741.

Kachuei, R., Hossein, Y.M., Sasan, R., Abdolamir, A., Naser, S., Farideh, Z. et al. 2009. Investigation of stored wheat mycoflora, reporting the *Fusarium* cf. *langsethiae* in three provinces of Iran during 2007. Annals of Microbiology 59(2): 383. DOI: 10.1007/BF03178344.

Kachapulula, P.W., Akello, J., Bandyopadhyay, R. and Cotty, P.J. 2017. *Aspergillus* section *Flavi* community structure in Zambia influences aflatoxin contamination of maize and groundnut. International Journal of Food Microbiology 261: 49–56. DOI: 10.1016/j.ijfoodmicro.2017.08.014

Kamika, I., Mngqawa, P., Rheeder, J.P., Teffo, S.L. and Katerere, D.R. 2014. Mycological and aflatoxin contamination of peanuts sold at markets in Kinshasa, Democratic Republic of Congo, and Pretoria, South Africa. Food Additives & Contaminants: Part B 7(2): 120–126. DOI: 10.1080/19393210.2013.858187.

Katsurayama, A.M., Martins, L.M., Iamanaka, B.T., Fungaro, M.H.P., Silva, J.J., Frisvad, J.C. et al. 2018. Occurrence of *Aspergillus* section *Flavi* and aflatoxins in Brazilian rice: From field to market. International Journal of Food Microbiology 266: 213–221. DOI: 10.1016/j.ijfoodmicro.2017.12.008.

Kjærbølling, I., Vesth, T., Frisvad, J.C., Nybo, J.L., Theobald, S., Kildgaard, S. et al. 2020. A comparative genomics study of 23 *Aspergillus* species from Section *Flavi*. Nature Communications 11: 1106. DOI: 10.1038/s41467-019-14051-y.

Klich, M.A., Lee, L.S. and Huizar, H.E. 1986. The occurrence of *Aspergillus flavus* in vegetative tissue of cotton plants and its relation to seed infection. Mycopathologia 95(3): 171–174. DOI: 10.1007/BF00437123.

Kornher, L. 2018. Maize markets in Eastern and Southern Africa (ESA) in the context of climate change. The State of Agricultural Commodity Markets. (SOCO) 2018. Rome. FAO: p. 58.

Kos, J., Janić Hajnal, E., Šarić, B., Jovanov, P., Mandić, A., Đuragić, O. et al. 2018. Aflatoxins in maize harvested in the Republic of Serbia over the period 2012–2016. Food Additives & Contaminants: Part B 11(4): 246–255. DOI: 10.1080/19393210.2018.1499675.

Kowalska, A., Walkiewicz, K., Kozieł, P. and Muc-Wierzgoń, M. 2017. Aflatoxins: Characteristics and impact on human health. Postepy higieny i medycyny doswiadczalnej 71: 315–327. DOI: 10.5604/01.3001.0010.3816.

Kumar, P., Mahato, D.K., Kamle, M., Mohanta, T.K. and Kang, S.G. 2017. Aflatoxins: A global concern for food safety, human health and their management. Frontiers in Microbiology 7: 2170. DOI: 10.3389/fmicb.2016.02170.

Kumeda, Y., Asao, T., Takahashi, H. and Ichinoe, M. 2003. High prevalence of B and G aflatoxin-producing fungi in sugarcane field soil in Japan: Heteroduplex panel analysis identifies a new genotype within *Aspergillus* section *Flavi* and *Aspergillus nomius*. FEMS Microbiology Ecology 45: 229–238 DOI: 10.1016/S0168-6496(03)00154-5.

Kurtzman, C.P., Horn, B.W. and Hesseltine, C.W. 1987. *Aspergillus nomius*, a new aflatoxin-producing species related to *Aspergillus flavus* and *Aspergillus tamarii*. Antonie van Leeuwenhoek 53(3): 147–158. DOI: 10.1007/BF00393843.

Lewis, L., Onsongo, M., Njapau, H., Schurz-Rogers, H., Luber, G., Kieszak, S. et al. and the Kenya Aflatoxicosis Investigation Group. 2005. Aflatoxin contamination of commercial maize products during an outbreak of acute aflatoxicosis in eastern and central Kenya. Environmental Health Perspectives 113: 1763–1767. DOI: 10.1289/ehp.7998.

Magan, N., Medina, A. and Aldred, D. 2011. Possible climate-change effects on mycotoxin contamination of food crops pre- and post-harvest: Mycotoxins and climate change. Plant Pathology 60(1): 150–163. DOI: 10.1111/j.1365-3059.2010.02412.x.

Makhlouf, J., Carvajal-Campos, A., Querin, A., Tadrist, S., Puel, O., Lorber S. et al. 2019. Morphologic, molecular and metabolic characterization of *Aspergillus* section *Flavi* in spices marketed in Lebanon. Scientific Reports 9(1): 5263. DOI: 10.1038/s41598-019-41704-1.

Mamo, F.T., Shang, B., Selvaraj, J.N., Wang, Y. and Liu, Y. 2018. Isolation and characterization of *Aspergillus flavus* strains in China. Journal of Microbiology 56(2): 119–127. DOI: 10.1007/s12275-018-7144-1.

Manizan, A.L. 2019. Evaluation de la contamination par les mycotoxines des céréales et oléagineux les plus consommes en Côte d'Ivoire: cas du circuit post-récolte de l'arachide (*Arachis hypogea* L.). Ph.D. Thesis, Université Nangui Abrogoua, Abidjan, Côte d'Ivoire.

Marin, S., Ramos, A.J. and Sanchis, V. 2012. Modelling *Aspergillus flavus* growth and aflatoxins production in pistachio nuts. Food Microbiology 32: 378–388. DOI: 10.1016/j.fm.2012.07.018

Martins, L.M., Sant'Ana, A.S., Fungaro, M.H.P., Silva, J.J., do Nascimento, M.D.S., Frisvad, J.C. et al. 2017. The biodiversity of *Aspergillus* section *Flavi* and aflatoxins in the Brazilian peanut production chain. Food Research International 94: 101–107. DOI: 10.1016/j.foodres.2017.02.006.

Massi, F.P., Vieira, M.L.C., Sartori, D., Penha, R.E.S., de Freitas Munhoz, C., Ferreira, J.M. et al. 2014. Brazil nuts are subject to infection with B and G aflatoxin-producing fungus, *Aspergillus pseudonomius*. International Journal of Food Microbiology 186: 14–21. DOI: 10.1016/j.ijfoodmicro.2014.06.006.

Mateo, E.M., Gil-Serna, J., Patiño, B. and Jiménez, M. 2011. Aflatoxins and ochratoxin A in stored barley grain in Spain and impact of PCR-based strategies to assess the occurrence of aflatoxigenic and ochratoxigenic *Aspergillus* spp. International Journal of Food Microbiology 149(2): 118–126. DOI: 10.1016/j.ijfoodmicro.2011.06.006.

Mauro, A., Battilani, P., Callicott, K.A., Giorni, P., Pietri, A. and Cotty, P.J. 2013. Structure of an *Aspergillus flavus* population from maize kernels in northern Italy. International Journal of Food Microbiology 162(1): 1–7. DOI: 10.1016/j.ijfoodmicro.2012.12.021.

Mazzani, C., Luzón, O., Chavarri, M., Fernández, M. and Hernández, N. 2008. *Fusarium verticillioides* y fumonisinas en maíz cosechado en pequeñas explotaciones y conucos de algunos estados de Venezuela. Fitopatología Venezolana 21(1): 18–22.

Mazen, M.B., El-Kady, I.A. and Saber, S.M. 1990. Survey of the mycoflora and mycotoxins of cotton seeds and cotton seed products in Egypt. Mycopathologia 110(3): 133–138. DOI: 10.1007/BF00437536.

Medina, A., Rodríguez, A., Sultan, Y. and Magan, N. 2015. Climate change factors and *Aspergillus flavus*: Effects on gene expression, growth and aflatoxin production. World Mycotoxin Journal 8 (2): 171–179. DOI: 10.3920/WMJ2014.1726.

Mohamad, R., Mohamed, M.S., Suhaili, N., Salleh, M.M. and Ariff, A.B. 2010. Kojic acid: Applications and development of fermentation process for production. Biotechnology and Molecular Biology Reviews 5(2): 24–37.

Mohamed, A. and Chala, A. 2014. Incidence of *Aspergillus* contamination of groundnut (*Arachis hypogaea* L.) in Eastern Ethiopia. African Journal of Microbiology Research 8(8): 759–765. DOI: 10.5897/AJMR12.2078.

Moore, F.C. and Lobell, D.B. 2015. The fingerprint of climate trends on European crop yields. Proceedings of the National Academy of Sciences 112(9): 2670–2675. DOI: 10.1073/pnas.1409606112.

Moore, G.G., Mack, B.M., Beltz, S.B. and Gilbert, M.K. 2016. Draft genome sequence of an aflatoxigenic *Aspergillus* species, *A. bombycis*. Genome Biology and Evolution 8: 3297–3300. DOI: 10.1093/gbe/evw238. DOI: 10.1093/gbe/evw238.

Moretti, A., Pascale, M. and Logrieco, A.F. 2019. Mycotoxin risks under a climate change scenario in Europe. Trends in Food Science & Technology 84: 38–40. DOI: 10.1016/j.tifs.2018.03.008.

Mphande, F.A., Siame, B.A. and Taylor, J.E. 2004. Fungi, aflatoxins, and cyclopiazonic acid associated with peanut retailing in Botswana. Journal of Food Protection 67(1): 96–102. DOI: 10.4315/0362-028X-67.1.96.

Nesci, A. and Etcheverry, M. (2002). *Aspergillus* section *Flavi* populations from field maize in Argentina. Letters in Applied Microbiology 34(5): 343–348. DOI: 10.1046/j.1472-765X.2002.01094.x.

Njobeh, P.B., Dutton, M.F., Koch, S.H., Chuturgoon, A., Stoev, S. and Seifert, K. 2009. Contamination with storage fungi of human food from Cameroon.

International Journal of Food Microbiology 135(3): 193–198. DOI: 10.1016/j. ijfoodmicro.2009.08.001.

Norlia, M., Jinap, S., Nor-Khaizura, M.A.R., Son, R. and Chin, C.K. 2018. Polyphasic approach to the identification and characterization of aflatoxigenic strains of *Aspergillus* section *Flavi* isolated from peanuts and peanut-based products marketed in Malaysia. International Journal of Food Microbiology 282 (3): 9–15. DOI: 10.1016/j.ijfoodmicro.2018.05.030

Okoth, S., De Boevre, M., Vidal, A., Di Mavungu, J.D. Landschoot, S., Kyallo, M. et al. 2018. Genetic and toxigenic variability within *Aspergillus flavus* population isolated from maize in two diverse environments in Kenya. Frontiers in Microbiology 9: 57. DOI: 10.3389/fmicb.2018.00057.

Onana, B., Essono, G., Bekolo, N., Ambang, Z. and Ayodele, M. 2013. Mycoflora associated with processed and stored cassava chips in rural areas of southern Cameroon. African Journal of Microbiology Research 7(43): 5036–5045. DOI: 10.5897/AJMR2013.6081.

Ortega-Beltran, A., Jaime, R. and Cotty, P.J. 2015. Aflatoxin-producing fungi in maize field soils from sea level to over 2000 masl: A three-year study in Sonora, Mexico. Fungal Biology 119(4): 191–200. DOI: 10.1016/j.funbio.2014.12.006.

Osamwonyi, U.O. and Wakil, S.M. 2012. Isolation of fungal species from fermentating pearl millet gruel and determination of their antagonistic activities against indicator bacterial species. Nigerian Food Journal 30(1): 35–42. DOI: 10.1016/S0189-7241(15)30011-4.

Pacheco, A.M., Lucas, A., Parente, R. and Pacheco, N. 2010. Association between aflatoxin and aflatoxigenic fungi in Brazil nut (*Bertholletia excelsa* HBK). Food Science and Technology 30(2): 330–334. DOI: 10.1590/S0101-20612010000200007

Pacin, A.M., Gonzalez, H.H.L., Etcheverry, M., Resnik, S.L., Vivas, L. and Espin, S. 2003. Fungi associated with food and feed commodities from Ecuador. Mycopathologia 156(2): 87–92. DOI: 10.1023/A: 1022941304447.

Pankaj, S.K., Shi, H. and Keener, K.M. 2018. A review of novel physical and chemical decontamination technologies for aflatoxin in food. Trends in Food Science & Technology 71: 73–83. DOI: 10.1016/j.tifs.2017.11.007.

Park, J.W., Choi, S.Y., Hwang, H.J. and Kim, Y.B. 2005. Fungal mycoflora and mycotoxins in Korean polished rice destined for humans. International Journal of Food Microbiology 103(3): 305–314. DOI: 10.1016/j.ijfoodmicro.2005.02.001.

Paterson, R.R.M. and Lima, N. 2011. Further mycotoxin effects from climate change. Food Res. Int. 44: 2555–2566. DOI: 10.1016/j.foodres.2011.05.038.

Perrone, G., Gallo, A. and Logrieco, A.F. 2014. Biodiversity of *Aspergillus* section *Flavi* in Europe in relation to the management of aflatoxin risk. Frontiers in Microbiology 5: 377. DOI: 10.3389/fmicb.2014.00377.

Peterson, S.W., Ito, Y., Horn, B.W. and Goto, T. 2001. *Aspergillus bombycis,* a new aflatoxigenic species and genetic variation in its sibling species, *A. nomius.* Mycologia 93: 689–703. DOI: 10.2307/3761823.

Pildain, M.B., Frisvad, J.C., Vaamonde, G., Cabral, D., Varga, J. and Samson, R.A. 2008. Two novel aflatoxin-producing *Aspergillus* species from Argentinean peanuts. International Journal of Systematic and Evolutionary Microbiology 58: 725–735. DOI: 10.1099/ijs.0.65123-0.

Pitt, J.I. and Hocking, A.D. 2006. Mycotoxins in Australia: Biocontrol of aflatoxin in peanuts. Mycopathologia 162(3): 233–243. DOI: 10.1007/s11046-006-0059-0.

Pitt, J.I. and Hocking, A.D. 2009. Fungi and Food Spoilage, 3rd ed., Springer US: New York, USA. ISBN 978-0-387-92206-5.

Pitt, J.I., Taniwaki, M.H. and Cole, M.B. 2012. Mycotoxin production in major crops as influenced by growing, harvesting, storage and processing, with emphasis on the achievement of Food Safety Objectives. Food Control 32(1): 205–215. DOI: 10.1016/j.foodcont.2012.11.023.

Prencipe, S., Siciliano, I., Contessa, C., Botta, R., Garibaldi, A., Gullino, M.L. et al. 2018. Characterization of *Aspergillus* section *Flavi* isolated from fresh chestnuts and along the chestnut flour process. Food Microbiology 69: 159–169. DOI: 10.1016/j.fm.2017.08.004.

Probst, C., Callicott, K.A. and Cotty, P.J. 2012. Deadly strains of Kenyan *Aspergillus* are distinct from other aflatoxin producers. European Journal of Plant Pathology 132: 419–429. DOI: 10.1007/s10658-011-9887-y

Probst, C., Bandyopadhyay, R. and Cotty, P.J. 2014. Diversity of aflatoxin-producing fungi and their impact on food safety in sub-Saharan Africa. International Journal of Food Microbiology 174: 113–122. DOI: 10.1016/j.ijfoodmicro.2013.12.010.

Purcell, S.L., Phillips, D.J. and Mackey, B.E. 1980. Distribution of *Aspergillus flavus* and other fungi in several almond-growing areas of California. Phytopathology 70(9): 926–929.

Rahimi, P., Sharifnabi, B. and Bahar, M. 2007. Detection of aflatoxin in *Aspergillus* species isolated from pistachio in Iran. Journal Phytopathology 156(1): 15–20. DOI: 10.1111/j.1439-0434.2007.01312.x.

Rasheed, U., Wu, H., Wei, J., Ou, X., Qin, P., Yao, X. et al. 2019. A polyphasic study of *Aspergillus* section *Flavi* isolated from corn in Guangxi, China – a hot spot of aflatoxin contamination. International Journal of Food Microbiology 310: 108307. DOI: 10.1016/j.ijfoodmicro.2019.108307.

Razzaghi-Abyaneh, M., Shams-Ghahfarokhi, M., Allameh, A., Kazeroon-Shiri, A., Ranjbar-Bahadori, S., Mirzahoseini, H. et al. 2006. A survey on distribution of *Aspergillus* section *Flavi* in corn field soils in Iran: Population patterns based on aflatoxins, cyclopiazonic acid and sclerotia production. Mycopathologia 161(3): 183–192. DOI: 10.1007/s11046-005-0242-8.

Reddy, K.R.N., Farhana, N.I., Wardah, A.R. and Salleh, B. 2010. Morphological identification of foodborne pathogens colonizing rice grains in South Asia. Pakistan Journal of Biological Sciences 13(16): 794–801.

Reis, T.A.D., Oliveira, T.D., Baquião, A.C., Gonçalves, S.S., Zorzete, P. and Corrêa, B. 2012. Mycobiota and mycotoxins in Brazil nut samples from different states of the Brazilian Amazon region. International Journal of Food Microbiology 159(2): 61–68. DOI: 10.1016/j.ijfoodmicro.2012.08.005.

Riba, A., Mokrane, S., Mathieu, F., Lebrihi, A. and Sabaou, N. 2008. Mycoflora and ochratoxin-A producing strains of *Aspergillus* in Algerian wheat. International Journal of Food Microbiology 122(1–2): 85–92. DOI: 10.1016/j.ijfoodmicro.2007.11.057.

Riba, A., Bouras, N., Mokrane, S., Mathieu, F., Lebrihi, A. and Sabaou, N. 2010. *Aspergillus* section *Flavi* and aflatoxins in Algerian wheat and derived products. Food and Chemical Toxicology 48(10): 2772–2777. DOI: 10.1016/j.fct.2010.07.005.

Rodrigues, P., Venâncio, A., Kozakiewicz, Z. and Lima, N. 2009. A polyphasic approach to the identification of aflatoxigenic and non-aflatoxigenic strains of *Aspergillus* section *Flavi* isolated from Portuguese almonds. International Journal of Food Microbiology 129(2): 187–193. DOI: 10.1016/j. ijfoodmicro.2008.11.023.

Rodrigues, P., Venâncio, A. and Lima, N. 2012. Aflatoxigenic fungi and aflatoxins in Portuguese almonds. The Scientific World Journal vol n pp DOI: 10.1100/2012/471926.

Rychlik, M., Humpf, H.U., Marko, D., Dänicke, S., Mally, A., Berthiller, F. et al. 2014. Proposal of a comprehensive definition of modified and other forms of mycotoxins including "masked" mycotoxins. Mycotoxin Research 30(4): 197–205. DOI: 10.1007/s12550-014-0203-5.

Saadullah, A. and Abdullah, S. 2015. Mycobiota and incidence of toxigenic fungi in dried fruits from Duhok Markets, North Iraq. Egyptian Academic Journal of Biological Sciences, G. Microbiology 7(1): 61–68.

Saberi-Riseh, R., Javan-Nikkhah, M., Heidarian, R., Hosseini, S. and Soleimani, P. 2004. Detection of fungal infectious agent of wheat grains in store-pits of Markazi province, Iran. Communications in Agricultural and Applied Biological Sciences 69(4): 541–544.

Saleem, M.J., Hannan, A. and Qaisar, T.A. 2012. Occurrence of aflatoxins in maize seed under different conditions. International Journal of Agriculture & Biology 14(3): 473–476.

Sales, A.C. and Takumi, Y. 2005. Updated profile of aflatoxin and *Aspergillus* section *Flavi* contamination in rice and its byproducts from the Philippines. Food Additives & Contaminants 22(5): 429–436. DOI: 10.1080/02652030500058387.

Samson, R.A. and K.A. Seifert. 1985. The ascomycete genus *Penicilliopsis* and its anamorphs. pp. 397–426. *In:* R.A. Samson and J.I. Pitt (eds.). Advances in *Penicillium* and *Aspergillus* Systematics. Plenum Press, New York and London.

Samson, R.A., Visagie, C.M., Houbraken, J., Hong, S.B., Hubka, V., Klaassen, C.H. et al. 2014. Phylogeny, identification and nomenclature of the genus *Aspergillus*. Studies in Mycology 78: 141–173. DOI: 10.1016/j.simyco.2014.07.004.

Savi, G.D., Piacentini, K.C., Tibola, C.S. and Scussel, V.M. 2014. Mycoflora and deoxynivalenol in whole wheat grains (*Triticum aestivum* L.) from Southern Brazil. Food Additives & Contaminants: Part B 7(3): 232–237. DOI: 10.1080/19393210.2014.898337.

Schatzmayr, G. and Streit, E. 2013. Global occurrence of mycotoxins in the food and feed chain: Facts and figures. World Mycotoxin Journal 6(3): 213–222. DOI: 10.3920/WMJ2013.1572

Sebők, F., Dobolyi, C., Hartman, M., Risa, A., Kritafon, C., Szoboszlay, S. et al. 2014. Occurrence of potentially aflatoxin producing *Aspergillus* species in maize fields of Abstract Book of the Meeting of the Hungarian Microbiological Society and the EU FP7 PROMISE Hungary. *In:* Regional Meeting, pp. 62–63.

Şenyuva, H., Gilbert, J., Samson, R.A., Özcan, S., Öztürkoğlu, Ş. and Önal, D. 2008. Occurrence of fungi and their mycotoxins in individual Turkish dried figs. World Mycotoxin Journal 1(1): 79–86. DOI: 10.3920/WMJ2008.x009

Sharfun-Nahar, S.N., Mushtaq, M. and Pathan, I.H. 2004. Seed-borne mycoflora of *Capsicum annuum* imported from India. Pakistan Journal of Botany 36(1): 191–198.

Sharma, S., Gupta, D. and Sharma, Y.P. 2013. Aflatoxin contamination in chilgoza pine nuts (*Pinus gerardiana* wall.) commercially available in retail markets of Jammu, India. International Journal of Pharma and Bio. Sciences 4: 751–759.

Singh, P. and Cotty, P.J. 2019. Characterization of Aspergilli from dried red chilies (Capsicum spp.): Insights into the etiology of aflatoxin contamination. International Journal of Food Microbiology 289: 145–153. DOI: 10.1016/j. ijfoodmicro.2018.08.025.

Singh, P., Orbach, M.J. and Cotty, P.J. 2018. *Aspergillus texensis*: A novel aflatoxin producer with S morphology from the United States. Toxins 10: 513. DOI:10.3390/toxins10120513.

Sinha, A.K. and Sinha, K.K. 1990. Insect pests, *Aspergillus flavus* and aflatoxin contamination in stored wheat: A survey at North Bihar (India). Journal of Stored Products Research 26(4): 223–226. DOI: 10.1016/0022-474X(90)90026–O.

Škrinjar, M.M., Miklič, V.J., Kocić-Tanackov, S.D., Jeromela-Marjanović, A.M., Maširević, S.N., Suturović, I.Z. et al. 2013. Xerophilic mycopopulations isolated from rapeseeds (*Brassica napus*). Acta Periodica Technologica 44: 115–124. DOI: 10.2298/APT1344115S.

Soares, C., Rodrigues, P., Peterson, S.W., Lima, N. and Venâncio, A. 2012. Three new species of *Aspergillus* section *Flavi* isolated from almonds and maize in Portugal. Mycologia 104: 682–697. DOI: 10.3852/11–088.

Stack, J. and Carlson, M. 2003. NF571 *Aspergillus flavus* and Aflatoxins in Corn. Historical Materials from University of Nebraska-Lincoln Extension.

Suárez-Bonnet, E., Carvajal, M., Méndez-Ramírez, I., Castillo-Urueta, P., Cortés-Eslava, J., Gómez-Arroyo, S. et al. 2013. Aflatoxin (B1, B2, G1, and G2) contamination in rice of Mexico and Spain, from local sources or imported. J. Food Sci. 78(11): T1822-T1829.

Sultan, Y. and Magan, N. 2010. Mycotoxigenic fungi in peanuts from different geographic regions of Egypt. Mycotoxin Research 26: 133–140. DOI: 10.1007/s12550-010-0048-5.

Takahashi, H., Kamimura, H. and Ichinoe, M. 2004. Distribution of aflatoxin-producing *Aspergillus flavus* and *Aspergillus parasiticus* in sugarcane fields in the southernmost islands of Japan. Journal of Food Protection 67(1): 90–95. DOI: 10.4315/0362-028X-67.1.90.

Taniwaki, M.H., Pitt. J.I., Iamanaka, B.T., Sartori, D., Copetti, M.V., Balajee, A. et al. 2012. *Aspergillus bertholletius* sp. *nov.* from Brazil Nuts. PLoS ONE 7: 42480. DOI: 10.1371/journal.pone.0042480.

Taniwaki, M.H., Frisvad, J.C., Ferranti, L.S., de Souza Lopes, A., Larsen, T.O., Fungaro, M.H.P. et al. 2017. Biodiversity of mycobiota throughout the Brazil nut supply chain: From rainforest to consumer. Food Microbiology 61: 14–22. DOI: 10.1016/j.fm.2016.08.002

Taniwaki, M.H., Pitt, J.I. and Magan, N. 2018. *Aspergillus* species and mycotoxins: Occurrence and importance in major food commodities. Current Opinion in Food Science 23: 38–43. DOI: 10.1016/j.cofs.2018.05.008.

TePaske, M.R., Gloer, J.B., Wicklow, D.T. and Dowd, P.F. 1992. Aflavarin and beta-aflatrem: New anti-insectan metabolites from the sclerotia of *Aspergillus flavus*. Journal of Natural Products 55: 1080–1086. http://dx.doi.org/10.1021/np50086a008.

Theumer, M.G., Henneb, Y., Khoury, L., Snini, S.P., Tadrist, S., Canlet, C. et al. 2018. Genotoxicity of aflatoxins and their precursors in human cells. Toxicology Letters 287: 100–107. DOI: 10.1016/j.toxlet.2018.02.007.

Tomar-Balhara, M., Kapoor, R. and Bhatnagar, A.K. 2006. Seed mycoflora and aflatoxin contamination in mustard during storage. pp. 321–338. *In*: Mukerji, K.G. and Manoharachary, C. (eds.). Current Concepts in Botany. I.K. International, New Delhi.

Tomar, D.S., Shastry, P.P., Nayak, M.K. and Sikarwar, P. 2012. Effect of seed borne mycoflora on cotton seed (JK 4) and their control. Journal Cotton Restitution of Developpment 26(1): 105–108.

Tran-Dinh, N., Kennedy, I., Bui, T. and Carter, D. 2009. Survey of Vietnamese peanuts, corn and soil for the presence of *Aspergillus flavus* and *Aspergillus parasiticus*. Mycopathologia 168(5): 257–268. DOI: 10.1007/s11046-009-9221-9.

Trung, T., Tabuc, C., Bailly, S., Querin, A., Guerre, P. and Bailly, J. 2008. Fungal mycoflora and contamination of maize from Vietnam with aflatoxin B1 and fumonisin B1. World Mycotoxin Journal 1(1): 87–94. DOI: 10.3920/WMJ2008. x010.

Turgay, E.B. and Ünal, F. 2009. Detection of seed borne mycoflora of sorghum in Turkey. The Journal of Turkish Phytopathology 38(1-3): 9–20. DOI:

Turner, P.C., Sylla, A., Gong, Y.Y., Diallo, M.S., Sutcliffe, A.E., Hall, A.J. et al. 2005. Reduction in exposure to carcinogenic aflatoxins by postharvest intervention measures in West Africa: A community based intervention study. Lancet 365: 1950–1956. DOI: 10.1016/S0140-6736(05)66661-5.

Umemura, M., Koike, H., Nagano, N., Ishii, T., Kawano, J., Yamane, N. et al. 2013. MIDDAS-M: Motif-independent de novo detection of secondary metabolite gene clusters through the integration of genome sequencing and transcriptome data. PLoS One 8: e84028. DOI: 10.1371/journal.pone.0084028.

Vaamonde, G., Patriarca, A., Pinto, V.F., Comerio, R. and Degrossi, C. 2003. Variability of aflatoxin and cyclopiazonic acid production by *Aspergillus* section *Flavi* from different substrates in Argentina. International Journal of Food Microbiology 88(1): 79–84. DOI: 10.1016/S0168-1605(03)00101-6.

Varga, J., Frisvad, J.C. and Samson, R.A. 2011. Two new aflatoxin producing species and an overview of *Aspergillus* section *Flavi*. Studies in Mycology 69: 57–80. DOI: 10.3114/sim.2011.69.05.

Varga, J., Baranyi, N., Chandrasekaran, M., Vágvölgyi, C. and Kocsubé, S. 2015. Mycotoxin producers in the *Aspergillus* genus: An update. Acta Biologica Szegediensis 59: 151–167.

Wallace, H.A.H. and Sinha, R.N. 1962. Fungi associated with hot spots in farm stored grain. Canadian Journal of Plant Science 42(1): 130–141. DOI: 10.4141/ cjps62-016.

Wartu, J.R., Whong, C.M.Z., Umoh, V.J. and Diya, A.W. 2015. Occurrence of aflatoxin levels in harvest and stored groundnut kernels in Kaduna State, Nigeria. Journal of Environmental Science, Toxicology and Food Technology 9(1): 62–66. DOI: 10.9790/2402-09126266.

Watarai, N., Yamamoto, N., Sawada, K. and Yamada, T. 2019. Evolution of *Aspergillus Oryzae* before and after domestication inferred by large-scale comparative genomic analysis. DNA Research 26: 465–472 DOI: 10.1093/ dnares/dsz024.

Weledesemaya, G.T., Gezmu, T.B., Woldegiorgis, A.Z. and Gemede H.F. 2016. Study on *Aspergillus* species and aflatoxin levels in sorghum (*Sorghum bicolor* L.) stored for different period and storage system in Kewet Districts, Northern Shewa, Ethiopia. JSOA Journal of Food Science and Nutrition 2(1): 1–8. DOI: 10.24966/FSN-1076/100010.

Wu, F. and Guclu, H. 2012. Aflatoxin regulations in a network of global maize trade. PloS One 7(9): 45151. DOI: 10.1371/journal.pone.0045151.

Yogendrarajah, P., Vermeulen, A., Jacxsens, L., Mavromichali, E., De Saeger, S., De Meulenaer, B. et al. 2016. Mycotoxin production and predictive modelling kinetics on the growth of *Aspergillus flavus* and *Aspergillus parasiticus* isolates in whole black peppercorns (*Piper nigrum* L). International Journal of Food Microbiology 228: 44–57. DOI: 10.1016/j.ijfoodmicro.2016.03.015.

Zhang, S., Monahan, J.B., Tkacz, J.S. and Berry, S. 2004. Indole diterpene gene cluster from *Aspergillus flavus*. Applied and Environmental Microbiology 70: 6875–6883. DOI: 10.1128/AEM.70.11.6875–6883.2004.

Zohri, A.A. and Abdel-Gawad, K.M. 1993. Survey of mycoflora and mycotoxins of some dried fruits in Egypt. Journal of Basic Microbiology 33(4): 279–288. DOI: 10.1002/jobm.3620330413.

Mycotoxins in Foods and Feeds in Morocco: Occurrence, Sources of Contamination, Prevention/Control and Regulation

Amina Bouseta[1*], Adil Laaziz[1], Hassan Hajjaj[2] and Rajae Belkhou[3]

[1] Laboratory of Biotechnology, Environment, Agri-food and Health (LBEAS) Faculty of Science Dhar Mahraz, University Sidi Mohamed Ben Abdallah, P.B. 1796 Atlas Fez, Morocco

[2] Laboratory of Biotechnologies and Development of Bio-resources, Cluster of Competency "Agri-food, Safety and Security", IUC VLIR-UOS, Moulay Ismail University, Marjane 2, BP 298, Meknes, Morocco

[3] Laboratory of Biotechnology, Environment, Agri-food and Health (LBEAS), High School of Technology, University Sidi Mohamed Ben Abdallah, P.B. 1796 Atlas Fez, Morocco

1. Introduction

Mycotoxins are secondary metabolites, which are very harmful to human and animal health. They are produced by molds mainly belonging to the genera *Aspergillus*, *Penicillium* and *Fusarium*. Park et al. (1999) reported that the Food and Agriculture Organization (FAO) of the United Nations estimated that at least 25% of the world's food crops are contaminated with mycotoxins. According to Eskola et al. (2019), the origin of the statement that the FAO estimated global food crop contamination with mycotoxins to be 25% is largely unknown. To assess the rationale for it, the authors examined the relevant literature and data of around 500,000 analyses from the European Food Safety Authority (EFSA) and large global survey for aflatoxins (AFs), fumonisins B (FB), deoxynivalenol (DON), T-2 toxin (T-2) and HT-2 toxin (HT-2), zearalenone (ZEA) and ochratoxin A (OTA) in cereals and nuts. This study seems to confirm the value of 25% estimated by FAO, although this figure underestimates, according to the study, the occurrence of mycotoxins above the detection limits (up to 60–80%). These figures only point out that, in terms

*Corresponding author: amina.bouseta@usmba.ac.ma

of health, contamination of food with mycotoxins is one of the main global health concerns. The European Union's rapid alert system for food and feed reports shows that mycotoxins are in first place according to the total number of danger notifications (RASFF 2018). In Morocco, the mycotoxin problem has been identified as one of the major challenges for food safety (Montet et al. 2020).

Several mycotoxins are thermally stable and are not easily eliminated during food processing or by physical and chemical treatments. Currently, more than 400 mycotoxins have been reported but only a limited number have toxic characteristics for humans and/or animals (Pitt et al. 2000, Shi et al. 2018). Certain mycotoxins can be produced by several species belonging to different genera. Likewise, a species can develop several mycotoxins. However, within the same species known to be toxigenic, not all strains have this capacity (Hussein and Brasel 2001, Reboux 2006). The number of contaminated products and emerging mycotoxins continues to increase due to the evolution of extraction and analysis techniques. Indeed, since the first detection of mycotoxins by TLC, several analysis techniques have been developed (ELISA, HPLC, GC, GC-MS, LC-MS, LC-MS/MS ...) leading to multi-detection and quantification of mycotoxins in foodstuffs.

The presence of mycotoxins has been strongly studied in various foods such as cereals and derivatives (Lee and Ryu 2017), coffee and tea drinks (Pallarès et al. 2017, García-Moraleja et al. 2015), vegetables (Dong et al. 2019), fruit juices and cooked foods (Sakuma et al. 2013, Carballo et al. 2018). In addition to the human and animal health problem, mycotoxins also cause significant economic losses (Pitt and Miller 2017, Wu and Mitchell 2016). Currently, the most important mycotoxins from a food safety and regulatory point of view are AFs, DON, T-2, HT-2, ZEA, FB, OTA, ergot alkaloids (EA), patulin (PAT) and citrinin (CIT) (Eskola et al. 2019).

2. Mycotoxins

2.1 Aflatoxins

Aflatoxins were first identified in 1960 as the causative agents of "Turkey X disease" killing 100,000 turkeys in England. In humans, the first aflatoxicoses were reported in India in 1974 causing the death of 100 people (Krishnamachari et al. 1975) and in Kenya in 2004, resulting in the death of 125 people (Muture and Ogana 2005). In both cases, the consumption of corn contaminated with high aflatoxins levels seems to have caused the epidemic. Aflatoxins are toxic metabolites synthesized mainly by the species of *Aspergillus parasiticus*, *A. flavus* and *A. nomius* (Van den Broek et al. 2001, Richard 2007). Four types of aflatoxins can be produced by these species: AFB1, AFB2, AFG1 and AFG2. The species *A. flavus* produces only aflatoxins B while the other two species can produce both aflatoxins B and G (Creppy 2002, Bennett and Klich 2003). The growth of *Aspergillus* strains and the production of four aflatoxins (AFB1, AFB2, AFG1 and AFG2) are highly dependent on environmental factors such as temperature, humidity, aeration and the nature of the environment. The major proliferation risk is linked to transport and storage conditions. Aflatoxin-producing fungi grow on a wide variety of foods such as cereals (corn, rice, barley, oats and sorghum), dried fruits (grapes and figs), peanuts, pistachios, almonds, nuts and cotton seeds (Bennett and Klich 2003, De Boevre 2012, Soubra et al. 2009). Milk can also be contaminated with aflatoxin M1 (AFM1) produced by hydroxylation of AFB1 by the hepatic microsomal cytochrome P450 in cows fed with a diet contaminated with AFB1 (Bennett and Klich 2003). AFM1 can also be detected in cheese with a higher concentration than that of raw milk. AFM1 is thermostable, can bind to casein and therefore is not affected by the cheese-making process (Barbiroli et al. 2007). AFs have carcinogenic, teratogenic, hepatotoxic, mutagenic and immunosuppressive properties, with AFB1 being the most toxic. According to the International Agency for Research on Cancer (IARC), AFB1 is classified in group 1 with high risks of hepatocellular carcinoma while AFM1 being less toxic, it is classified in group 2B (possible carcinogen for humans). AFB1 forms DNA adducts by covalent binding to N7-guanine, resulting in persistent DNA lesions and AFB1 induces oxidative stress including modulation of antioxidant (EFSA 2020). The pharmacokinetics of aflatoxin is not yet fully understood (Dohnal et al. 2014). The liver is the primary site of aflatoxin metabolism, where they are converted to the 8,9-epoxide form by cytochrome P450 (CYP) enzymes (Wild and Turner 2002, Verma 2004). In fact, these enzymes oxidize AFB1 in the liver to form AFB1-8,9-exoepoxide and AFB1-endo-epoxide. AFB1-8,9-exoepoxide can bind to DNA mainly forming the adduct (AFB1- N7-Gua) which is responsible for the mutagenic properties of AFB1. The endo-epoxide AFB1 cannot bind to nucleic acids, it is less toxic (Wild and Turner 2002, Verma 2004, Dohnal et al. 2014).

2.2 Ochratoxine A

Ochratoxin A (OTA) is a secondary metabolite produced by several species of mold belonging to the genera *Aspergillus* and *Penicillium* (Abarca et al. 2003, Varga et al. 2003). OTA was first identified in South Africa in cereals (Van der Merwe et al. 1965), but the cereal sector is not the only one affected by this mycotoxin.

Indeed, the presence of OTA has been reported in many other products such as vegetables, coffee, beer, wine, grape juice, raisins as well as cocoa products, nuts and spices (EFSA 2006). It has also been found in the blood and tissues of animals and in the serum of people who have eaten contaminated food (EFSA 2006, Marquardt and Frohlich 1992). The kidney is the main target organ of exposure to OTA, in fact it causes nephrotoxicity in animals. OTA has also been associated with Endemic Balkan Nephropathy (Krogh et al. 1987), although the causality of OTA in human nephropathy remains unclear. OTA has carcinogenic, teratogenic, immunotoxic and possibly neurotoxic properties; IARC (1993) classified it in group 2B as a possible human carcinogen. Recently, Ostry et al. (2017) reported that new data on the formation of OTA-DNA adducts, on the role of OTA in oxidative stress and the identification of epigenetic factors involved in OTA carcinogenesis could lead to the reclassification of OTA.

2.3 Trichothecenes

Trichothecenes belong to a family of around 200 structurally related mycotoxins. These are cyclic sesquiterpenoids characterized by a stable C12-C13 epoxy cycle and a double bond between C9 and C10. Trichothecenes are divided into four groups (A–D) according to their functional groups (acetoxy and hydroxyl). Type A is represented by toxin HT-2 and toxin T-2, and type B is often represented by deoxynivalenol (DON) and nivalenol (NIV) (Marin et al. 2013). Types C and D include some lesser important trichothecenes (Marin et al. 2013). Trichothecenes type A and B are produced by several *Fusarium* species (Nielsen and Thrane 2001), but also by certain *Trichoderma* species (Nielsen et al. 2005). The most important T-2 and HT-2 producing species are *F. sporotrichioides*, *F. langsethiae*, *F. acuminatum* and *F. poae* while the main DON producers are *F. graminearum*, *F. culmorum* and *F. cerealis* (Marin et al. 2013). In addition to DON, its acetylated derivatives, 3-acetyl-deoxynivalenol (3-Ac-DON) and 15-acetyl-deoxynivalenol (15-Ac-DON) can be produced by these fungi and have been detected with DON but at levels generally 10% lower than those reported for DON (FAO/WHO 2011). These fungi that grow on crops in fields are phytopathogens and can develop in temperate climates (Marin et al. 2013). Some species are responsible for *Fusarium* wilt (FHB), one of the most serious plant diseases that results in quality loss and reduced grain yield (Krnjaja et al. 2011). Trichothecenes are generally very stable compounds during storage, grinding, cooking of food and are not degraded by high temperatures (EFSA 2011). Animal exposure to TCs leads to *in vitro* and *in vivo* apoptosis in several organs such as lymphoid organs, hematopoietic tissues, liver, intestinal crypts, bone marrow and thymus (Pestka 2007, EFSA 2011). They interact with ribosomes and mitochondria causing inhibition of protein synthesis, their cytotoxic effects on cell membranes and inside the cell are facilitated by their amphiphilic nature (Pace et al. 1988). Acute high dose toxicity of trichothecenes is characterized by diarrhea, vomiting, leukocytosis, hemorrhage, circulatory shock and death, while chronic low dose toxicity is characterized by anorexia, weight gain suppression, neuroendocrine and immunological changes (Pestka 2007, Larsen et al. 2004). With regard to human exposure, cereals and cereal-based products were the main sources of DON, while for farm and companion animals, cereals, cereal by-products and feed corn were the main sources of DON and contribute the most to their exposure (EFSA 2017).

The main sources of exposure to T-2 and HT-2 toxins are cereals and cereal-based foods, especially bread, fine bakery products, cereal millings and breakfast cereals (EFSA 2011). The T-2 toxin is rapidly metabolized into a large number of products, the HT-2 toxin being its major metabolite. Estimates of chronic human dietary exposure to the sum of T-2 and HT-2 toxins based on the occurrence data are below the tolerable daily intake (TDI), these toxins do not constitute a health problem (EFSA 2011).

2.4 Zearalenone

Zearalenone is a mycotoxin produced by species of the genus *Fusarium* and in particular *F. graminearum, F. culmorum, F. semitectum, F. equiseti,* and *F. verticillioides* (EFSA 2011). It was first isolated in 1962 from corn contaminated with *Giberella zea* (Stob et al. 1962). It is a natural contaminant of grains, especially corn, but can also be found in other crops such as wheat, barley, sorghum and rye. ZEA is considered a field mycotoxin but its production can also take place under poor storage conditions. It is thermostable and withstands a temperature of 120°C for 4 h (Yiannikouris et al. 2002). ZEA is a macrocyclic lactone derived from resorcyclic acid. It is described as a mycotoxin which induces obvious estrogenic effects in humans and animals, given its structural similarity to natural estrogens (Bennett and Klich 2003). The reduction of zearalenone to zearalenol, a key step in its bioactivation, is catalyzed by hepatic 3-α-hydroxysteroid dehydrogenase. Zearalenone and its derivatives will bind competitively to estrogen receptors (ERα and ERβ) in various animal species and promote the synthesis of RNA, proteins, as well as cell proliferation, increasing the mass of organs and causing changes and lesions in the female animal reproductive system (Hussein and Brasel 2001). Thus, the public health concern regarding the ZEA is associated mainly with its strong estrogenic activity. IARC (1993) has classified this mycotoxin in Group 3 (not classifiable as their carcinogenicity to humans).

2.5 Fumonisins

Fumonisins are natural contaminants of corn and corn products; there are currently four types of fumonisins (FA, FB, FC and FP) and the most important of which are type B (FB1, FB2 and FB3). These are mycotoxins produced by species of the genus *Fusarium* and mainly *Fusarium verticillioides* (syn. *Fusarium moniliforme*) and *Fusarium proliferatum*. Other fungal species, including *F. napiforme, F. dlamini* and *F. nygamai,* also produce fumonisins (EFSA 2005). The putative fumonisin biosynthesis gene cluster and the production of FB2 and FB4 have been demonstrated in certain strains of *Aspergillus* sect. Nigri (Frisvad et al. 2011, Mogensen et al. 2010). Only *A. niger* and *A. welwitschiae* species are able to produce FB (Frisvad et al. 2007, 2011, Mogensen et al. 2010) and can contribute to the accumulation of FB2 in corn kernels. Recently, Ferrara et al. (2020) have developed a rapid test based on loop-mediated isothermal amplification (LAMP) for the rapid and selective detection of *Aspergillus* strains producing FB (*A. niger* and *A. welwitschiae*) among the non-producing strains. This rapid molecular test is based on the detection of the *fum10* gene, a structural gene of the fumonisin cluster in the toxigenic species of *Aspergillus.* Their results showed that very small quantities of conidia are necessary to detect the presence of the *fum10* gene giving information on the presence of *Aspergillus* species producing FB2 and on the possible contamination of fumonisins in corn.

FBs can react during food processing, resulting in the formation of modified Maillard-like forms. FBs can strongly interact by non-covalent bond with macro constituents of the matrix, giving rise to the so-called hidden FBs. Hidden forms can release unchanged parent forms of FB into the gastrointestinal tract. Marin et al. (2013) reported that certain heat treatments or extrusions reduce the presence of FB in food and that in general the levels of FB in products intended for direct human consumption such as corn flakes are low (Marin et al. 2013). According to the IARC, fumonisins are classified in category 2B (possible carcinogenicity). They produce a wide range of toxic effects in animals such as encephalitis (or leukoencephalomalacia), a serious and generally fatal disease in horses, pulmonary edema in pigs, an often-fatal disease (Bolger et al. 2001, Haschek et al. 2001) as well as nephrotoxicity and liver cancer in rats (EFSA 2005).

This mycotoxin has a cytotoxic effect; it inhibits the synthesis of proteins and DNA and promotes oxidative stress. It also induces DNA fragmentation and stops the cell cycle (Abado-Becognee et al. 1998). In humans, there is a link between high consumption of fumonisin-contaminated corn and the development of esophageal cancer in some parts of the world, and fumonisins have been reported as potential risk factors for malformations of the neural tube, craniofacial and other congenital anomalies originating from neural crest cells (Marasas et al. 2004). Fumonisins are structural analogues of sphingoid bases and they inhibit ceramide synthase. This induces a disturbance in the metabolism of sphingolipids and pathological changes (Riley and Merrill 2019).

2.6 Patulin

Patulin (PAT) is a mycotoxin produced by certain species of the genera *Penicillium*, *Aspergillus*, *Paecilomyces* and *Byssochlamys* (Steiman et al. 1989). It has often been associated with *Penicillium expansum*, but other fungal species of *Penicillium* are able to produce PAT (Morales et al. 2007, Frisvad et al. 2004). It mainly contaminates apples and derivatives (Beltrán et al. 2014, Zhong et al. 2018, Saleh and Goktepe 2019). It has also been detected in other fruits and derivatives, in vegetables and in cereals and cereal products (Shephard et al. 2010). The concentrations of PAT observed in Europe are generally low whereas certain samples of products from other countries have shown higher concentration levels (Vidal et al. 2019). Although classified by the IARC in group 3 as non-carcinogenic, recent reviews on PAT have reported that this mycotoxin has been linked to neurological, gastrointestinal and immunological adverse effects, mainly causing liver damage and renal problems (Puel et al. 2010, Pal et al. 2017). Puel et al. (2010) reported that the affinity of patulin for sulfhydryl groups explains its inhibitory effect on many enzymes. In addition, the WHO considers patulin as a possible genotoxic compound (WHO 2005). Recently, the harmful effects of mycotoxins in general and of patulin in particular on the sensitive structures of the intestines have been widely studied and the toxicity of patulin on the function of the intestinal barrier has been demonstrated (Akbari et al. 2017).

2.7　Citrinin

Citrinin (CIT) is a polyketide mycotoxin first described in the species *Penicillium citrinum* (Hetherington and Raistrick 1931). Since then, it has been isolated from several species of the genera *Aspergillus* (*A. terreus*, *A. carneus* and *A. niveus*), *Penicillium* (*P. verrucosum* and *P. expansum*) and *Monascus* (*M. ruber* and *M. purpureus*) (Blanc et al. 1995, Hajjaj 2000a, Doughari 2015, de Oliveira Filho et al. 2017). Some of the citrinin-producing fungi are also able to produce other mycotoxins such as OTA, PAT or AFs leading to co-occurrence in commodities of CIT and these mycotoxins (Blanc et al. 1995, EFSA 2012, Doughari 2015). Although described for its antibiotic activities against Gram positive bacteria (*Bacillus, Streptococcus*) but also against *Pseudomonas, Saccharomyces* and *Candida*, its use for this purpose has been rejected because of its toxicity, despite positive results in the treatment of tropical skin diseases (Bastin 1949).

　　Citrinin is generally formed after harvest under storage conditions and it occurs mainly in grains. Stored grain samples (wheat, oats, barley and rye) associated with lung problems in farmers and silo operators were contaminated by citrinin (up to 80 mg/kg) and ochratoxin A (Scott et al. 1972). The occurrence of CIT has been detected in various commodities worldwide such as cereals, beans, fruits, fruit and vegetable juices, herbs and spices and also in spoiled dairy products (EFSA 2012, Zaied et al. 2012, Doughari et al. 2015). Analysis of biological fluids showed the presence of CIT in urine and plasma (Martins et al. 2019, Ouhibi et al. 2020). CIT has also been isolated from fungal ferment species of the genus *Monascus*, used for industrial purposes for the production of a red food coloring and commonly used in Asia (Blanc et al. 1995, Bennett and Klich 2003). It has been shown that the addition of short chain fatty acids (C8, C10 and C12) or methyl ketones (2-heptanone, 2-nonanone or 2-undecanone) in the culture medium in *Monascus ruber*, lowers the titer of the mycotoxin (Hajjaj et al. 2000b). Oxygen is not only the final electron acceptor of the respiratory chain, but also a substrate in certain reactions of secondary metabolism. It is indeed established that oxygen is used thanks to monooxygenases in the biosynthesis of pentacetides (Turner 1971). Discontinuous fermenter cultures carried out in *M. ruber* showed that when the oxygen is in limiting concentration, the production of citrinin is weak and the stoichiometric balance is far from being balanced with the important production of ethanol (fermentative metabolism) (Hajjaj et al. 2015).

　　Conversely, under conditions of no oxygen limitation, the production of citrinin is greater (Hajjaj et al. 1999). The nature of the amino acids has been shown to modify the concentration of citrinin in *M. ruber*, with high levels in the presence of glutamate, alanine or proline as a source of nitrogen (Hajjaj et al. 2012).

　　Citrinin is acutely nephrotoxic at relatively high doses in mice and rats, rabbits, pigs and poultry, causing swelling and eventual necrosis of the kidneys and affecting the liver function at a lesser extent. Based in the available data, the International Agency for Research on Cancer (IARC) concluded that there is limited evidence for carcinogenicity in animals (EFSA 2012). Recently, Sun et al. (2020) reported that exposure of citrinin at 30 μM disrupts organelle distribution and functions in mouse oocytes.

3. Occurrence and Source of Mycotoxins in Food and Feed in Morocco

Morocco is an agricultural North African country with agricultural GDP of about 14%. Agricultural employment contributes to 38.8% of total employment and 73.7% of rural employment. Cereal grains and associated products rank the first in usable agricultural area (59%) while only 16%, 5% and 3% of the total usable agricultural area are dedicated to plantations, fodder and market gardening, respectively. However, cereal production varies strongly according to rainfall; a decrease of 49% of production in 2018/2019 compared to 2017/2018 were reported (MAPMDREF 2019). Indeed, an assessment highlighted that Morocco was vulnerable in terms of cereal imports for climatic factors between the period of 2001 to 2017 (FAO et al. 2018).

Fungal colonization is a major cause of quality deterioration of food crops, leading to economic losses and constituting a potential human health hazard through synthesis of toxic secondary metabolites: mycotoxins. In addition, given the large national cereals consumption (around 185 kg/inhabitant/year), Morocco imports annually 41 to 76 million quintals of cereals (2009–2019) from different countries (France, Canada, Ukraine, Russia and United States) (HCP 2018, ONICL 2020) which represented 35–66% of the total consumption. Mycotoxins and toxigenic fungi can also contaminate these cereals, if conditions are favorable. The occurrence of mycotoxins in other commodities such as grapes and derivatives, fruits and milk have been reported by several studies and, in some cases, the co-detection of several mycotoxins in the same product has been demonstrated.

3.1 Cereals and Derived Products

Morocco's food consumption model is still largely dominated by cereals, mainly common wheat (FAO 2011). A national survey on Moroccan household consumption and expenditure with a sample of 16,000 households between 2013 and 2014 showed that cereals represent almost 25% of food expenditure. Flour is the main cereal commodity consumed by households (62.5% of cereals; 98.5 kg/person/year) followed by bought bread (10.4%; 19.3 kg/person/year) and the non-transformed grains (10%, 18.5 kg/person/year), represented mainly by rice. The consumption of couscous is estimated at 3.6% of the cereals consumed (6.7 kg/person/year) (HCP 2018).

Since the 1960s, the presence of mycotoxins and toxigenic fungi in cereals and their derivatives have been widely studied worldwide. However, only about fifteen papers were published on mycotoxins in commodities in Morocco and studies on toxigenic fungi are still rare. The occurrence of mycotoxins in cereals and derived products are summarized in Table 1. Tantaoui et al.

Table 1: Occurrence of mycotoxins in cereals and derived products in Morocco

Foodstuff	Mycotoxin	n/N	Range (mean samples) µg/kg	Year of study	Analysis method	References	[1]Maximum levels (µg/kg)
Barley	AFB1	0/75	ND	1991–1992	TLC	Tantaoui-Elaraki et al. (1994)	2
	OTA	3/75	1.13–2.83	1991–1992	TLC		3
	OTA	11/20	Up to 0.80			Zinedine et al. (2006)	3
Barley semolina couscous	AFB1	0/8	ND	2014–2015	LC-MS/MS	Zinedine et al. (2017)	2
	AFB2	1/8	2.5 (0.31)				ΣAFs: 4
	AFG1	2/8	2.7–28.6 (3.9)				
	AFG2	2/8	4.2–8.3 (1.6)				
	OTA	0/8	ND				3
	FB1	0/8	ND				-
	FB2	0/8	ND				-
	FB3	0/8	ND				-
	ZEA	2/8	7.6–543.3 (68.9)				-
	DON	0/8	ND				-
	NIV	2/8	182.5–240 (52.8)				-
	T-2	0/8	ND				-

Food	Mycotoxin	Incidence	Range (mean)	Year	Method	Reference	Limit
Corn	HT-2	0/8	ND				-
	AFB1	1/50	18 (0.36)	1991–1992	TLC	Tantaoui-Elaraki et al. (1994)	2
	OTA	8/20	Up to 7.22	-	HPLC	Zinedine et al. (2006)	3
	FB1	10/20	Up to 5960				B1+B2:1000
	ZEA	3/20	Up to 17				100
Corn flour	AFB1	16/20	0.23–11.2	-	HPLC-FLD	Zinedine et al. (2007a)	2
	AFB2	16/20	0.03–1.05				∑AFs: 4
	AFG1	16/20	0.21–0.41				
	AFG2	16/20	0.1–0.24				
Corn semolina couscous	AFB1	1/6	31.1 (5.18)	2014–2015	LC-MS/MS	Zinedine et al. (2017)	2
	AFB2	1/6	2.1 (0.35)				∑AFs: 4
	AFG1	0/6	ND				
	AFG2	2/6	2.8–17.5 (10.15)				3
	OTA	0/6	ND				
	FB1	5/6	253.3–847.9 (464.7)				B1+B2:1000
	FB2	5/6	41.9–343.4 (168.1)				
	FB3	5/6	72.9–356.7 (181.1)				-

(Contd.)

Table 1: (*Contd.*)

Foodstuff	Mycotoxin	n/N	Range (mean samples) μg/kg	Year of study	Analysis method	References	[1]Maximum levels (μg/kg)
	ZEA	0/6	ND				-
	DON	0/6	ND				-
	NIV	0/6	ND				-
	T-2	1/6	0.8 (0.13)				-
	HT-2	0/6	ND				-
Corn product	Fumonisin	1/1	19.53	2014	ELISA	El Madani et al. (2016)	B1+B2:1000
Wheat	AFB1	0/107	ND	2014–2015	LC-MS/MS	Tantaoui-Elaraki et al. (1994)	2
	OTA	2/17	Up to 30.6 (3.5)	-	HPLC-FLD	Hajjaji et al. (2006)	3
	OTA	8/20	Up to 1.73	-	HPLC-FLD	Zinedine et al. (2006)	
	FB1	0/80	ND	-	LC-MS/MS	Blesa et al. (2014)	-
	FB2	0/80	ND				-
	FB3	0/80	ND				-
	ZEA	0/80	ND				-
	DON	9/81	321–1,340	-	LC	Ennouari et al. (2013)	-

Food	Toxin	Incidence	Concentration	Year	Method	Reference	Notes
	NIV	7/17	Up to 128 (27.1)	-	HPLC-UV	Hajjaji et al. (2006)	-
		4/80	121–1480 (33)	-	LC-MS/MS	Blesa et al. (2014)	-
	T-2	0/80	ND	-	LC-MS/MS	Blesa et al. (2014)	-
	HT-2	0/80	ND	-			-
		0/80	ND				-
	PAT	0/15	ND	2012	HPLC-UV, TLC	Mansouri et al. (2014)	-
Durum wheat Soft wheat	CIT	(N=8) (N=8)	Up to 750 600–1400	2011–2012	TLC, HPLC	Mansouri (2018)	-
Wheat Flour	Aflatoxin	4/5	103.56 – 152.65 (90.73)	2014	ELISA	El Madani et al. (2016)	ΣAFs: 4
	AFB1	3/17	0.03–0.15			Zinedine et al. (2007a)	2
	AFB2	3/17	ND				
	AFG1	3/17	ND				
	AFG2	3/17	ND				
	Fumonisin	1/6	484.78 (80.79)	2014	ELISA	El Madani et al. (2016)	-
	DON	1/6	91.52 (15.25)	2011–2012	TLC, HPLC	Mansouri (2018)	-
	CIT	(N=3)	1700				-
Wheat semolina couscous	AFB1	0/84	ND	2014–2015	LC-MS/MS	Zinedine et al. (2017)	2

(Contd.)

Table 1: (*Contd.*)

Foodstuff	Mycotoxin	n/N	Range (mean samples) µg/kg	Year of study	Analysis method	References	[1]Maximum levels (µg/kg)
	AFB2	1/84	2.6 (0.03)				ΣAFs: 4
	AFG1	2/84	1.0–2.5 (0.04)				
	AFG2	7/84	1.6–5.5 (0.27)				3
	OTA	0/84	ND				-
	FB1	0/84	ND				-
	FB2	0/84	ND				-
	FB3	0/84	ND				-
	ZEA	29/84	22.0–132.1 (33.4)				-
	DON	18/84	20.6–106.6 (14.9)				-
	NIV	13/84	52.4–462.2 (37.9)				-
	T-2	2/84	5.3–5.8 (0.14)				-
	HT-2	0/84	ND				-
Semolina	Fumonisin	1/2	42.79 (21.39)	2014	ELISA	El Madani et al. (2016)	-
	DON	2/2	45.78–58.45 (52.11) 2250				
	CIT	(N=3)		2011–2012	TLC, HPLC	Mansouri (2018)	
Wheat bran	AFB1	0/54	ND	1991–1992	TLC	Tantaoui-Elaraki et al. (1994)	2

Food	Mycotoxin	Incidence	Range (mean)	Year	Method	Reference	
Bread	Aflatoxin	21/23	42.07–139.26 (74.17)	2014	ELISA	El Madani et al. (2016)	4
	OTA	48/100	0.14–149	2006	LC-FLD	Zinedine et al. (2007b)	3
	Fumonisin	14/23	5.63–133.77 (45.44)	2014	ELISA	El Madani et al. (2016)	-
	DON	6/23	7.09–70.11 (8.40)	2014	ELISA	El Madani et al. (2016)	-
Biscuits	Aflatoxin	4/4	113.13–125.55 (118.90)	2014	ELISA	El Madani et al. (2016)	4
	Fumonisin	4/4	25.43–188.71 (95.24)				-
	DON	4/4	47.66–55.41 (51.65)				-
Rice	OTA	18/20	0.02–32.4	2005	LC-FLD	Zinedine et al. (2007c)	3
		26/100	0.08–47	2006	LC-FLD	Juan et al. (2008a)	-
Pasta	ZEA	55/106	0.5–3	2016–2017	LC-MS/MS	Bouafifssa et al. (2018)	-
	DON	43/106	16–900				-
	T-2	17/106	4–50				-
Breakfast cereals	OTA	4/48	5.1–224.6	-	LC-FLD	Zinedine et al. (2010)	2
	FB1	17/48	Up to 152	-	LC-MS/MS	Mahnine et al. (2012)	-
	FB2	18/48	Up to 62.3				-

(Contd.)

Table 1: (*Contd.*)

Foodstuff	Mycotoxin	n/N	Range (mean samples) μg/kg	Year of study	Analysis method	References	[1]Maximum levels (μg/kg)
	FB3	17/48	Up to 13.9				-
Infant cereals	OTA	0/20	-	-	LC-FLD	Zinedine et al. (2010)	0.5
	FB1	1/20	2 (0.1)	-	LC-MS/MS	Mahnine et al. (2012)	-
	FB2	2/20	1.2–2.3 (0.17)				-
	FB3	0/20	ND				-
Poultry feeds	AFB1	315	20–5625	1989–1991	ELISA, TLC	Kichou and Walser. (1993)	
	AFB1	14/21	0.05–5.38	-	HPLC-FLD	Zinedine et al. (2007a)	
	AFB2	14/21	0.03–0.58				
	AFG1	0/21	ND				
	AFG2	0/21	ND				
	OTA	19/62	0.24–26.8	2013–2014	HPLC-FLD	Sifou et al. (2016)	
Beer	OTA	0/5	ND	-	HPLC-FLD	Filali et al. (2001)	3

[1]Moroccan Regulation (Arrêté Conjoint MAPM/MS 2016)

(1994) published the first study within the framework of a vast investigation on the contamination of foodstuffs by mycotoxins. The 336 samples of food commodities including cereals (N=232) and barn (N=54) collected in Morocco from 1991 to 1992 have been investigated using thin layer chromatography (TLC). The results showed that aflatoxin B1 was detected only in one corn sample out of 50 (18 µg/kg), and one sample of peanuts was contaminated by aflatoxins (820 µg/kg of AFs). No wheat samples were contaminated with the analyzed mycotoxins and OTA was detected only in three barley samples (1.13–2.83 µg/kg). OTA analysis in barley samples (N=20) using the HPLC-FLD system also showed that this mycotoxin was detected in 55% of barely samples with maximum level of 0.80 µg/kg (Zinedine et al. 2006). Using the multi-detection LC-MS/MS method and the QuEChERS extraction procedure, samples of Barely semolina couscous (N=8), corn semolina couscous (N=6) and wheat semolina couscous (N=84) were also analyzed for their content in several mycotoxins (Zinedine et al. 2017). Aflatoxin B1, OTA, FB1-3, DON, T-2 and HT-2 toxins were not detected in all samples of barely semolina couscous. However, one sample was contaminated by AFB2 (2.5 µg/kg) and AFG1 and AFG2 were detected in two samples. Although the AFs levels are high in one of the samples, the means of AFB2, AFG1 and AFG2 for all the samples are 0.3, 3.9 and 1.6 µg/kg, respectively. NIV and ZEA were also detected in two samples of corn semolina couscous; mean levels were 52.8 and 68.9 µg/kg.

In corn semolina couscous, all samples were free of AFG1, OTA, ZEA, DON, NIV and HT-2 toxin. Only one sample was contaminated by AFB1 and AFB2 with levels of 31.1 and 2.1 µg/kg, respectively (mean values were 5.2 µg/kg for AFB1 and 0.3 µg/kg for AFB2). AFG2 was detected in two samples with a mean concentration of 4.1 µg/kg. The incidence of fumonisins contamination was of 83.3% and means were of 464.7 µg/kg for FB1 and 168.1 µg/kg for FB2, sum of FB1 + FB2 remains below MRL. Although the concentrations of aflatoxins are high in one or two samples of barely or corn semolina couscous, sample size was low (6 to 8 samples). According to Whitaker (2001), for small sample sizes, sampling is usually the largest source of variability.

Taking into account the consumption of raw materials for the manufacture of wheat, corn or barely couscous (HCP 2018); consumption of barley or corn semolina couscous seems to be very low compared to the wheat semolina couscous. In the same study, analysis of a large sample of wheat semolina couscous (N=84) collected from various areas showed that this cereal product is less contaminated by AFs and OTA than the two other types of couscous. So, AFB1, FB1-3, OTA and HT-2 toxin were not detected in any sample of wheat semolina couscous. However, one wheat couscous sample was contaminated with AFB2 with a level of 2.6 µg/kg. Two samples were contaminated with AFG1 with a total mean value of 0.04 µg/kg (maximum level of 2.5 µg/kg) and seven were contaminated with AFG2 with a range of 1.6 to 5.5 µg/kg mean value of 0.27 µg/kg. Authors also reported that sum of AFs in three samples (incidence of 3.57%) exceeded the value of 4 ng/g, MRL fixed by European authorities (EC 2006).

Mycotoxins have also been reported to contaminate wheat, and derivatives such as flour, bread, pasta and cookies. Analysis of wheat samples collected from different regions of Morocco showed that average contamination levels in positive samples varied according the study from 0.42 to 29.4 μg/kg for OTA (Hajjaji et al. 2006, Zinedine et al. 2006) and 65.9 to 502.1 μg/kg for DON (Hajjaji et al. 2006; Ennouari et al. 2013, Blesa et al. 2014). Fumonisins, ZEA, DON, NIV, T-2, HT-2 and PAT were not detected in wheat samples (Blesa et al. 2014, Mansouri et al. 2014). The contamination of durum and soft wheat from the field and storage as well as derivatives (semolina and flour) by citrinin have been reported for the first with high levels (1.4-2.25 mg/kg) in semolina, storage soft wheat and flour (Mansouri 2018). The occurrence of AFs (AFB1, AFB2, AFG1 and AFG2) in flour (N=17) was also reported by Zinedine et al. (2007a). Their results showed that the incidence of aflatoxins in corn flour and wheat flour was 80%, and 17.6%, respectively. The mean levels of AFB1, AFB2, AFG1 and AFG2 in positive samples of corn flour are 1.57, 0.16, 0.31 and 0.17 μg/kg, respectively. For wheat flour, three samples were contaminated by AFB1 with a range from 0.03 to 0.15 μg/kg (mean of 0.07 μg/kg), while AFB2, AFG1 and AFG2 were not detected in all wheat flour samples. In a study carried out by El Madani et al. (2016) on 36 cereal samples collected from Fez city popular market and supermarket were analyzed by ELISA method. For wheat flours (n=6), a range of 103.6 to 152.7 μg/kg of AFs was detected in four samples. Incidence of AFs contamination of biscuits was 100% with a range of 113.1 to 125.6 μg/kg for biscuits, indicating high risk of aflatoxins contamination beyond the maximum tolerable level by the Moroccan regulation (4 μg/kg). In the same study, AFs, Fumonisin and DON also contaminated biscuits and bread. The four samples of biscuits were highly contaminated by AFs in the range of 113.1 to 125.6 μg/kg (mean value of 118.9 μg/kg). Mean levels of Fumonisin and DON were of 95.2 and 51.7 μg/kg (incidence of 100%). In bread, 21 out of the 23 bread samples (91.3%) were contaminated by AFs with an average for positive samples of 81.2 μg/kg. Levels range of Fumonisin and DON contamination were 5.6–133.8 and 7.1–70.1 μg/g, respectively. Zinedine et al. (2007b) reported that bread (N=100) also contaminated by OTA with incidence of 48% and a range of 0.14 to 149 μg/kg; mean value of positive samples of 13.00 μg/kg (26% of total samples were above the MRL of 3 μg/kg set by EU regulation). For orientation purposes, the authors also attempted to determine daily intake of OTA from bread for an adult (60 kg body weight) based on their results and published bread consumption in Morocco of 210 kg/person/year (Chaoui et al. 2003). The estimated daily intake was 126 ng/kg bw/day, higher than the tolerable daily intake of 17.1 ng/kg bw/day set by the EFSA (2006).

The highest levels of OTA were also reported in rice; the incidence and range according to the studies were of 26 to 90% and 0.02 to 47 μg/kg, respectively (Juan et al. 2008a, Zinedine et al. 2007c). Recently, the multi-occurrence of 20 mycotoxins in pasta samples collected in different cities in Morocco was assessed using QEChERS extraction procedure and LC-MS/MS, GC-MS/MS methods (Bouafifssa et al. 2018). Their results showed that 99 out of 106 total samples (93.4%) were contaminated with at least one mycotoxin. AFB1 detected only in two samples with levels of 0.01 µg/kg for Rabat sample and 0.25 µg/kg for Agadir sample. Mycotoxins ZEA, DON, T-2 and HT-2 were detected in the different studied cities with range of 0.5–3, 16–900, 4-50 and 4–419 µg/kg, respectively.

OTA analysis of 68 samples of breakfast and infants cereals products collected from different supermarkets and pharmacies in Rabat (Morocco) were done using pressurized liquid extraction and liquid chromatography (Zinedine et al. 2010). Authors reported all infant cereal samples were free of OTA while four samples of breakfast cereals (incidence of 5.8%) were contaminated with level range of 5.1 to 224.6 µg/kg.

In poultry feed, Kichou and Walser (1993) analyzed 315 samples for the occurrence of AFB1, the levels of this mycotoxin ranged from 20 to 5625 µg/kg. Occurrence of AFB1, OTA and ZEA in poultry feed was also evaluated in 70 samples and results showed that only one sample was contaminated with 1.4 ppb of AFB1 (Benkerroum and Tantaoui-Elaraki 2001). Later, Zinedine et al. (2007a) evaluated the level of total aflatoxins (AFB1, AFB2, AFG1 and AFG2) in poultry feed samples and reported that 14 out of 21 samples were contaminated by AFB1 and AFB2 with a range from 0.05 to 5.38 µg/kg and from 0.03 to 0.58 µg/kg, respectively. AFG1 and AFG2 were not detected in all analyzed samples. OTA was also detected in poultry feeds with an incidence about 30% and a range of 0.24–26.8 µg/kg (Sifou et al. 2016).

Beer contamination with mycotoxins such as OTA and thricothecenes has been the subject of numerous studies. The main source of these metabolites in beer seems to derive from contaminated feedstocks: barley and malt (Grajewski et al. 2019). Analysis of ten samples of beer produced in Morocco showed that OTA was not detected (Filali et al. 2001).

3.2   Fruits, Dried Fruits and Fruit Juices

Several surveys on the incidence of mycotoxins in fruits and derivatives worldwide have been published; however similar studies remains very limited in Morocco. The consumption of fruits and derivatives represents 7.3% of food expenditure with a share of 43% reserved for pome fruits including grapes. 27.5% of fruit expenditure were dedicated to the fresh stone fruit, dried fruit and oilseeds (plums, peaches, apricots, dates, prunes, raisins …) and the consumption of fruits increased significantly from 39.1 in 2001 to 67.5 kg/person/year in 2014 (HCP 2018). Fruits and dried fruits are also used in the food industry for several preparations such as bakery products, breakfast cereals and infant cereals. Results published on the natural occurrence of mycotoxins in fruits, dried fruits and fruit juices in Morocco are summarized in Table 2. Tantaoui-Elaraki et al. (1994) and Juan et al. (2008b) have examined aflatoxins contamination of dried fruits; their results showed that maximum levels of AFB1 ranged from 0.28 to 2500 µg/kg. For peanuts, Tantaoui-Elaraki et al. (1994) reported that one of six analyzed samples by TLC was contaminated by high level of 820 µg/kg (total AFs): 250 µg/kg for AFB1 and AFG1 and 160 µg/kg for AFB2 and AFG2. Using HPLC-FLD, Juan

Table 2: Occurrence of mycotoxins in fruits, dried fruits and fruit juices in Morocco

Foodstuff	Mycotoxin	n/N	Range (mean samples) µg/kg	Year of study	Analysis method	References	[1]Maximum levels (µg/kg)
Peanuts	Aflatoxins	1/6	820 (136.66)	1991–1992	TLC	Tantaoui-Elaraki et al. (1994)	4
	AFB1	1/20	0.17 (0.008)	2006	HPLC-FLD	Juan et al. (2008b)	2
	OTA	5/20	0.10–2.36	2005	LC-FLD	Zinedine et al. (2007c)	-
Pistachio	AFB1	9/20	Up to 1430	2006	HPLC-FLD	Juan et al. (2008b)	12
	OTA	0/20	ND	2005	LC-FLD	Zinedine et al. (2007c)	-
Walnuts	AFB1	6/20	Up to 2500	2006	HPLC-FLD	Juan et al. (2008b)	5
	OTA	7/20	0.04–0.23	2005	LC-FLD	Zinedine et al. (2007c)	-
Dried figs	AFB1	1/20	0.28	2006	HPLC-FLD	Juan et al. (2008b)	6
	OTA	13/20	0.03–1.42	2005	LC-FLD	Zinedine et al. (2007c)	-
Grapes	OTA	13/22	Up to 4 (0.86)	2006	HPLC-FLD	Selouane et al. (2009)	-
Raisins	AFB1	4/20	Up to 13.9	2006	HPLC-FLD	Juan et al. (2008b)	2
	OTA	6/20	0.05–4.95	2005	LC-FLD	Zinedine et al. (2007c)	-
Wine	OTA	30/30	0.028–3.24 (0.65)	2001	HPLC-FLD	Filali et al. (2001)	-
Fruit juices	OTA	1/14	1.16 (0.08)	2001	HPLC-FLD	Filali et al. (2001)	-
Black olive	AFB1	4/10	0.6–5 (0.97)	-	HPLC-FLD	El Adlouni et al. (2006)	-
	CIT	8/10	0.45– 0.52 (0.06)				
	OTA	10/10	0.31–1.02 (0.4)				
	OTA	9/25	0.62–4.8	-	HPLC-FLD	Idrissi et al. (2004)	-

[1]Moroccan Regulation (Arrêté Conjoint MAPM/MS 2016)

et al. (2008b) also reported AFB1 contamination of one peanut sample out of the twenty analyzed, but at low level (0.17 µg/kg). The highest levels of AFB1 were recorded for pistachio and walnuts with maximum levels of 1430 and 2500 µg/kg, respectively. Dried figs presented a lower level of 0.28 µg/kg in one sample and four samples of raisins were contaminated with maximum level of 13.9 µg/kg (Juan et al. 2008b). AFB1 was not detected in the 44 grape samples analyzed by Tantaoui-Elaraki et al. 1994). The occurrence of OTA in dried fruits was also reported by Zinedine et al. (2007c). This study showed that the incidence of OTA in peanuts, dried grapes, walnuts and dried figs was 25%, 30%, 35% and 65%, respectively, while no OTA was detected in any pistachio samples. The mean level of OTA in positive samples of peanuts, dried figs, dried grapes and walnuts are 0.68, 0.33, 0.96 and 0.11 µg/kg, respectively. A study of grapes from Morocco showed that OTA was detected in 13 of the 22 samples analyzed. The concentrations ranged from 0.08 to 4.00 µg/kg and mean level in positive samples was 1.48 µg/kg (Selouane et al. 2009a). In fruit juices, the results of Filali et al. (2001) showed that only one grape fruit juice sample from 14 various fruit juices (cocktail, pineapple, mango, peach, orange, grapefruit and clementine) was contaminated by OTA with a concentration of 1.16 µg/L. However, all the thirty analyzed wine samples were contaminated by OTA at levels range of 0.028 to 3.24 µg/L.

The natural occurrence of OTA in black olives have been investigated by Idrissi et al. (2004) and the authors found that 36% of the analyzed samples were contaminated by OTA with a range from 0.62 to 4.8, average of positive samples was of 1.43 µg/kg. Indeed, in a previous contamination by mycotoxins, levels of total aflatoxins and OTA were found in all samples (incidence of 100%) with a maximum level of 0.5 µg/kg and 1.02 µg/kg respectively. The same authors reported that, 8 out of 10 black olives samples were contaminated by CIT with a maximum concentration of 0.52 µg/kg.

3.3 Other Commodities: Milk, Spices and Herbal Green Tea

Although several dairy products are consumed in Morocco such as milk, yogurts, lben and cheeses, only three studies on mycotoxins have been carried out on milk (Table 3). Zinedine et al. (2007d) reported that 48 samples of pasteurized milk were contaminated by AFM1 with a range of 0.001 to 0.117 µg/L and incidence of 88.9%. El Marnissi et al. (2012) reported the presence of AFM1 in 13 out of 48 raw milk samples with a maximum level of 0.1 µg/L and an average for positive samples of 0.043 µg/L. Further, in a more recent study of Alahlah et al. (2020), the incidence of AFM1 in powder milk and UHT milk was 100% and 35%, respectively. The mean values of AFM1 in powder milk and UHT milk samples were 0.0255 and 0.0148 µg/kg, respectively.

Natural occurrence of aflatoxins in 55 spices samples commercialized in Morocco was reported by Zinedine et al. (2006). This study showed that the incidence of aflatoxins in all samples of spices was 83.63%, the higher levels of contamination were found in red paprika and ginger, with an average level of 5.23 µg/kg and 1.47 µg/kg for total aflatoxins, respectively, and the maximum concentration for red paprika was 9.68 µg/kg and 9.10 µg/kg for ginger. In

Table 3: Occurrence of mycotoxins in milk, spices and herbal green tea in Morocco

Foodstuff	Mycotoxin	n/N	Range (mean samples) µg/kg or µg/L	Year of study	Analysis method	References	[1]Maximum levels (µg/kg or µg/L)
Pasteurized milk	AFM1	48/54	0.001–0.117 (0.018)	2006	LC-FLD	Zinedine et al. (2007d)	-
Raw milk		13/48	0.01–0.1 (0.011)	2009-2010	HPLC-FLD	El Marnissi et al. (2012)	-
Powder milk		7/7	0.0152–0.0399 (0.0255)	-	HPLC-FLD	Alahlah et al. (2020)	-
UHT milk		14/40	0.0051–0.0444 (0.0148)				-
Black pepper	Aflatoxins	14/15	0.04–0.55		HPLC-FLD	Zinedine et al. (2006)	10
Ginger		10/12	0.03–9.10				10
Red paprika		14/14	1.30–9.68				10
Cumin		8/14	0.01–0.18				-
Herbal green tea	AFB1	[2]76/126	1.8–41.8	2018	LC-MS/MS	Mannani et al. (2020)	-
	AFB2		1.6–15.4				-
	AFG1		8–30.2				-
	AFG2		1.3–76				-

[1]Moroccan Regulation (Arrêté Conjoint MAPM/MS 2016)
[2]76 of 126 samples contaminated by at least one aflatoxin

this survey, incidence of total aflatoxins in cumin and black pepper was lower than in red paprika and ginger.

In herbal green tea commercialized in Morocco, Mannani et al. (2020) reported the occurrence of AFB1, AFB2, AFG1 and AFG2 (Table 3). This study showed that the incidence of total aflatoxins in 126 herbal green tea samples was 60.31%. The mean levels of AFB1, AFB2, AFG1 and AFG2 in positive samples are 10.3, 10.83, 18.66 and 12.06 µg/kg, respectively, and the higher concentrations were 41.8, 15.4, 30.2 and 76 µg/kg, respectively.

4. Sources of Mycotoxin Contamination

Mold is a ubiquitous filamentous fungus that occurs naturally in soil and can contaminate air, water, food and feed around the world. Mold-contaminated food and feed have been known to be associated to the occurrence of mycotoxins, health hazard fungal secondary metabolites. The sources of exposure to mycotoxins as well as the admissible daily doses are not yet established in Morocco. However, according to consumption data and the multiple occurrence of several mycotoxins, cereals and their derivatives are probably the main sources of exposure to mycotoxins. The mycotoxin-producing fungi identified in Moroccan food and feed are mainly species of the *Fusarium*, *Aspergillus*, and *Penicillium* genera. Table 4 summarizes the toxigenic fungal species, their isolation source and the produced mycotoxins. The study of Hajjaji et al. (2006) isolated strains from wheat grains samples stored in farm warehouses in the region of Beni Mellal, representing seven genera and the most important of them were *Aspergillus* (37.4%), *Penicillium* (14.1%) and *Fusarium* (9%). Among *Aspergillus*, identification of isolated revealed that the main species were *A. niger*, *A. flavus*, *A. terreus*, *A. nidulans*, *A. ochraceus* and *A. alliaceus*; strains of *A. alliaceous* (100%), *A niger* (47%) and *A. ochraceus* (17%) were able to synthesize OTA on CYA medium. However, none of *Penicillium* isolates can produce this toxin. Authors also reported that the optimal conditions for the growth and OTA production were different. Optimal conditions of growth for *A. alliaceus* and *A. terreus* were 30°C/0.98 a_w and 25°C/0.93–0.95 a_w for *A. niger* while the optimal production of OTA was observed at 30°C/0.93 a_w and 30°C/0.99 a_w for *A. alliaceus* and *A. niger* respectively. In the Meknes region, fungal flora contaminating the wheat and barley grain in the field, the storage silo, and products of transformation (flour, semolina) were evaluated (Mansouri et al. 2014, Mansouri 2018). Macroscopic and microscopic characters enabled to identify over 140 isolates belonging to several genera (mainly of *Aspergillus, Penicillium, Fusarium, Cladosporium, Alternaria, Ulocladium, Rhizopus, Mucor* and *Trichoderma*) and their proportion varied according to the type of commodities. Using the TLC and HPLC analysis method, 15.7% of *Penicillium* isolates produced patulin, including six identified as *Penicillium expansum*. Four of these *Penicillum expansum* strains produced both patulin and citrinin. Mansouri et al. (2017) studied the effect of carbon, nitrogen and physicochemical factors on patulin production by *Penicillium expansum* isolated Moroccan cereal have. Their results showed that

Table 4: Mycotoxins and toxigenic fungal species isolated from Moroccan commodities

Fungi		Mycotoxins	Foodstuffs and feed	References
Aspergillus	*A. niger*	OTA	Grapes, wheat, durum wheat, olive, olive cake	Selouane et al. (2009a, 2009b), Ennouari et al. (2018), Hajjaji et al. (2006), Roussos et al. (2006)
	A. carbonarius	OTA	Grapes	Selouane et al. (2009a, 2009b)
	A. tubingensis	OTA	Grapes, durum wheat	Selouane et al. (2009a, 2009b), Ennouari et al. (2018)
	A. niger aggregate		Grapes	Selouane et al. (2009a, 2009b)
	A. alliaceus	OTA	Wheat	Hajjaji et al. (2006)
	A. ochraceus	OTA	Wheat	Hajjaji et al. (2006)
	A. ochraceus Wilhelm	OTA	Poultry feeds	Benkerroum and Tantaoui-Elaraki (2001)
	A. flavus	AFB1	Poultry feeds, olive, olive cake	Benkerroum and Tantaoui-Elaraki (2001), Roussos et al. (2006)
		AFB1, AFB2, AFG1, CPA	Cumin, paprika, white pepper	
	A. minisclerotigenes	AFB1, AFG1, CPA	Cumin, paprika, white pepper	El Mahgubi et al. (2013)
	A. section Flavi	AFB1, AFB2, AFG1, CPA	Cumin, paprika, white pepper	El Mahgubi et al. (2013)
Penicillium	*P. expansum*	PAT, CIT	Durum wheat, barley, semolina, flour, apple	Mansouri (2018), Mansouri et al. (2014), Rharmitt et al. (2016)
	P. verrucosum	OTA	Poultry feeds	Benkerroum and Tantaoui-Elaraki (2001)
Fusarium	*F. graminearum*	DON	Durum wheat	Ennouari et al. (2018)
	F. equiseti	DON	Durum wheat	Ennouari et al. (2018)

F. ecuminatum	DON	Durum wheat	Ennouari et al. (2018)
F. sporotrichioides	DON	Durum wheat	Ennouari et al. (2018)
F. verticillioides	DON	Durum wheat	Ennouari et al. (2018)
F. coeruleum	DON	Durum wheat	Ennouari et al. (2018)
F. oxysporum	DON	Durum wheat	Ennouari et al. (2018)

pH and temperature had great influence on patulin production and carbon, nitrogen as well as their ratio (carbon/nitrogen) influenced patulin synthesis with L-glutamate as an important nitrogen source in synthetic nutrient media.

Producing-patulin of *Penicillium expansum* strains were also isolated from apples by Rharmitt et al. (2016). Recently, Ennouari et al. (2018) isolated species of *Aspergillus* and *Fusarium* from samples of durum wheat collected in seven areas in Morocco and studied their capacity to produce OTA and DON. The results of biodiversity showed that the most abundant species were belonging to *Alternaria* spp. (86.2%), *Fusarium* spp. (71.1%), and *Aspergillus* spp. (14%). Results of LC analysis indicated also that 2.75 % of *Fusarium* spp. isolates produced DON while OTA was produced by 18.6% of *Aspergillus* section *Nigri* isolates. Molecular identification of the isolates showed the presence of the main species of *F. graminearum, F. equiseti, F. ecuminatum, F. sporotrichioides, F. verticillioides, F. coeruleum* and *F. oxysporum*. The identification of *Aspergillus* species showed the abundance of *A. flavus, A. niger, A. tubingensis* and *A. niger* aggregate. In poultry feeds samples, analysis of fungal contamination showed that molds belonged to 10 different genera; *Penicillium* (35.7%) and *Aspergillus* (20.4%) were the most represented genera. The other genera identified were *Fusarium, Alternaria, Trichoderma, Cladosporium, Verticillium, Mucor, Rhizopus* and *Ulocladium*. Isolates of *A. flavus* (30%) produced aflatoxin B1 on rice and *A. ochraceus* Wilhelm (25%) and *P. verrucosum* (14.3%) produced OTA on wheat (Benkerroum and Tantaoui-Elaraki 2001).

In Moroccan grapes, Selouane et al. (2009a) showed that the most abundant species were member of *Aspergillus niger* aggregate (82.5%) whereas *A. carbonarius* isolates were recovered only occasionally and in small numbers (3.3%). Considering the abundance of *A. niger* aggregate species and their capacity to produce OTA (42.8% of isolates), these species are the most probable source of OTA in Moroccan grapes while *A. carbonarius* has been described as the main source of OTA in European grapes and derived products (Battilani et al. 2006). The complex interaction of several factors (temperature, water activity, culture media, incubation period) affect fungal growth and OTA production by these toxicogenic species (Selouane et al. 2009b, 2014), which makes it difficult to control this exposure.

During the 2003 and 2004 olive oil production campaigns in Morocco, strains belonging to ten genera were isolated from olive and olive cake, represented mainly by *Penicillium* (32.3%) and *Aspergillus* (26.9%). Seven of the nine tested *A. flavus* strains produced AFB1 and 27 of the 36 *A. niger* strains produced OTA (Roussos et al. 2006), which probably explains the contamination of olives by AFs and OTA.

As shown below, El Mahgubi et al. (2013) reported the contamination of the spices (paprika, cumin, black pepper and white pepper) by AFB1. From the same samples, authors isolated fungal flora from these spices and tested their capacity to produce AFs and CPA. They observed a widespread contamination of spices with *Aspergilli* section *Flavi* and 57% of isolates were found to be toxigenic. The most frequent chemotypes were able to produce both aflatoxin B and cyclopiazonic acid (25%) followed by producers of aflatoxins

B only (16%). Toxigenic isolates (3/36) that produce sclerotia in culture were identified as *Aspergillus minisclerotigenes* and were able to synthesis aflatoxin G1.

5. Prevention and Control of Mycotoxins in Morocco

Consumption of products contaminated with mycotoxins leads to harmful health risks for human and animal health, since several Moroccan products are contaminated. Therefore, the prevention and reduction strategies of these toxins before their entering the food chain must be the subject of more and continuous attention. As known, molds are ubiquitous and heterotrophic microorganisms that can colonize various substrates and some species produced mycotoxins. The production of these toxins depends on different factors such as temperature, moisture content, water activity, substrate type, nutrient composition, climate and insect's damages.

Generally, pre-harvest prevention strategies, such as good agricultural practices (GAP) and good manufacturing practices (GMP), are the best way to prevent the synthesis of mycotoxins in agricultural and agri-food products. However, once mycotoxins are present, this approach may not eliminate them (Karlovsky et al. 2016). Therefore, other post-harvest detoxification procedures should be considered (Jarda et al. 2011, Temba et al. 2016, Zhu et al. 2016). For control strategies, innovative chemical, biological and other methods have been proposed. GAP and GMP included crop rotation, selection of resistant varieties; management of pests and harvest management are the most effective solutions to prevent fungal growth and the mycotoxins production (FAO/WHO 2013, Magan et al. 2004, Magan and Aldred 2007). Hazard analysis and critical control points (HACCP) also play an important role in the prevention and management of mycotoxins, these systems include prevention, control and good hygiene and manufacturing practices at all stages of field management, storage monitoring, segregation and cleaning procedures (FAO 2002, Stove 2013). Preventing contamination of grape berries, cereal seeds, coffee during harvesting and new harvested fruit, depends on several factors such as harvest time, the good performance/cleanliness of the agricultural and harvesting equipment and elimination of the soiled grains (FAO/WHO 2013, FAO 2009, Bucheli and Taniwaki 2002). The control of storage conditions (temperature, humidity of warehouses and darkness) turn out essential to limit and prevent/reduce fungal growth and mycotoxins production (Peraica et al. 2002, Lanyasunya et al. 2005). Corbett (2003) has reported that a long period of storage of apples leads to a dramatic increase in patulin content in apple juice.

In Morocco, the prevention and reduction of mycotoxins in food and feed is included in the country's overall agricultural strategy, namely "Green Morocco Plan" launched in 2008 for the first period 2009–2019 and renewed for the second period 2020–2030. Indeed, considerable efforts have been made and will continue to increase the agricultural areas and for the improvement of productivity, quality, food safety and food security. In this context, national agricultural research system including several institutions has developed different resistant varieties of seeds and plants. To reduce losses and improve the quality of agricultural products, several training and support programs for farmers have been carried out in different areas (management of pesticide use, field management, etc.). In research, the rare published data focused on chemical and biological control.

5.1 Chemical Control

The use of synthetic fungicides is one of the most widely used strategies to control fungal growth and are often considered essential for securing the global food supply (Strange and Scott 2005). Many substances with very different chemical constituents are used as fungicides (Gupta 2018). According to national office for food safety (ONSSA "Office National de Sécurité Sanitaire des produits Alimentaires"), there are 1055 commercial specialty pesticides and 375 active ingredients in Morocco reported in Phytosanitary Index in 2015 (UNEP-FAO-RC-SHPF 2015). The statistics and prognoses done by the Ministry of Agriculture, Fisheries, Rural Development and Water and Forests (MAFRDWF) that agricultural development via the Green Morocco Plan will increase the amount of pesticides used by 200% at 2020 compared to the reference year 2008–2009 (UNEP-FAO-RC-SHPF 2015). The absence of phytosanitary product manufacturing units in our country means that 95% of these phytosanitary products are imported. In 2008, insecticides dominated the market, followed by fungicides and herbicides. However, fungicides predominated with a proportion of around 42.5% in 2011 and 2013. On the other hand, 35% to 45% of what is imported is repackaged in small packages adapted to meet the needs of small farmers (UNEP-FAO-RC-SHPF 2015).

Studies on the effectiveness of different fungicides have shown that these chemicals reduce fungal growth but, in some cases, stimulate mycotoxins production (Zouhair et al. 2014, Qjidaa 2015). The antifungal activity of benomyl, thiabendazole and pyrimethanil was tested *in vitro* and *in vivo* on the development of species of *P. expansum*, *A. niger* and *A. fumigatus* isolated from conserved apples in Morocco. They have shown that the fungicide pyrimethanil was more effective against fungal growth (Attrassi et al. 2007). However, thiabendazole was more effective than benomyl and azoxystrobin against the surface development of *Fusarium oxysporum* and the deep development of *Aspergillus fumigatus* and *Alternaria tenuissima* (Bouigoumane et al. 2008). Studies on the effect of five fungicides on fungal growth and OTA production by *A. carbonarius*, *A. niger*, *A. tubingensis* and *A. foetidus* isolated from Moroccan grapes showed that thiabendazole, hexaconazole, benomyl and pyrimethanil significantly reduce mycelial growth with total inhibition at concentrations higher than 0.5×R.D, but stimulation of OTA production were observed in some cases. However, the reduction in radial growth was less marked for azoxystrobin and varied from 20 to 51.6% depending on the strain and the azoxystrobin concentration (Qjidaa et al. 2014, Zouhair et al. 2014,

Qjidaa 2015, Laaziz et al. 2017). Thus, fungicides must therefore be applied with caution and their use must take into account both their ability to inhibit fungal growth but also for their effect in terms of biosynthesis of mycotoxins.

5.2 Biological Control

The development of fungicide resistance in many fungal pathogens and the growing public concern about the risks associated with pesticide use have generated interest in the development of alternative methods against these fungal pathogens, especially environmentally friendly methods. Biological control using microbial antagonists, alone or as part of an integrated control strategy to reduce the use of pesticides, is a promising approach for the control of mycotoxins in crops before and after harvest. Several microorganisms, including bacteria, yeasts and non-toxigenic fungi belonging to the genus *Aspergillus* (Xu et al. 2013), have been tested for their ability to reduce contamination by mycotoxins. Thus, lactic acid bacteria with their antifungal properties seem to be potentially very effective in preventing the formation of mycotoxins. Culture of *Lactobacillus casei* CCM 1825 has been reported to reduce citrinin production by 73.2% in *P. citrinum* (Gourama, 1997). Guimarães et al. (2017) selected a strain of *Lactobacillus plantarum* for inhibiting the growth of *Aspergillus flavus*. Further, they reported that cell-free supernatant (CFS) of this strain inhibits the aflatoxins production by 91%. This inhibition was dependent on CFS pH, increased with increasing doses of CFS, and was independent of fungal growth, which was inhibited only by 32%. Another study by Fuchs et al. (2008) showed that lactic acid bacteria are capable of absorbing patulin and that *Lactobacillus plantarum* has the best absorption capacity. Furthermore, the capacity of lactic acid bacteria to adsorb OTA varied between 8% and 28%, and no degradation product was detected, which suggests that elimination was achieved by a binding process (Del Prete et al. 2007). In addition, the use of *B. licheniformis* isolated from fermented soybeans from Thailand effectively eliminated OTA, with a removal efficiency of 92.5% for a 48 h treatment at 37°C (Petchkongkaew et al. 2008). Studies have confirmed that *Brevibacterium* species degrade OTA by hydrolyzing its amide bond (Rodriguez et al. 2011, Moreno et al. 2013).

The use of natural substances and their extracts such as essential oils (EOs) is another recommended alternative to reduce chemical fungicides. Indeed, natural plant extracts such as eugenol and essential oils of cinnamon, oregano, onions, lemongrass, turmeric and mint are known to prevent both mold growth and mycotoxin production (Kabak et al. 2006). The composition of EOs is very complex, a single EO contains several volatile compounds (Bohme et al. 2014), known to have fungicidal and bactericidal activities (Combrinck et al. 2011, Teixeira et al. 2013). These activities can be linked to single compounds or groups of compounds. Like other Mediterranean countries, Morocco is very rich in medicinal and aromatic plants, used mainly in traditional medicine (Ghourri et al. 2013, Benlamdini et al. 2014, Fakchich and Elachouri 2014, El Hamsas-El Youbi et al. 2016).

Although the antibacterial and antioxidant activity of EOs and plant extracts are widely studied, their antifungal activity, mainly against toxigenic species are scarce. In 2003, Chebli et al. reported that *Origanum compactum* and *Thymus glandulosus EOs* greatly inhibited the fungal growth of *Botrytis cinerea* (up to 100% at 100 ppm) while *Mentha pulegium* EO exhibited moderate activity of 58.5% at 250 ppm. El Ouadi et al. (2017) investigated the biological activity of EOs against post-harvest phytopathogenic fungi in apples and reported that growth inhibition of *P. expansum* reaches 100% at 1 µL/mL of *Melissa officinalis* EO, whereas it takes 2 µL/mL to have the total inhibition of *Rhizopus stolonifera*.

According to the study of Bouddine et al. (2012), EOs of oregano and thyme are more efficient against *A. niger* than those of clove and rosemary. They also demonstrated that mycelial growth of *A. niger* was completely inhibited by carvacrol, thymol and eugenol at the concentrations of 0.025, 0.025 and 0.05%, respectively. Qjidaa (2015) reported that *Mentha pulegium* EO has a significant antifungal activity against *A. tubingensis* and *A. foetidus* (up to 100%, depending on the strain and the EO concentration). At 200 µL/L, the inhibition rate was between 87.8 and 100%. OTA was not detected in most cases for ochratoxigenic strains, which suggests its anti-ochratoxinogenic effect. Qjidaa et al. (2018) demonstrated that trans-resveratrol inhibits fungal growth but the average inhibition rates are between 8 and 24% and it also seems to inhibit the production of OTA. On the other hand, Laaziz et al. (2019) demonstrated that essential oil of thymus has pronounced effect on fungal growth and OTA production of/by two strains of *Aspergillus niger*. The growth inhibition rates increased as the concentration of thyme EO increased and reached 41% when thyme EO concentration was 200 µL/L and 100% at 500 µL/L. HPLC-FLD analysis of culture medium extracts showed that the application of thyme EO at a dose of 100 and 200 µL/L stimulated the OTA production.

6. Moroccan Regulations and Surveillance Program

Aware of the impact of mycotoxins on human and animal health, many countries have adopted regulations relating to these mycotoxins in food and feed products in order to preserve human health and animals, as well as the economic interests of producers, processors and traders. Establishing regulations for these mycotoxins is a complex operation in which many factors, scientific and socio-economic, can affect the establishment of limit values and regulation for these secondary metabolites. Availability of toxicological data and those on the presence of mycotoxins in various products are the first two factors that provide the information necessary for the assessment of the risk and the exposure and constitutes the main scientific basis for the establishment of regulations (FAO 2001, 2002). Other factors are also important such as knowledge of the distribution of mycotoxin concentrations in a batch, availability of analytical methods, legislation of countries with which commercial contacts exist and need for an adequate food supply.

Regarding Moroccan legislation, in relation to mycotoxin contaminants, it protects consumers through two following measures: (a) Establishment of maximum levels for mycotoxins in food and feed to ensure that they do not harm human or animal health and (b) Keeping mycotoxin levels as low as reasonably achievable by application of the preventive measures (good agricultural, storage and food processing practices).

For all primary products and food products, these maximum limits as well as the alert thresholds apply to the edible part of the product concerned. However, for products intended for infants and young children, the maximum limits set apply to the product as it is consumed. A Morocco joint decree of the Minister of Agriculture and Maritime Fisheries and the Minister of Health n° 1643-16 of 23 Chaabane 1437 (May 30, 2016) setting the maximum authorized limits of contaminants in primary products and food products was published in the Official Bulletin on 03/11/2016 (Arrêté Conjoint MAPM/MS 2016). In this joint decree, a contaminant is defined as any substance which is not intentionally added to the primary product or to the food product, but which remains present in the product concerned as a residue. Two indicators have been defined for these contaminants:

- Maximum limit: the concentration, above which, for a given pair of a primary product or a food product in relation to the contaminant, the risk to human health is sufficient to consider the product non-compliant; and
- Alert threshold: the maximum contaminant concentration limit beyond which an investigation aimed at understanding the methods and kinetics of product contamination is carried out.

Any exceedance of the alert threshold noted before its application deadline does not lead to the product not being in conformity, but gives rise to an investigation by the monitoring agent, aimed at understanding the methods and kinetics of contamination of the product. After its application deadline, the alert threshold can be transformed into a maximum limit. The period granted until the deadline for applying the alert threshold must allow the sectors concerned to adapt to the alert threshold provided for and to put in place any measure to understand the causes and reduce the probability of a breach of said threshold. After this date, it is decided to modify this alert threshold, to maintain it, to delete it or to transform it into a maximum limit of the same level or of a different level. However, in the event that scientific data shows the need to modify the alert threshold in order to ensure maximum protection of the health of the consumer, the latter may be modified before its application deadline.

Since the discovery of aflatoxins in the 1960s, many countries have adopted regulations to protect consumers from the harmful effects of this family of mycotoxins. Regulatory limits for AFB1 or for the sum of AFB1, AFB2, AFG1 and AFG2 in human and/or animal food products have been developed since 1995. Most of the existing regulations on mycotoxins in Africa also refer to aflatoxins. Morocco was considered the country with the most precise regulations on mycotoxins (Shephard 2004). Subsequently, other mycotoxins were added to the list, namely AFM1, OTA, PAT, DON, ZEA, FB1 and FB2. The primary product and the food product listed in the table of Moroccan regulation on mycotoxin for which an alert threshold and its application deadline are set, must be regularly monitored by the competent services of the national office food safety "ONSSA" until to the deadline for applying the alert threshold which is May 30, 2020.

The maximum limits for mycotoxins reported by the Moroccan authorities in this decree are those adopted by EU countries (European Commission 2006). Other maximum limits for mycotoxins (T-2, HT-2 and CIT) are still absent in Moroccan law and require increased monitoring. Moreover, studies have shown the presence of citrinin in some foodstuffs, we can cite the presence of this mycotoxin in black olives or rice, bought at retailers and at the supermarket in Morocco (Gourama and Bullerman 1988, El Adlouni et al. 2006). Citrinin was also detected in samples of durum wheat and common wheat from the field, storage and in transformation products (flour and semolina), with contents varying between 0.6 and 2.55 mg/kg (Mansouri 2018). In general, very few articles reported to the presence of CIT in original food products or marketed in Morocco in comparison with OTA or aflatoxins. This would not exclude the presence of citrinin or strains producing this mycotoxin. As the phenomenon of OTA-CIT synergy has been demonstrated and as a precautionary principle, it becomes more prudent to systematically survey CIT in foodstuffs marketed especially when the presence of ochratoxin has been demonstrated.

6.1   National Office of Food Safety: Supervisory Body

The National Office of Food Safety (ONSSA) is a Moroccan public establishment created in 2009, placed under the supervision of the ministry in charge of agriculture and endowed with the moral personality and the financial autonomy. It exercises, on behalf of the State, the powers relating to the preservation of animal and plant health and health security along the entire food chain. Indeed, the law n° 25.08 establishing the ONSSA entrusted to the latter a multitude of powers, all relating to the protection of the health of the consumer and the preservation of the health of animals and plants. ONSSA is thus responsible for ensuring government policy relating to the protection of consumer health through the control of the entire food chain. As such, the Office intervenes in animal health, plant health, sanitary control and compliance of food products, control of inputs, phytosanitary products and veterinary drugs, as well as in import and export.

7. Conclusion

Mycotoxins are toxic secondary metabolites contaminated foodstuffs and feeds all over the globe. This review summarizes the status of major mycotoxin contaminations in food and feed in Morocco. The overall range of mycotoxin contamination levels based on the published information is summarized in Tables 1–3. Analysis of the results published in the databases shows that the foodstuffs produced or marketed in Morocco are contaminated with mycotoxins and/or toxigenic fungi. Average contents rarely exceed the maximum limits authorized by both Moroccan and European regulations. Does this study reflect that there is no risk associated with mycotoxins in Morocco?

The available data on the presence of mycotoxins in food cannot answer this much more complex question. As can be seen, the number of publications on mycotoxins in Morocco is low compared to other countries, mainly the Euro-Mediterranean countries. In addition, most of the studies were carried out with reduced and unrepresentative sampling. However, sometimes the existence of one or a few samples with contents higher than the authorized limits and the multi-detection of several mycotoxins in the same sample showed that mycotoxins constitute a significant health risk. Furthermore, the results of a collective expertise on societal challenges in Africa and in which three Moroccan universities participated (USMBA, UCA, UAE) demonstrated that mycotoxins constitute a societal challenge and a risk in the field of food safety (Montet et al. 2020) and also highlighted the practical approach to control and prevent them (Montet et al. 2019). To reduce the risk of mycotoxins in Morocco, it is necessary to set up means of control as well in pre-harvest as during storage and processing, favoring rather preventive methods and development of decontamination techniques for uncontrollable cases. To achieve these objectives, collective expertise to develop strategies for the evaluation, prevention and control of mycotoxins can be used, promoting training, communication and synergy between the various stakeholders. As known, the exposure to mycotoxins depends on the level of these contaminants in different foods and on their intake (Kroes et al. 2002). Moreover, several mycotoxins in foods co-occur frequently at low concentrations. The challenges to prevent fungal growth and mycotoxin contamination in Morocco can include these following points:

- Training of agricultural staff in different area of agriculture, including the use of agricultural inputs and their impact on health and the environment;
- GAP/GMP and quality assurance systems for agro-industries, essential to prevent the fungal growth—mainly toxigenic fungi and prevent the consequent production of mycotoxins along the production chain;
- Impact of effect of environmental factors and climate change on fungal contamination and the production of mycotoxins in the various sectors;
- Collection of a database of predominant fungi and mycotoxins in Morocco;
- Collection database on the consumption of foods which requires care for specific groups of high consumption of some foods;
- Collection database on biomarkers measurements of serum or urinary in humans;
- Development of detoxification method.

References

Abado-Becognee, K., Mobio, T.A., Ennamany, R., Fleurat-Lessard, F., Shier, W.T., Badria, F. et al. 1998. Cytotoxicity of fumonisin B1: Implication of lipid peroxidation and inhibition of protein and DNA syntheses. Archives of Toxicology 72: 233–236. DOI: 10.1007/s002040050494.

Abarca, M.L., Accensi, F., Bragulat, M.R., Castellá, G. and Cabañes, F.J. 2003. *Aspergillus carbonarius* as the main source of ochratoxin-A contamination in dried vine fruits from the Spanish market. Journal of Food Protection 66: 504–506. DOI: 10.4315/0362-028x-66.3.504.

Akbari, P., Braber, S., Varasteh, S., Alizadeh, A., Garssen, J. and Fink-Gremmels, J. 2017. The intestinal barrier as an emerging target in the toxicological assessment of mycotoxins. Archives of Toxicology 91(3): 1007–1029. DOI: 10.1007/s00204-016-1794-8.

Alahlah, N., El Maadoudi, M., Bouchriti, N., Triqui, R. and Bougtaib, H. 2020. Aflatoxin M1 in UHT and powder milk marketed in the northern area of Morocco. Food Control.vol n pp DOI: 10.1016/j.foodcont.107262.

Arrêté Conjoint MAPM/MS. 2016. Arrêté Conjoint du Ministre de l'Agriculture et de la Pêche Maritime et du Ministre de la Santé n°1643-16 du 23 chaabane 1437 (30 mai 2016) fixant les limites maximales autorisées des contaminants dans les produits primaires et les produits alimentaires. Bulletin Officiel n°6514 du 03/11/2016. P. 1681. (Available in: http://www.onssa.gov.ma/images/reglementation/transversale/ARR.1643-16.FR.c1.pdf).

Attrassi, K., Benkirane, R., Attarassi, B., Badoc, A. and Douira, A. 2007. Efficacité de deux fongicides benzimidazolés et de l'anilinopyrimidine sur la pourriture des pommes en conservation. Bulletin de la Société de pharmacie de Bordeaux 146: 195–210.

Barbiroli, A., Bonomi, F., Benedetti, S., Mannino, S., Monti, L., Cattaneo, T. et al. 2007. Binding of aflatoxin M1 to different protein fractions in ovine and caprine milk. Journal of Dairy Sciences 90: 532–540. DOI: 10.3168/jds.S0022-0302(07)71536-9.

Bastin, R. 1949. La citrinine: sa production, son utilisation thérapeutique. Bulletin de la Société de chimie biologique 31(3/4): 865–875.

Battilani, P., Magan, N. and Logrieco, A. 2006. European research on ochratoxin-A in grapes and wine. International Journal of Food Microbiology 111(Supplement 1): S2-S4. DOI: 10.1016/j.ijfoodmicro.2006.02.007.

Beltrán, E., Ibáñez, M., Sancho, J.V. and Hernández, F. 2014. Determination of patulin in apple and derived products by UHPLC–MS/MS. Study of matrix effects with atmospheric pressure ionisation sources. Food Chemistry 142: 400–407. DOI: 10.1016/j.foodchem.2013.07.069.

Benkerroum, S. and Tantaoui-Elaraki, A. 2001. Study of toxigenic moulds and mycotoxins in poultry feeds. Revue de Médecine Vétérinaire 152: 335–342.

Benlamdini, N., Elhafian, M., Rochdi, A. and Zidane, L. 2014. Étude floristique et ethnobotanique de la flore médicinale du Haut Atlas oriental (Haute Moulouya). Journal of Applied Biosciences 78: 6771–6787. DOI: 10.4314/jab.v78i1.17.

Bennett, J.W. and Klich, M. 2003. Mycotoxins. Clinical Microbiology Reviews 16: 497–516. DOI: 10.1128/CMR.16.3.497-516.2003.

Blanc, P.J., Loret, M.O., Santerre, A. and Goma, G. 1995. Production of citrinin by various species of *Monascus*. Biotechnology Letters 17: 291–294. DOI: 10.1007/BF01190639.

Blesa, J., Moltó, J.C., El Akhdari, S., Mañes, J. and Zinedine, A. 2014. Simultaneous determination of *Fusarium* mycotoxins in wheat grain from Morocco by liquid chromatography coupled to triple quadrupole mass spectrometry. Food Control 46: 1–5. DOI: 10.1016/j.foodcont.2014.04.019.

Böhme, K., Barros-Velázquez, J., Calo-Mata, P. and Aubourg, S.P. 2014. Antibacterial, antiviral and antifungal activity of essential oils: Mechanisms and applications. *In:* Villa Veiga-Crespo, T.P. (eds). Antimicrobial Compounds. Springer, Berlin, Heidelberg.

Bolger, M., Coker, R.D., Dinovi, M., Gaylor, D., Gelderblom, M.O., Paster, N. et al. 2001. Fumonisins. pp. 103–279. *In:* Safety Evaluation of Certain Mycotoxins in Food. Food and Agriculture Organization of the United Nations, Paper n° 74. World Health Organization Food Additives, series 47. Geneva: WHO Food Additives.

Bouafifssa, Y., Manyes, L., Rahouti, M., Mañes, J., Berrada, H., Zinedine, A. et al. 2018. Multi-occurrence of twenty mycotoxinsin pasta and a risk assessment in the Moroccan population. Toxins 10: 432. DOI: 10.3390/toxins10110432.

Bouddine, L., Louaste, B., Achahbar, S., Chami, N., Chami, F. and Remmal, A. 2012. Comparative study of the antifungal activity of some essential oils and their major phenolic components against *Aspergillus niger* using three different methods. African Journal of Biotechenology 11: 14083–14087. DOI: 10.5897/AJB11.3293.

Bouigoumane, I., Selmaoui, K., Ouazzani-Touhami, A. and Douira, A. 2008. Efficacité *in vitro* et *in vivo* de différents fongicides sur le développement de la pourriture des poires dans la chambre froide d'Oulmès (Maroc). Reviews in Biology and Biotechnology 7(2): 37–47.

Bucheli, P. and Taniwaki, M.H. 2002. Research on the origin, and on the impact of post-harvest handling and manufacturing on the presence of ochratoxin-A in coffee. Food Additives and Contaminants 19: 655–665. DOI: 10.1080/02652030110113816.

Carballo, D., Pinheiro-Fernandes-Vieira, P., Tolosa, J., Font, G., Berrada, H. and Ferrer, E. 2018. Dietary exposure to mycotoxins through fruits juice consumption. Revista-de-Toxicologia 35: 2–6.

Chaoui, A., Faid, M. and Belhcen, R. 2003. Effect of natural starters used for sourdough bread in Morocco on phytate biodegradation. Eastern Mediterranean Health Journal 9: 141–147.

Chebli, B., Achouri, M., Idrissi-Hassani, L.M. and Hmamouchi, M. 2003. Chemical composition and antifungal activity of essential oils of seven Moroccan Labiatae against *Botrytis cinerea* Pers: Fr. Journal of Ethnopharmacology 89(1): 165–169. DOI: 10.1016/s0378-8741(03)00275-7.

Combrinck, S., Regnier, T. and Kamatou, G.P.P. 2011. *In vitro* activity of eighteen essential oils and some major components against common postharvest fungal pathogens of fruit. Industrial Crops and Products 33: 344–349. DOI: 10.1016/j.indcrop.2010.11.011.

Corbett, D. 2003. Patulin – U.K. Producers Perspective. *In:* Patulin Technical Symposium. February 18–19, 2003; Kissimmee, Florida, USA. National Center for Food Safety and Technology, Summit, III.

Creppy, E.E. 2002. Update of survey, regulation and toxic effects of mycotoxins in Europe. Toxicology Letters 127(1-3): 19–28. DOI: 10.1016/S0378-4274(01)00479-9.

De Boevre, M., Mavungu, J.D., Landshchoot, S., Audenaert, K., Eeckhout, M. and Maene, P. 2012. Natural occurrence of mycotoxins and their masked forms in food and feed products. World Mycotoxin Journal 5: 207–219. DOI: 10.3920/WMJ2012.1410.

de Oliveiri Filho, J.W.G., Islam, M.T., Ali, E.S., Uddin, S.J., Santos, J.V.O., de Alencar, M.V.O.B. et al. 2017. A comprehensive review on biological properties of citrinin. Food and Chemical Toxicology 110: 130–141. DOI: 10.1016/j.fct.2017.10.002.

Del Prete, V., Rodriguez, H., Carrascosa, A.V., De las Rivas, B., Garcia-Moruno, E. and Munoz, R. 2007. *In vitro* removal of OTA by wine lactic acid bacteria. Journal of Food Protection 70(9): 2155–2160. DOI: 10.4315/0362-028X-70.9.2155.

Dohnal, V., Wu, Q. and Kuča, K. 2014. Metabolism of aflatoxins: Key enzymes and interindividual as well as interspecies differences. Archives of Toxicology 88(9): 1635–1644. DOI: 10.1007/s00204-014-1312-9.

Dong, H., Xian, Y., Xiao, K., Wu, Y., Zhu, L. and He, J. 2019. Development and comparison of single-step solid phase extraction and QuEChERS clean-up for the analysis of 7 mycotoxins in fruits and vegetables during storage by UHPLC-MS/MS. Food Chemistry 274: 471–479. DOI: 10.1016/j.foodchem.2018.09.035.

Doughari, J.H. 2015. The occurrence, properties and significance of citrinin mycotoxin. Journal of Plant Pathology and Microbiology 6(11). pp DOI: 10.4172/2157-7471.1000321

El Adlouni, C., Tozlovanu, M., Naman, F., Faid, M. and Pfohl-Leszkowicz, A. 2006. Preliminary data on the presence of mycotoxins (ochratoxin A, citrinin and aflatoxin B1) in black table olives "Greek style" of Moroccan origin. Molecular Nutrition and Food Research 50(6): 507–512. DOI: 10.1002/mnfr.200600055.

El Hamsas-El Youbi, A., El Mansouri, L., Boukhira, S., Daoudi, A. and Bousta, D. 2016. *In vivo* anti-inflammatory and analgesic effects of aqueous extract of *Cistus ladanifer* L. from Morocco. American Journal of Therapeutics 23(6): 1554–1559. DOI: 10.1097/MJT.0000000000000419.

El Madani, H., Taouda, H. and Aarab, L. 2016. Evaluation of contamination of wheat and bread by fungi and mycotoxins in Fez region of Morocco. European Journal of Advanced Research in Biological and Life Sciences 4(2): 44–52. DOI: 10.13140/RG.2.1.1287.1288.

El Mahgubi, A., Puel, O., Bailly, S., Tadrist, S., Querin, A., Ouadia, A. et al. 2013. Distribution and toxigenicity of *Aspergillus* section Flavi in spices marketed in Morocco. Food Control 32: 143–148. DOI: 10.1016/j.foodcont.2012.11.013.

El Marnissi, B., Belkhou, R., Morgavi, D., Bennani, L. and Boudra, H. 2012. Occurrence of aflatoxin M1 in raw milk collected from traditional dairies in Morocco. Food Chemistry and Toxicology 50: 2819–2821. Doi.org/10.1016/j.fct.2012.05.031.

El Ouadi, Y., Manssouri, M., Bouyanzer, A., Majidi, L., Bendaif, H., Elmsellem, H. et al. 2017. Essential oil composition and antifungal activity of *Melissa officinalis* originating from north-East Morocco, against postharvest phytopathogenic fungi in apples. Microbial Pathogenesis 107: 321–326. DOI: 10.1016/j.micpath.2017.04.004.

Ennouari, A., Sanchis, V., Marin, S., Rabouti, M. and Zinedine, A. 2013. Occurrence of deoxynivalenol in durum wheat from Morocco. Food Control 32: 115–118. DOI: 10.1016/j.foodcont.2012.10.036.

Ennouari, A., Sanchis, V., Rahouti, M. and Zinedine, A. 2018. Isolation and molecular identification of mycotoxin producing fungi in durum wheat from Morocco. Journal of Materials and Environmental Sciences 9 (7): 1470–1479. DOI: 10.26872/jmes.2018.9.5.161.

Eskola, M., Kos, G., Elliott, C.T., Hajšlová, J., Mayar, S. and Krska, R. 2019. Worldwide contamination of food-crops with mycotoxins: Validity of the widely cited 'FAO estimate' of 25%. Critical Reviews in Food Science and Nutrition 1-17. DOI: 10.1080/10408398.2019.1658570.

European Commission. 2006. Commission Regulation No. 1881/2006 of December 19 setting maximum levels of certain contaminants in foodstuffs. Official Journal of the European Union No. L364/5 of December 20, 2006.

European Commission. 2007. Commission Regulation No. 1126/2007 of September 28 amending regulation (EC) No 1881/2006 setting maximum levels contaminants in foodstuffs as regards *Fusarium* toxins in maize and maize products. Official Journal of European Union L255/14 of September 29, 2006.

European Commission. 2013. Commission Recommendation of 27 March 2013 on the presence of T-2 and HT-2 toxin in cereals and cereal products (2013/165/EU). Official Journal of European Union L91/12 of April 4, 2013.3.4.2013.

EFSA (European Food Safety Authority). 2005. Opinion of the Scientific Panel on Contaminants in Food Chain on a request from the Commission related to fumonisins as undesirable substances in animal feed. The EFSA Journal 235: 1–32.

EFSA (European Food Safety Authority). 2006. Opinion of the scientific panel on contaminants in the food chain on a request from the Commission related to ochratoxin A in food. The EFSA Journal 365: 1–56.

EFSA (European Food Safety Authority). 2011. Scientific opinion on the risks for animal and public health related to the presence of T-2 and HT-2 toxin in food and feed. EFSA Journal 9(12): 2481.

EFSA (European Food Safety Authority). 2012. Scientific opinion on the risks for public and animal health related to the presence of citrinin in food and feed. EFSA Journal 10(3): 2605.

EFSA (European Food Safety Authority). 2017. Risks to human and animal health related to the presence of deoxynivalenol and its acetylated and modified forms in food and feed. EFSA Journal 15(9): 4718.

EFSA (European Food Safety Authority). 2018. Risks for animal health related to the presence of fumonisins, their modified forms and hidden forms in feed. EFSA Journal 16(5): 5242.

EFSA (European Food Safety Authority). 2020. Risk assessment of aflatoxins in food. EFSA Journal 18(3): 6040.

Fakchich, J. and Elachouri, M. 2014. Ethnobotanical survey of medicinal plants used by people in Oriental Morocco to manage various ailments. Journal of Ethnopharmacology 154 (1): 76–87. DOI: 10.1016/j.jep.2014.03.016.

FAO. 2001. Safety evaluation of certain mycotoxins in food. Document prepared by the Joint FAO/WHO Expert Committee on Food Additives (JECFA) at its fifty-sixth meeting.

FAO. 2002. Food and drug administration investigative. Operations Manual. Available at http://www.fda.gov/ora/inspect_ref/iom/Contents/ch4_TOC.html.

FAO. 2002. Manual on the application of the HACCP system in mycotoxin prevention and control. Rome, Italy: Joint FAO/WHO Food Standards Programme FAO.

FAO. 2009. Code of practice for the prevention and reduction of ochratoxin-a contamination in coffee. CAC/RCP 69–2009.

FAO. 2011. Profil nutritionnel de pays. Royaume du Maroc. Organisation des Nations Unies pour l'Alimentation et l'Agriculture. Available at http://www.fao.org/3/a-bc635f.pdf.

FAO, FIDA, OMS, PAM and UNICEF. 2018. L'État de la sécurité alimentaire et de la nutrition dans le monde 2018. Renforcer la résilience face aux changements climatiques pour la sécurité alimentaire et la nutrition. Rome, FAO.

FAO/WHO. 2011. Safety evaluation of certain food additives and contaminants. Deoxynivalenol. WHO Food Additives Series 63: 317–485.

FAO/WHO. 2013. Codex Alimentarius Commission – Proposed draft annex for the prevention and reduction of aflatoxins and ochratoxin a contamination in sorghum (code of practice for the prevention and reduction of mycotoxin contamination in cereals (cac/rcp 51- 2003).

Ferrara, M., Logrieco, A.F., Moretti, A. and Susca, A. 2020. A loop-mediated isothermal amplification (LAMP) assay for rapid detection of fumonisin producing *Aspergillus* species. Food Microbiology 90. DOI. 10.1016/j.fm.2020.103469.

Filali, A., Ouammi, L., Betbeder, A.M., Baudrimont, I., Soulaymani, R., Benayada, A. et al. 2001. Ochratoxin-A in beverages from Morocco: A preliminary survey. Food Additives and Contaminants 18(6): 565–568. DOI: 10.1080/02652030117365.

Frisvad, J.C., Smedsgaard, J., Larsen, T.O. and Samson, R. 2004. Mycotoxins, drugs and other extrolites produced by species in *Penicillium* subgenus *Penicillium*. Studies in Mycology 49, 201–241.

Frisvad, J.C., Smedsgaard, J., Samson, R.A., Larsen, T.O. and Thrane, U. 2007. Fumonisin B2 Production by *Aspergillus niger*. Journal of Agricultural and Food Chemistry 55(23): 9727–9732. Doi.org/10.1021/jf0718906.

Frisvad, J.C., Larsen, T.O, Thrane, U., Meijer, M., Varga, J. Samson, R.A. et al. 2011. Fumonisin and ochratoxin production in industrial *Aspergillus niger* strains. PLoS One 6(8). pp DOI: 10.1371/journal.pone.0023496.

Fuchs, S., Sontag, G., Stidl, R., Ehrlich, Font, V., Kundi, M. and Knasmüller, S. 2008. Detoxification of patulin and OTA, two abundant mycotoxins, by lactic acid bacteria. Food and Chemical Toxicology 46(4): 1398–1407. DOI: 10.1016/j.fct.2007.10.008.

García-Moraleja, A.G., Mañes, J. and Ferrer, E. 2015. Analysis of mycotoxins in coffee and risk assessment in Spanish adolescents and adults. Food and Chemical Toxicology 86: 225–233. DOI: 10.1016/j.fct.2015.10.014.

Ghourri, M., Zidane, L. and Douira, A. 2013. Usage des plantes médicinales dans le traitement du Diabète Au Sahara marocain (Tan-Tan). Journal of Animal and Plant Sciences 17(1): 2388–2411. DOI: MMM-05-2010-4-3-1957-2557-101019-201002287.

Gourama, H. 1997. Inhibition of growth and mycotoxin production of *Penicillium* by *Lactobacillus* species. LWT - Food Science and Technology 30(3): 279–283. DOI: 10.1006/fstl.1996.0183.

Gourama, H. and Bullerman L.B. 1988. Mycotoxin production by molds isolated from 'Greek-style' black olives. International Journal of Food Microbiology 6(1): 81–90. DOI: 10.1016/0168-1605(88)90087-6.

Govaris, A., Roussi, V., Koidis, P.A. and Botsoglou, N.A. 2001. Distribution and stability of aflatoxin M1 during processing, ripening and storage of Telemes cheese. Food Additives and Contaminants 18: 437–443. DOI: 10.1080/02652030120550.

Grajewski, J., Kosicki, R., Twaruzek, M. and Błajet-Kosicka, A. 2019. Occurrence and risk assessment of mycotoxins through Polish beer consumption. Toxins 11(254). pp DOI: 10.3390/toxins11050254.

Guimarães, A., Santiago, A., José, A., Teixeira-Venâncio, A. and Abrunhosa, L. 2017. Anti-aflatoxigenic effect of organic acids produced by *Lactobacillus plantarum*. International Journal of Food Microbiology 264: 31–38. DOI: 10.1016/j.ijfoodmicro.2017.10.025.

Gupta, P.K. 2018. Toxicity of fungicides. pp. 569-580. *In:* Veterinary Toxicology (3rd ed) Elsevier Inc. DOI: 10.1016/B978-0-12-811410-0.00045-3.

Hajjaj, H., Klaebe, A., Loret, M.O., Goma, G., Blanc, P.J. and Francois, J. 1999. Biosynthetic pathway of citrinin in the filamentous fungus *Monascus ruber* as revealed by 13C nuclear magnetic resonance. Applied and Environmental Microbiology 65: 311–314.

Hajjaj, H., Blanc, P.J., Groussac, E., Goma, G., Uribelarrea, J.L. and Loubiere, P. 2000a. Improvement of red pigment/citrinin production ratio as a function of environmental conditions by *Monascus ruber*. Biotechnology and Bioengineering 64(4): 497–501. DOI: 10.1002/(sici)1097-0290(19990820)

Hajjaj, H., Klaebe, A., Goma, G., Blanc, P.J., Barbier, E. and Francois, J. 2000b. Medium chain fatty acids affect citrinin production in the filamentous fungus *Monascus ruber*. Applied and Environmental Microbiology 66(3): 1120–1125. DOI: 10.1128/AEM.66.3.1120-1125.2000.

Hajjaj, H., François, J.M., Goma, G. and Blanc, P.J. 2012. Effect of amino acids on red pigments and citrinin production in *Monascus ruber*. Journal of Food Science 77(3): M156-M159. DOI: 10.1111/j.1750-3841.2011.02579.x.

Hajjaj, H., Goma, G. and François, J.M. 2015. C/N ratio and cultivation mode effects on citrinin and red pigment production from *Monascus ruber*. International Journal of Food Science & Technology 50: 1731–1736. DOI: 10.1111/ijfs.12803.

Hajjaji, A., El Otmani, M., Bouya, D., Bouseta, A., Mathieu, F., Collin, S. et al. 2006. Occurrence of mycotoxins (ochratoxin-A, deoxynivalenol) and toxigenic fungi in Moroccan wheat grains: Impact of ecological factors on the growth and ochratoxin-A production. Molecular Nutrition and Food Research 50: 494–499. DOI: 10.1002/mnfr.200500196.

Haschek, W.M., Gumprecht, L.A., Smith, G., Tumbleson, M.E. and Constable, P.D. 2001. Fumonisin toxicosis in swine: An overview of porcine pulmonary edema and current perspectives. Environmental Health Perspectives 109: 251–257. DOI: 10.1289/ehp.01109s2251.

HCP (Haut-Commissariat au Plan). 2018. Enquête nationale sur la consommation et les dépenses des ménages 2013/2014. Rapport de synthèse. Available: https://

www.hcp.ma/downloads/Enquete-Nationale-sur-la-Consommation-et-les-Depenses-des-Menages_t21181.html.

Hetherington, A.C. and Raistrick, H. 1931. Studies in the Biochemistry of Micro-organisms. Part XIV. On the production and chemical constitution of a new yellow colouring matter, citrinin, produced from glucose by *Penicillium*. DOI: 10.1098/rstb.1931.0025.

Hussein, H.S. and Brasel, J.M. 2001. Toxicity, metabolism, and impact of mycotoxins on humans and animals. Toxicology 167: 101–134. DOI: 10.1016/S0300-483X(01)00471-1.

IARC (International Agency for Research on Cancer), 1993. Ochratoxin-A. *In:* IARC monographs on the evaluation of carcinogenic risks to humans: Some naturally occurring substances; food items and constituents, heterocyclic aromatic amines and mycotoxins. 56: 489–521.

Idrissi, L., Betbedez, A.M., Benlemlih, M., Faid, M., Creppy, E.E. and Zinedine, A. 2004. Ochratoxin-A, determination in dried fruits and black olives from Morocco. Alimentaria: Revisita de tecnologia e hygiene de los alimentos, 73–76.

Jarda, G., Liboz, T., Mathieua, F., Guyonvarc'h, A. and Lebrihi, A. 2011. Review of mycotoxin reduction in food and feed: From prevention in the field to detoxification by adsorption or transformation. Food Additives & Contaminants 28(11): 1590–1609. DOI: 10.1080/19440049.2011.595377.

Juan, C., Zinedine, A., Idrissi, I. and Manes, J. 2008a. Ochratoxin-A in rice on the Moroccan retail market. International Journal of Food Microbiology 126: 83–85. DOI: 10.1016/j.ijfoodmicro.2008.05.005.

Juan, C., Zinedine, A., Soriano, J.M., Moltó, J.C., Idrissi, L. and Mañes, J. 2008b. Aflatoxins levels in dried fruits and nuts available in Rabat-Salé area, Morocco. Food Control 19: 849–853. DOI: 10.1016/j.foodcont.2007.08.010.

Kabak, B., Dobson, A.D.W. and Var, I. 2006. Strategies to prevent mycotoxin contamination of food and animal feed: A review. Critical Reviews in Food Science and Nutrition 46(8): 593–619. DOI: 10.1080/10408390500436185.

Karlovsky, P., Suman, M., Berthiller, F., De Meester, J., Eisenbrand, G., Perrin, I. et al. 2016. Impact of food processing and detoxification treatments on mycotoxin contamination. Mycotoxin Research 32: 197–205. DOI: 10.1007/s12550-016-0257-7.

Kichou, F. and Walser, M.M. 1993. The natural occurrence of aflatoxin B1 in Moroccan poultry feeds. Veterinary and Human Toxicology 35: 105–108.

Krishnamachari, K., Bhat, R.V., Nagarajan, V. and Tilac, T. 1975. Investigations into an outbreak of hepatitis in Western India. Indian Journal of Medical Research 63: 1036–1048.

Krnjaja, V., Levic, J., Stankovic, S. and Stepanic, A. 2011. *Fusarium* species and their mycotoxins in wheat grain. Zbornik Matice Srpske Za Prirodne Nauke 120: 41–48. DOI: 10.2298/ZMSPN1120041K.

Krogh, P.1987. Ochratoxins in food. pp. 97–121. *In:* Krogh P. (ed.). Mycotoxins in Food. Academic Press, London.

Kroes, R., Müller, D., Lambe, J., Löwik, M.R.H, van Klaveren, J., Kleiner, J. 2002. Assessment of intake from the diet. Food and Chemical Toxicology 40(2–3): 327-385. DOI: 10.1016/S0278-6915(01)00113-2.

Laaziz, A., El Hammoudi, Y., Hajjaji, A., Qjidaa, S. and Bouseta, A. 2019. Evaluation of the control ability of *Thymus satureioides* essential oil against *Aspergillus niger* growth and ochratoxin-A production. Colloque International sur la Recherche en Agroalimentaire February 05–06, Larache, Morocco. Available in: http://www.erbgb.ma/Proceeding_2019.pdf

Laaziz, A., Qjidaa, S., El Hammoudi, Y., Hajjaji, A. and Bouseta, A. 2017. Chemical control of fungal growth and ochratoxin-A production by *Aspergillus* isolated from Moroccan grapes. South Asian Journal of Experimental Biology 7(2): 84–91.

Lanyasunya, T.P., Wamae, L.W., Musa, H.H., Olowofeso, O. and Lokwaleput, I.K. 2005. The risk of mycotoxins contamination of dairy feed and milk on small holder dairy farms in Kenya. Pakistan Journal of Nutrition 4: 162–169. DOI: 10.3923/pjn.2005.162.169.

Larsen, J.C., Hunt, J., Perin, I. and Ruckenbauer. P. 2004. Workshop on trichothecenes with a focus on DON: Summary report. Toxicology Letters 153: 1–22. doi. org/10.1016/j.toxlet.2004.04.020.

Lee, H.J. and Ryu, D. 2017. Worldwide occurrence of mycotoxins in cereals and cereal-derived food products: Public health perspectives of their co-occurrence. Journal of Agricultural and Food Chemistry 65(33): 7034–7051. DOI: 10.1021/acs.jafc.6b04847.

Magan, N., Sanchis, V. and Aldred, D. 2004. Role of spoilage fungi in seed deterioration. pp. 311–323. *In:* Aurora D.K. (ed.). Fungal Biotechnology in Agricultural, Food and Environmental Applications. Marcell Decker, New York.

Magan, N. and Aldred, D. 2007. Post-harvest control strategies: Minimizing mycotoxins in the food chain. International Journal of Food Microbiology 119: 131–139. DOI: 10.1016/j.ijfoodmicro.2007.07.034.

Mannani, N., Tabarani, A., Abdennebi, E.H. and Zinedine, A. 2020. Assessment of aflatoxin levels in herbal green tea available on the Moroccan market. Food Control 108: 106882. DOI: 10.1016/j.foodcont.2019.106882.

Mansouri, A., Hafidi, M., Mazouz, H., Zouhair, R., El Karbane, M. and Hajjaj, H. 2014. Mycoflora and patulin-producing strains of cereals in North-Western Morocco. South Asian Journal of Experimental Biology 4(5): 276–282.

Mansouri, A., Elkarbane, M., Ben Aziz, M., Nait M'Barek, H., Hafidi, M. and Hajjaj, H. 2017. Effect of carbon, nitrogen and physico-chemical factors on patulin production in *Penicillium expansum*. South Asian Journal of Experimental Biology 7(2): 57–63.

Mansouri A, 2018. Study of mycological diversity and risk assessment of mycotoxins in the cereal sector in the Meknes region. Thesis. April 28. Moulay Ismail University, Meknes.

MAPMDREF (Ministère de l'Agriculture, de la Pêche Maritime, du Développement Rural et des Eaux et Forêts). 2019. Agriculture en chiffres 2018. (Available in: http://www.agriculture.gov.ma/sites/default/files/19-00145-book_agricultures_en_chiffres_def.pdf).

Marasas, W.F., Riley, R.T., Hendricks, K.A., Stevens, V.L., Sadler, T.W., Gelineau-van Waes, J. et al. 2004. Fumonisins disrupt sphingolipid metabolism, folate transport, and neural tube development in embryo culture and *in vivo*: A potential risk factor for human neural tube defects among populations consuming fumonisin-contaminated maize. The Journal of Nutrition 134(4): 711–716. DOI: 10.1093/jn/134.4.711.

Marin, S., Ramos, A.J., Cano-Sancho, G. and Sanchis, V. 2013. Mycotoxins: Occurrence, toxicology, and exposure assessment. Food and Chemical Toxicology 60: 218–237. DOI: 10.1016/j.fct.2013.07.047.

Marquardt, R.R. and Frohlich, A.A. 1992. A review of recent advances in understanding ochratoxicosis. Journal of Animal Science 70: 2968–3988. DOI: 10.2527/1992.70123968x.

Martins, C., Vidal, A., De Boevre, M., De Saeger, S., Nunes, C., Torrese, D. et al. 2019. Exposure assessment of Portuguese population to multiple mycotoxins: The human biomonitoring approach. International Journal of Hygiene and Environmental Health 222: 913–925. DOI: 10.1016/j.ijheh.2019.06.010.

Mogensen, J.M., Larsen, T.O. and Nielsen, K.F. 2010. Widespread occurrence of the mycotoxin fumonisin b(2) in wine. Journal of Agricultural and Food Chemistry 58(8): 4853–4857. DOI: 10.1021/jf904520t.

Montet, D., Hazm, J.E., Ouadia, A., Chichi, A., Mbaye, M.S., Diop, M.B. et al. 2019. Contribution of the methodology of collective expertise to the mitigation of food safety hazards in low- or medium-income countries. Food Control 99: 84–88. DOI: 10.1016/j.foodcont.2018.12.009.

Montet, D., Hazm, J.E., Ouadia, A., Chichi, A., Mbaye, M.S., Diop, M.B. et al. 2020. Use of collective expertise as a tool to reinforce food safety management in Africa. Journal of Food Research 9(3): 1–18. DOI: 10.5539/jfr.v9n3p9.

Morales, H., Sanchis, V., Rovira, A., Ramos, A.J. and Marin, S. 2007. Patulin accumulation in apples during postharvest: Effect of controlled atmosphere storage and fungicide treatments. Food Control 18(11): 1443–1448. DOI: 10.1016/j.foodcont.2006.10.008.

Moreno, M.R., Lopez, H.R., Ray, B.D., Pojan, I.M., Moruno, E.G. and Doria, F. 2013. Biological degradation of OTA into ochratoxin. US Patent Application Publication. No. 20130209609 A1.

Muture, B.N. and Ogana, G. 2005. Aflatoxin levels in maize and maize products during the 2004 food poisoning outbreak in Eastern Province of Kenya. East African Medical Journal 82(6): 275–279. DOI: 10.4314/eamj.v82i6.9296.

Nielsen, K.F., Gräfenhan, T., Zafari, D. and Thrane, U. 2005. Trichothecene production by *Trichoderma brevicompactum*. Journal of Agricultural and Food Chemistry 53(21): 8190–8196. DOI: 10.1021/jf051279b.

Nielsen, K.F. and Thrane, U. 2001. Fast methods for screening of trichothecenes in fungal cultures using gas chromatography–tandem mass spectrometry. Journal of Chromatography A 929(1-2): 75–87. DOI: 10.1016/s0021-9673(01)01174-8.

ONICL. 2020. Office National Interprofessionnel des Céréales et des Légumineuses. Statistiques, Maroc. Production-importation céréales principales. Available in: https://www.onicl.org.ma/portail/situation-du-march%C3%A9/ statistiques.

Ostry, V., Malir, F., Toman, Y.G. and Grosse Y. 2017. Mycotoxins as human carcinogens – the IARC Monographs classification. Mycotoxin Research 33(1): 65–73. DOI: 10.1007/s12550-016-0265-7.

Ouhibi, S., Vidal, A., Martins, C., Gali, R., Hedhili, A., De Saeger, S. et al. 2020. LC-MS/MS methodology for simultaneous determination of patulin and citrinin in urine and plasma applied to a pilot study in colorectal cancer patients. Food and Chemical Toxicology 136. DOI: 10.1016/j.fct.2019.110994.

Pace, J.G., Watts, M.R. and Canterbury, W.J. 1988. T-2 mycotoxin inhibits mitochondrial protein synthesis. Toxicon 26(1): 77–85. DOI: 10.1016/0041-0101(88)90139-0.

Pal, S., Singh, N. and Ansari, K.M. 2017. Toxicological effects of patulin mycotoxin on the mammalian system: An overview. Toxicology Research 6(6): 764–771. Doi: 10.1039/c7tx00138j.

Pallarés, N., Font, G., Mañes, J. and Ferrer, E. 2017. Multimycotoxin LC–MS/MS analysis in tea beverages after Dispersive Liquid–Liquid Microextraction (DLLME). Journal of Agricultural and Food Chemistry 65(47): 10282–10289. Doi: 10.1021/acs.jafc.7b03507.

Park, D.L., Njapau, H. and Boutrif, E. 1999. Minimizing risks posed by mycotoxins utilizing the HACCP concept. FAO Food. Nutrition and Agriculture Journal 23: 49–56.

Peraica, M., Domijan, A.M., Jurjevic, Z. and Cvjetkovic, B. 2002. Prevention of exposure to mycotoxins from food and feed. Archives of Industrial Hygiene and Toxicology 53: 229–237.

Pestka, J.J. 2007. Deoxynivalenol: Toxicity, mechanisms and animal health risks. Anim. Feed Science Technology 137(3-4): 283–298. Doi: 10.1016/j.anifeedsci.2007.06.006.

Petchkongkaew, A., Taillandier, P., Gasaluck, P. and Lebrihi, A. 2008. Isolation of *Bacillus* spp. from Thai fermented soybean (Thua-nao). Screening for aflatoxin B1 and OTA detoxification. Journal of Applied Microbiology 104: 1495–1502. Doi: 10.1111/j.1365-2672.2007.03700.x.

Pitt, J.I., Basilico, J.C., Abarca, M.L. and Lopez, C. 2000. Mycotoxins and toxigenic fungi. Medical Mycology 38 (Supplement 1), 41–46. Doi: 10.1080/mmy.38.s1.41.46.

Pitt, J.I. and Miller, J.D. 2017. A Concise History of Mycotoxin Research. Journal of Agricultural and Food Chemistry 65 (33): 7021–7033. Doi: 10.1021/acs.jafc.6b04494.

Puel, O., Galtier, P. and Oswald, I.P. 2010. Biosynthesis and toxicological effects of patulin. Toxins 2(4): 613–631. Doi: 10.3390/toxins2040613.

Qjidaa, S. 2015. Effet des fongicides, du resvératrol et de l'huile essentielle de *Mentha pulegium* sur la croissance fongique et la production de l'OTA par des souches d'*Aspergillus tubingensis* et *A. foetidus* isolées à partir du raisin marocain. Université Sidi Mohammed Ben Abdellah Faculté des Sciences Dhar El Mahraz-Fès.

Qjidaa, S., Laaziz, A., Hajjaji, A. and Bouseta, A. 2018. Effet du trans-resvératrol sur la croissance et la production de l'ochratoxine A de certaines souches d'*Aspergillus* marocains. International Journal of Biological and Chemical Sciences 12: 1345–1355.

Qjidaa, S., Selouane, A., Zouhair, S., Bouya, D., Decock, C. and Bouseta, A. 2014. *In vitro* effect of pyrimethanil on the fungal growth of ochratoxigenic and no-ochratoxigenic species of *Aspergillus tubingensis* and *Aspergillus foetidus* isolated from Moroccan grapes. South Asian Journal of Experimental Biology 4: 76–84.

RASFF (The Rapid Alert System for Food and Feed). 2018. Available in: https://

ec.europa.eu/food/sites/food/files/safety/docs/rasff_annual_report_2018.pdf

Reboux, G. 2006. Mycotoxins: Health effects and relationship to other organic compounds. Revue Française d'Allergologie et d'Immunologie Clinique 46(3) : 208–212. doi.org/10.1016/j.allerg.2006.01.036.

Rharmitt, S., Hafidi, M., Hajjaj, H., Scordino, F., Giosa, D. Giuffrè, L. et al. 2016. Molecular characterization of patulin producing and non-producing *Penicillium* species in apples from Morocco. International Journal of Food Microbiology 217: 137–140. DOI: 10.1016/j.ijfoodmicro.2015.10.019.

Richard, J.L. 2007. Some major mycotoxins and their mycotoxicoses – An overview. International Journal of Food Microbiology 119(1–2): 3–10. DOI: 10.1016/j.ijfoodmicro.2007.07.019.

Riley, R.T. and Merrill, A.H. Jr. 2019. Ceramide synthase inhibition by fumonisins: A perfect storm of perturbed sphingolipid metabolism, signaling and disease. Journal of Lipid Research 60(7): 1183–1189. DOI: 10.1194/jlr.S093815.

Rodriguez, H., Reveron, I., Doria, F., Costantini, A., De Las, R.B. and Munoz, R. 2011. Degradation of OTA by *Brevibacterium* species. Journal of Agricultural and Food Chemistry 59: 10755–10760. DOI: 10.1021/jf203061p.

Roussos, S., Zaouia, N., Salih, G., Tantaoui-Elaraki, A., Lamrani, K., Cheheb, M. et al. 2006. Characterization of filamentous fungi isolated from Moroccan olive and olive cake: Toxinogenic potential of *Aspergillus* strains. Molecular Nutrition and Food Research 50(6): 500–506. DOI: 10.1002/mnfr.200600005.

Sakuma, H., Watanabe Y., Furusawa, H., Yoshinari, T., Akashi, H., Kawakami, H. et al. 2013. Estimated dietary exposure to mycotoxins after taking into account the cooking of staple foods in Japan. Toxins 5(5): 1032–1042. DOI: 10.3390/toxins5051032.

Saleh, I. and Goktepe, I. 2019. The characteristics, occurrence, and toxicological effects of patulin. Food and Chemical Toxicology 129: 301–311. DOI: 10.1016/j.fct.2019.04.036.

Scott, P.M., van Walbeek, W., Kennedy, B. and Anyeti, D. 1972. Mycotoxins (ochratoxin-A, citrinin, and sterigmatocystin) and toxigenic fungi in grains and agricultural products. Journal of Agricultural and Food Chemistry 20: 1103–1109. DOI: 10.1021/jf60184a010.

Selouane, A., Zouhair, S., Bouya, D., Lebrihi, A. and Bouseta A. 2009a. Natural occurrence of ochratoxigenic *Aspergillus* species and ochratoxin A in Moroccan grapes. World Applied Sciences Journal 7 (3): 297–305.

Selouane, A., Bouya, D., Lebrihi, A. and Bouseta, A. 2009b. Impact of some environmental factors on growth and production of ochratoxin-A of/by *Aspergillus tubingensis*, *A. niger* and *A. carbonarius* isolated from Morrocan grapes. The Journal of Microbiology 411–419. DOI: 10.1007/s12275-008-0236-6.

Selouane, A., Qjidaa, S., Zouhair, S., Lebrihi, A., Bouya, D. and Bouseta, A. 2014. Effect of culture medium, temperature and incubation time on growth and ochratoxin-A production by *Aspergillus tubingensis* and *A. niger* isolated from Moroccan vineyards. South Asian Journal of Experimental Biology 4(6): 290–299.

Shephard, G.S. 2004. Mycotoxins worldwide: Current issues in Africa. pp. 81–88. *In:* Barug, D., Van Egmond, H.P., López Garciá, R., van Osenbruggen,

W.A. and Visconti, A. (eds.). Meeting the Mycotoxin Menace. Wageningen Academic Publishers, The Netherlands.

Shephard, G.S., van der Westhuizen, L., Katerere, D.R., Herbst, M. and Pineiro, M. 2010. Preliminary exposure assessment of deoxynivalenol and patulin in South Africa. Mycotoxin Research 26: 181–185. DOI: 10.1007/s12550-010-0052-9.

Shi, H., Li, S., Bai, Y., Prates, L.L., Lei, Y. and Yu, P. 2018. Mycotoxin contamination of food and feed in China: Occurrence, detection techniques, toxicological effects and advances in mitigation technologies. Food Control 91: 202–215. Doi.org/10.1016/j.foodcont.2018.03.036.

Sifou, A., Mahnine, N., Manyes, L., El Adlouni, C., El Azzouzi, M. and Zinedine, A. 2016. Determination of ochratoxin-A in poultry feeds available in Rabat area (Morocco) by high performance liquid chromatography. Journal of Materials and Environmental Science 7(6): 2229–2234.

Soubra, L., Sarkis, D., Hilan, C. and Verger, P. 2009. Occurrence of total aflatoxins, ochratoxin-A and deoxynivalenol in food stuffs available on the Lebanese market and their impact on dietary exposure of children and teenagers in Beirut. Food Additives and Contaminants 26(2): 189–200. DOI: 10.1080/02652030802366108.

Steiman, R., Seigle-Murandi, F., Sage, L. and Krivobok, S. 1989. Production of patulin by micromycetes. Mycopathologia 105(3): 129–133. DOI: 10.1007/BF00437244.

Stob, M., Baldwin, R.S., Tuite, J., Andrews, F.N. and Gillette, K.G. 1962. Isolation of an anabolic uterotropic compound from corn infested with Gibberellazeae. Nature 196: 1318. DOI: 10.1038/1961318a0.

Stove, S.D. 2013. Food safety and increasing hazard of mycotoxin occurrence in foods and feeds. Critical Reviews in Food Science and Nutrition 53(9): 887–901. DOI: 10.1080/10408398.2011.571800.

Strange, R.N. and Scott, P.R. 2005. Plant disease: A threat to global food security. Annual Review Phytopathology 43: 83–116. Doi.org/10.1146/annurev. phyto.43.113004.133839.

Sun, M.H., Li, X.H., Xu, H., Xu, Y., Pan, Z.N. and Sun, S.C. 2020. Citrinin exposure disrupts organelle distribution and functions in mouse oocytes. Environmental Research 185. DOI: 10.1016/j.envres.2020.109476.

Tantaoui-Elaraki, A., Benabdellah, I., Majdi, M., Elalaoui, M.R. and Dahmani, A., 1994. Recherche de mycotoxines dans des denrées alimentaires distribuées au Maroc. Actes de l'Institut Agronomique et Vétérinaire 14: 11–16.

Teixeira, B., Marques, A., Ramosa, C., Nengc, N.R., Nogueirac, J.M.F., Saraiva, J.A. et al. 2013. Chemical composition and antibacterial and antioxidant properties of commercial essential oils. Industrial Crops and Products 43: 587–595. DOI: 10.1016/j.indcrop.2012.07.069.

Temba, B.A., Sultanbawa, Y., Kriticos, D.J., Fox, G.P., Harvey, J.J.W. and Fletcher, M.T. 2016. Tools for defusing a major global food and feed safety risk: Nonbiological postharvest procedures to decontaminate mycotoxins in foods and feeds. Journal of Agricultural and Food Chemistry 64(47): 8959–8972. DOI: 10.1021/acs.jafc.6b03777.

Turner, WB. 1971. Fungal Metabolites. Academic Press, London, New York.

UNEP-FAO-RC-SHPF. 2015. Etude sur le suivi de l'effet des pesticides sur la santé humaine et l'environnement. (http://www.pic.int/Portals/5/download. aspx?d=UNEP-FAO-RC-SHPF-Morocco-Report-20151127.Fr.pdf).

Van den Broek, P., Pittet, A. and Hajjaj, H. 2001. Aflatoxin genes and the aflatoxinogenic potential of Koji molds. Applied Biotechnology and Microbiology 57(1-2): 192–199. DOI: 10.1007/s002530100736.

Van der Merve, K.J., Steyn, P.S. and Fourie, L. 1965. Mycotoxins. Part II: The constitution of chratoxins A, B and C, metabolites of *Aspergillus ochraceus* Wilh. Journal of Chemical Society vol 7083–7088. DOI: 10.1039/jr9650007083.

Varga, J., Rigó, K., Tóth, B., Téren, J. and Kozakiewicz, Z. 2003. Evolutionary relationships among *Aspergillus* species producing economically important mycotoxins. Food Technology and Biotechnology 41(1): 29–36.

Verma, R.J. 2004. Aflatoxin cause DNA damage. International Journal of Human Genetics 4(4): 231–236. DOI: 10.1080/09723757.2004.11885899.

Vidal, A., Ouhibi, S., Ghali, R., Hedhili, A., De Saeger, S. and De Boevre, M. 2019. The mycotoxin patulin: An updated short review on occurrence, toxicity and analytical challenges. Food and Chemical Toxicology 129: 249–256. DOI: 10.1016/j.fct.2019.04.048.

Whitaker, T.B. 2001. Sampling techniques. pp. 11–24. *In:* Trucksess, M.W. and Pohland, A.E. (eds.). Mycotoxin Protocols. Vol. 157. Humana Press Inc., Totowa, NJ.

Wild, C.P. and Turner, P.C. 2002. The toxicology of aflatoxins as a basis for public health decisions. Mutagenesis 17(6): 471–481.

Wu, F. and Mitchell, N.J. 2016. How climate change and regulations can affect the economics of mycotoxins. World Mycotoxin Journal 9(5): 653–663. DOI: 10.3920/WMJ2015.2015.

Xu, D., Wang, H., Zhang, Y., Yang, Z. and Sun, X. 2013. Inhibition of non-toxigenic *Aspergillus niger* FS10 isolated from Chinese fermented soybean on growth and aflatoxin B1 production by *Aspergillus flavus*. Food Control 32(2): 359–365. DOI: 10.1016/j.foodcont.2012.12.013.

Yiannikouris, A. and Jouany, J.P. 2002. Mycotoxins in feeds and their fate in animals: A review. Animal Research 51: 81–99. DOI: 10.1093/mutage/17.6.471.

Zaied, C., Zouaoui, N., Bacha, H. and Abid, A. 2012. Natural occurrence of citrinin in Tunisian wheat grains. Food Control 28(1): 106–109. DOI: 10.1016/j.foodcont.2012.04.015.

Zhu, Y., Hassan, Y., Watts, C. and Zhou, T. 2016. Innovative technologies for the mitigation of mycotoxins in animal feed and ingredients—A review of recent patents. Animal Feed Science and Technology 216: 19–29. DOI: 10.1016/j.anifeedsci.2016.03.030.

Zinedine, A., Brera, C., Elakhdari, S., Catano, C., Debegnach, F., Angelini, S. et al. 2006. Natural occurrence of mycotoxins in cereals and spices commercialized in Morocco. Food Control 17(11): 868–874. DOI: 10.1016/j.foodcont.2005.06.001.

Zinedine, A., Juan, C., Soriano, J.M., Molto, J.C., Idrissi, I. and Manes, J. 2007a. Limited survey for the occurrence of aflatoxins in cereals and poultry feed from Rabat, Morocco. International Journal of Food Microbiology 115(1): 124–127. DOI: 10.1016/j.ijfoodmicro.2006.10.013

Zinedine, A., Juan, C., Idrissi, I. and Manes, J. 2007b. Occurrence of ochratoxin-A in bread consumed in Morocco. Biochemical Journal 87: 154–158. DOI: 10.1016/j.microc.2007.07.004.

Zinedine, A., Soriano, J.M., Juan, C., Mojemmi, B., Moltó, J.C., Bouclouze, A. et al. 2007c. Incidence of ochratoxin-A in rice and dried fruits from Rabat and Salé area, Morocco. Food Additives and Contaminants 24(3): 285–291. DOI: 10.1080/02652030600967230.

Zinedine, A., Gonzalez-Osnaya, L., Soriano, J.M., Molto, J.C., Idrissi, I. and Manes, J. 2007d. Presence of aflatoxin M1 in pasteurized milk from Morocco. International Journal of Food Microbiology 114(1): 25–29. DOI: 10.1016/j.ijfoodmicro.2006.11.001.

Zinedine, A., Blesa, J., Mahnine, N., El Abidi, A., Montesano, D. and Mañes, J. 2010. Pressurized liquid extraction coupled to liquid chromatography for the analysis of ochratoxin-A in breakfast and infants cereals from Morocco. Food Control 21(2): 132–135. DOI: 10.1016/j.foodcont.2009.04.009.

Zinedine, A., Fernández-Franzón, M., Mañes, J. and Manyes, L. 2017. Multi-mycotoxin contamination of couscous semolina commercialized in Morocco. Food Chemistry 214: 440–446. DOI: 10.1016/j.foodchem.2016.07.098.

Zhong, L., Carere, J., Lu, Z., Lu, F. and Zhou, F. 2018. Patulin in apples and apple-based food products: The burdens and the mitigation strategies. Toxins 10(475). pp DOI: 10.3390/toxins10110475.

Zouhair, S., Qjidaa, S., Selouane, A., Bouya, D., Decock, C. and Bouseta, A. 2014. Effect of five fungicides on growth and ochratoxin-A production by two *Aspergillus carbonarius* and *Aspergillus niger* isolated from Moroccan grapes. South Asian Journal of Experimental Biology 4 (3): 118–126.

Status and Management of Aflatoxins in Kenya

Ruth Nyaga
Nairobi 19441 00202.

1. Introduction

Aflatoxin is a poisonous substance, which when taken in large quantities exposes the consumer(s) to different digestive cancers, more so when the exposure is for a long period of time. Management of aflatoxin levels continue to be a topical issue as the substance is found in both cultivated and uncultivated land. Unfortunately, crops and animals that derive their nutrients and food from high aflatoxin infested lands tend to contain high levels of the substance, which in return is transferred to the community through consumption and in the process exposes a larger part of the community to in particular liver cancer (Hoffman et al. 2018).

Unfortunately, less developed countries such as Kenya are the most affected by liver cancer in the world that is mostly fueled by being exposed to high levels of aflatoxin. As indicated by the WHO (2015) assessment report, aflatoxin is transferred to the community through consumed food items. Unlike in the developed countries, measures employed to control aflatoxin transfer by fungi from the soil to animals and crops deliver better results as compared to less developed countries. In addition, the dietary habits of less developed countries due to limited resources has resulted in high reliance on grains such as maize and groundnuts, which have high aflatoxin levels in comparison to other crops and animal products (Hoffman et al. 2019). The findings by Hoffman et al. (2019) underscore the danger aflatoxin pose to less developed countries.

This chapter's focus therefore is limited to the situation of aflatoxin in Kenya, foods that are affected, local regulation and control means applied. This is in the belief that like in the developed worlds, the mycotoxins impact on the health of the communities in the country could be minimized.

Email: nyagahr@gmail.com

1.1 Background of Cancer Situation in Kenya

Cancer may not be considered a pandemic, but in Kenya it could as well be. Just before the COVID-19 issue it had come close to being declared a national disaster. As revealed by UICC (2019), there are 47,000 new cancer cases and 33,000 death cases annually. Narrowing down to aflatoxins and the consequences they pose in form of liver cancer, it accounts for 2.8% of new cases and deaths per annum and the percentage continues to increase (IARC 2018). This has seen the calls for cancer being declared a national disaster becoming the order of the day (Oketch 2020).

The same article (Oketch 2020) noted that as the number of cases continue to rise, the country due to economic limitations cannot adequately mobilize the required resources. The rallying point is to declare cancer a national disaster and aim at minimizing chances of exposure. This is in the belief that, if uprooted from the source, cancer will not overburden the country's health system. However, as noted by IARC (2018), there are different types of cancers and their causes differ. This therefore calls for identifying all the major types of cancer in the country, what prevailing circumstances expose Kenyans to each type of cancer. For the purposes of this chapter, focus is limited to aflatoxin and the impact it has on growth of liver cancer cases in the country.

1.2 Issue of Concern/Problem Statement

Kenya, a country of about 50 million people experienced about 47,000 new cases and 33,000 cancer deaths per annum (UICC 2019). Treatment of cancer is an expensive undertaking, which Kenya as a developing nation can ill afford (Oketch 2020). In addition, as noted by the same author (Oketch 2020), the country does not currently have adequate resources in the form of oncologists, hospital facilities and medication to treat cancer cases. This is compounded by limited access to health facilities by the poor majority with less than 1% of the county residents having health insurance cover (UNDP 2020).

The statistics as stated in the previous paragraph lead to several crucial conclusions. First and foremost, cancer in the country is a serious course for concern as the numbers continue to increase. Secondly, the country does not have adequate resources to access adequate health, a significant majority of the population are poor and cannot afford proper medical care, and finally, different types of cancer are caused by differing circumstances.

The focus of this exercise is limited to liver cancer, which is increasingly related to the level of aflatoxins Kenyans are exposed to. The issues explored will include; affected foods, local regulations and means applied to manage aflatoxin levels in the country.

1.3 Objectives of the Exercise

The following objectives were formulated to ensure that the undelaying challenges as stated under the "Issues under Focus" are addressed.

1. Farm products and their aflatoxin levels;

2. Analysis of aflatoxin regulations and management practices in Kenya; and

3. Technologies for aflatoxin control in crops in Kenya.

2. Literature Review

2.1 Farm Products and Their Aflatoxin Levels in Kenya

Little work on the prevalence of aflatoxins in Kenya was published before the 21st century. Early studies found high incidence of contamination with aflatoxins in the majority (93%) of the main meals and local alcoholic brew—in the Murang'a District of Kenya. High levels of aflatoxin continue to be found in crops such as maize, millet, sorghum, pigeon peas and yam components, which happen to be the staple food for Kenyans. Several studies undertaken in the 21st century have consistently shown aflatoxins in a variety of foodstuffs and from various regions in the country. Of interest from several of these publications are the alarmingly high proportions (Table 1) of food commodities that surpass the Kenyan regulatory threshold of 10 µg/kg set for total aflatoxins and 5 µg/kg set for aflatoxin B_1 content (KEBS, 2018a). Notably, the extremely high levels were recorded in Kenya's main staple starch, maize, peanuts and in animal feed. These statistics, coupled with the regular consumption of sizeable portions of maize products across diverse age groups provide insights into the high chronic aflatoxins exposure rates in the country, of about 67% of the population (Mutegi et al. 2018).

2.2 Analysis of Aflatoxin Regulations and Management Practices in Kenya

Many lives have been and continue to be lost from exposure to high levels of aflatoxins in staple foods in Kenya (Mutegi et al. 2018). Based on this, increased public and private sector participation in mitigation measures have been put in place as highlighted below:

(i) **The aflatoxins task force of the Kenya government**

This is an inter-departmental/inter-ministerial team instituted to spearhead surveillance of aflatoxins in maize as well as advise the government on looming outbreaks and containment measures. Operating under the Ministry of Agriculture, it has representatives from the Ministry of Health, the Kenya Bureau of Standards (KEBS), Kenya Plant Health Inspectorate Services (KEPHIS) and public universities (Mutegi et al. 2018).

(ii) **National Food Safety Coordinating Committee**

The National Food Safety Coordinating Committee (NFSCC) was constituted in February 2006 with broad consultations among stakeholders. The Secretariat is the Ministry of Health, while the chairmanship lies with the Ministry of Agriculture. Its membership is drawn from ministries/ state departments/Semi-Autonomous Government Agencies (SAGA),

Table 1: Selected data on aflatoxins prevalence in Kenya (1960-date) from published sources (Mutegi et al. 2018)

Subject (Origin)	Range[a]	Above threshold (%)[b]	n	References
Maize products (µg/kg)				
Makueni, Kitui	LOD–48,000 (9.1 Gm)	35[c]	716	Daniel et al. 2011
Nairobi (Korogocho; Dagoretti) g	0–88.83 (6.7); 0–20 (2.97)	16	99; 87	Kiarie et al. 2016
Kitui, Makueni, Machakos, Thika	1.0–46,400 (20.53)	55[c]	342	Lewis et al. 2005
Eastern, Nyanza	0.01–9,091.8 (46.9)	50.3	789	Mahuku 2018 unpublished data
Western Kenya	LOD–710	15	985	Mutiga et al. 2015
Upper and Lower eastern	LOD–4,839	39	1,500	Mutiga et al. 2014
Makueni	0.0–13,000	35.5[c]	104	Mwihia et al. 2008
Nairobi	0.11–4,593	83	144	Okoth and Kola 2012
Kwale, Isiolo, Tharaka Nithi, Kisii, Bungoma	<1.0–1,137	26[d]	497	Sirma et al. 2016
Peanut products (µg/kg)				
Busia, Kisii	0.1–591.1	48.8[d]	204	Menza et al. 2015
Nyanza, Western, Nairobi	LOD–32,328	37	1,161	Mutegi et al. 2013
Busia, Homabay	0.0–7,525	7.5[c]	769	Mutegi et al. 2009
Nairobi, Nyanza	LOD–2,377	43	82	Ndung'u et al. 2013
Eldoret and Kericho towns	0.0–2,345	–	228	Nyirahakizimana et al. 2013

(Contd.)

Table 1: *(Contd.)*

Subject (Origin)	Range ()[a]	Above threshold (%)[b]	n	References
Sorghum (µg/kg)				
Nairobi (Korogocho; Dagoretti)[g]	0.2–194.41 (8.07); 0.1–14.47 (2.59)	11	53; 36	Kiarie et al. 2016
Kwale, Isiolo, Tharaka Nithi, Kisii, Bungoma	<1.0–91.7	11[d]	164	Sirma et al. 2016
Millet (µg/kg)				
Kwale, Isiolo, Tharaka Nithi, Kisii, Bungoma	<1.0–1,658.2	10[d]	205	Sirma et al. 2016
Medicinal herbs (µg/kg)				
Eldoret and Mombasa towns	<0.25–24	–	100	Keter et al. 2017
Milk products (ng/kg)				
Eldoret, Machakos, Nyeri, Machakos, Nakuru, Nairobi	5.8–600	20[e]	613	Kang'ethe and Lang'a 2009
Makueni	1.4–152.7 (0.83)	22.2 (detected)	18	Kang'ethe et al. 2017
Nandi	0.5–0.8 (0.06)	9.5 (detected)	21	Kang'ethe et al. 2017
Nairobi (Korogocho; Dagoretti)[g]	0.002–2.56 (0.132); 0.007–0.64 (0.093)	63[e]	76; 52	Kiarie et al. 2016
Nairobi	LOD–1,675	55[e]	190	Kirino et al. 2016
Bomet	LOD–2.93	43.8[e]	156	Langat et al. 2016
Kwale, Isiolo, Tharaka Nithi, Kisii, Bungoma	<2–6,999 (3.2 Gm)	10.4[e]	512 (farmers)	Senerwa et al. 2016

Animal feed products (μg/kg)				
Eldoret, Machakos, Nyeri, Machakos, Nakuru, Nairobi	–	67[d]	830	Kang'ethe and Lang'a 2009
Nairobi	5.13–1,123	95	72	Okoth and Kola 2013
Kwale, Isiolo, Tharaka Nithi, Kisii, Bungoma	<1.0–4,682 (9.8 Gm)	61.8[d]	102 (feed manufacturers)	Senerwa et al. 2016
Kwale, Isiolo, Tharaka Nithi, Kisii, Bungoma	<1.0–1,198 (25.6 Gm)	90.3[d]	31 (feed retailers)	Senerwa et al. 2016
Human exposure (pg/mg) AFB_1-lysine adduct level				
Various	0.05–0.417 (AFB-gual)	12.6 (detected)	830	Autrup et al. 1987
Tharaka Nithi and Meru Counties	4.18–10.46 (7.82)	100 (detected)[f]	884	Leroy et al. 2015
Nyanza, Coast, eastern, Rift Valley	LOD–211 (2.01 Gm)	78 (detected)	597	Yard et al. 2013

[a] Mean values in brackets; in some instances the authors did not present mean values and in others, Arithmetic mean was not differentiated from geometric mean (Gm).
[b] Percent beyond Kenyan regulatory threshold (10 µg/kg).
[c] Percentage based on the then Kenyan regulatory threshold of 20 µg/kg.
[d] Percentage samples is based on the KEBS regulatory threshold for aflatoxin B_1 (5 µg/kg).
[e] Percentage based on a threshold of 0.05 µg/kg.
[f] Sample size comprised of women.
[g] Korogocho is a slum neighbourhood of Nairobi and Dagoretti is one of the eight divisions of Nairobi; aflatoxin data are provided for both areas, respectively.

Key: gm – grams; LOD – Limit of detection

Ministry of Health, Trade, academia, and the Council of Governors of the government of Kenya. The Food and Agriculture Organization of the United Nations (FAO), World Health Organization (WHO), World Food Program (WFP), Kenya Association of Manufacturers (KAM) and Cereal Millers Association (CMA) are also co-opted members. Its mission is to protect consumers' health by ensuring that food produced, distributed, marketed, and consumed meets required standards of food safety. Among its roles are to coordinate formulation of food safety policies and supportive legal framework, harmonize and coordinate the implementation of food control activities including food and feed analysis, inspection, enforce and coordinate food safety information, education and communication (IEC) and monitor and evaluate food safety programs. The committee is involved in annual testing for contamination by aflatoxins in maize, during and after harvest (Mutegi et al. 2018).

3. Regulatory Compliance through National Standards and Regulatory Bodies

Following the death of numerous dogs after eating commercial dog feed in the 1980s, the standards and regulatory agency (KEBS) developed and drafted a standard for dog feeds in 1985. Standards for maize grain and other food grains and products existed before this date and had been set at 20 µg/kg. This standard was thereafter revised by KEBS to 10 µg/kg for total aflatoxins and 5 µg/kg for AFB1 in 2007, after conducting a risk assessment based on dietary exposure, frequency of consumption of highly prone food items particularly maize, methodologies available for aflatoxins testing, and available literature on prevalence and exposure. Enforcement of standards has been a major challenge as regulatory entities have focused on the organized formal sector, leaving out the informal sector through which over 90% of the food supply in Kenya is transacted. Amongst those served by the unregulated informal sector include upper middle, lower middle, low-income populations, who subsequently become exposed to aflatoxins (Mutegi et al. 2018).

4. Collaborations between Government Ministries (Agriculture and Health) and FAO

Following the acute aflatoxicosis incidence in 2004 which caused numerous deaths in Kenya, FAO supported the Ministries of Agriculture and Health to identify causative factors for the outbreak as well as introduce measures that could be employed to mitigate future occurrences. The immediate action taken was to develop an extensive awareness raising campaign about aflatoxins through public and private sector stakeholders' consultations both at the national and county levels. Other measures included supply of moisture meters and aflatoxins testing equipment to extension agents to assist farmers in monitoring moisture content in grain and undertake surveillance. Long-term solutions proposed included development of early warning systems, revision of aflatoxins standards to ensure optimal safety for the end consumer while ensuring fair trade; establishment of aflatoxins prevalence data to identify areas requiring immediate intervention; testing/developing aflatoxins management options; continuous monitoring and surveillance programs; capacity building and strengthening aflatoxins-related policy and regulation (Mutegi et al. 2018).

5. Other Public-Private Partnership Initiatives

(a) The Aflatoxin Proficiency Testing and Control in Africa (APTECA) program is being piloted in various African countries, including Kenya, where Cereal Millers Association (CMA), use the Biosciences eastern and central Africa (BECA)-International Livestock Research Institute (BecA-ILRI hub) platform, and several government departments have partnered to advance a quality systems approach for managing aflatoxins. The Cereal Growers Association (CGA) continuously collaborates with private and public-sector players to educate farmers on aflatoxins mitigation and test ready to scale technologies in their mandated maize growing regions. The Mexican government recently partnered with the public sector including KALRO and the local universities, as well as private sector to advance nixtamalization (alkaline cooking) as a postharvest intervention to manage aflatoxins in maize grain (Mutegi et al. 2018).

(b) The Agricultural Cooperative Development International – Volunteers in Overseas Cooperative Assistance (ACDI-VOCA), an international not-for-profit, non-governmental organization has assisted commercial maize farmers in improving the quality of their grain through sound postharvest practices. ACDI-VOCA has also partnered with other private sector players, government, and the Consultative Group for the International Agricultural Research (CGIAR) centers to raise awareness about aflatoxins among farmers and extension officers through dissemination workshops. The recently concluded AflaSTOP project (Storage and Drying for Aflatoxin Prevention Project), a collaboration between ACDI-VOCA, Agribusiness Systems International, Meridian Institute and several public-sector institutions, has generated important information on comparative cost effectiveness of existing hermetic storage technologies and commercialization feasibility of the EasyDry M500 portable dryer (Mutegi et al. 2018).

(c) The World Food Programme (WFP) also set up the Purchase for Progress (PFP) process that ensures that international humanitarian organization purchases safe and quality grain from farmers after independent testing and this has been of significant benefit to various cereal value chains and farmers. Other important joint public-sector initiatives, such as the Safe Dairy Project, an effort of the University of Nairobi, Agrifood Research Finland (MTT), Finnish Food Safety Authority (EVIRA), KALRO and Egerton University in Kenya also are in place (Mutegi et al. 2018).

(d) Lastly, the Partnership for Aflatoxin Control in Africa (PACA) has played an important role in the mitigation of aflatoxins in Africa, through the leadership of the AU and supported by a steering committee with representation across sectors and disciplines. This has ensured that there is platform for engaging and deliberating solutions to aflatoxins in the continent aimed at safeguarding consumer health and facilitation of trade. Kenya has been a beneficiary of some of the regional initiatives that have been supported through PACA, including infrastructure development for advancing technologies such as the use of biological control for mitigation of aflatoxins and capacity building through infrastructural support (Mutegi et al. 2018).

In summary, the initiatives described above demonstrate the effectiveness of public-private partnerships that optimize on each partner's complementary strengths such as existing and trusted in-country mechanisms; infrastructure and regulation within the public sector to support information and technology dissemination; strong technical support from private sector players; pooling of resources from partners; mainstreaming proposed initiatives within government structures; and utilizing public and privates sector platforms for information dissemination (Mutegi et al. 2018).

6. Economic Drivers for Mitigation of Aflatoxins

The motivation to invest in management efforts is driven by aspects of the intervention that affect perception, uptake, buy-in and scaling. A major push for seeking solutions to address contamination by aflatoxins in Kenyan produce is the historical negative impacts on the health of the population. The cost of treating such health conditions is a burden to the affected families and the government health institutions providing treatment services. Moreover, the cost of technology directly affects adoption and upscaling efforts. Costs must make economic sense to the end user and the supplier of the technology. Several technologies have been advanced for use at smallholder level, but few studies have determined the economic incentives to promote uptake. Household income is another factor that determines the success of uptake of a technology (Mutegi et al. 2018).

7. Technologies for Aflatoxin Control in Crops in Kenya

7.1 Pre-harvest

In the United States at the pre-harvest stage, biocontrol products have been used since the 1990s which allows for the production of aflatoxin-compliant crops. In Africa, biocontrol products have been developed by the International Institute of Tropical Agriculture (IITA) and partners under the trade name Aflasafe (Aflasafe.com). Usage of Aflasafe products in Senegal (6-year data;

Aflasafe SN01) and Nigeria (12-year data; Aflasafe) in multiple agro-ecologies and under diverse cropping systems has resulted in aflatoxin reductions in maize and groundnut of 82–89% (95% confidence interval) compared to untreated plots after storage or simulated storage conditions. The product developed for use in Nigeria was also found to be effective to decrease aflatoxin content in chili peppers. Data on efficacy of Aflasafe products developed and used in other countries (i.e., Kenya, Tanzania, Burkina Faso, Ghana, Senegal, Malawi, Mozambique, and Zambia) is available but has not been published in scientific journals. Significant technical support may be required for successful use of this technology among smallholders in less organized farming systems and IITA and International Food Policy Research Institute (IFPRI) have an ongoing assessment to establish the level of farmer training required to achieve aflatoxin reduction goals (Hoffmann et al. 2018).

7.2 Drying

Good post-harvest practices, including adequate drying and controlled storage, have long been the standard recommendation for controlling aflatoxin in crops. Providing groundnut farmers with training on aflatoxin control and plastic sheets on which they could dry their crops was shown to reduce aflatoxin by approximately 50% in Ghana, compared to farmers who were trained but not given drying sheets. In a separate study in Kenya, maize farmers were given drying sheets as well as access to other technologies. Drying sheets were the most widely used technology; the stored maize of farmers who reported drying their maize on these and not using any of the other technologies offered was 79% less contaminated with aflatoxin three months after harvest than that of farmers in villages where no intervention was offered. Overall, aflatoxin levels in intervention villages were 53% lower than in control. A biomass drying technology is slightly more effective, leading to an 85% reduction in contamination after three months of storage in an experimental study. This approach was developed into a mobile drying technology, the EasyDry M500, by ACDI-VOCA (Hoffmann et al. 2018).

7.3 Storage

Hermetic storage bags, designed to reduce insect pest losses, have been shown to reduce aflatoxin contamination by between 40 and 66% after three months of storage, depending on the starting moisture content, level of contamination, and other environmental factors (Hoffmann et al. 2018)

8. Summary of Findings, Recommendations and Conclusions

The high levels of aflatoxins in food and feed commodities and related deaths gives credible evidence that the issue of aflatoxin mitigation must be managed sooner than later. Various studies and efforts have come to a common conclusion that, to effectively implement management of aflatoxins in Kenya, capacity development in the form of human resource base and infrastructure is critical. The human capacity to address various facets of aflatoxins mitigation is still low. Up-to-date research facilities for mycotoxin research, for food commodities and human and animal exposure, is required particularly in public institutions. The relevance of social learning and networks in promoting aflatoxins mitigation efforts amongst smallholder farmers need to be considered, not ignoring the importance of provision of diagnostic tools and testing equipment that must be properly targeted. Their application needs to ruminate on cost, availability, rapidity in the decision-making process on the testing outcome, the scale of use, whether the grain is for home or market consumption, as well as the capacity of the end user to competently and appropriately use the testing methods. Alongside, sustained public awareness is necessary to develop a population that is conscious of the benefits of consuming safe food and consequently demand for it. Finally, government and private sector can play a crucial role in strengthening policies that impact on food safety, as well as support risk assessment initiatives to ensure that well thought out standards for mycotoxins are in place. Scarce resources available to advance management efforts must be utilized well by proper targeting and ensuring that duplication of efforts is minimized. Unlike in several other countries where trade primarily drives the impacts of aflatoxins and attempts to get solutions, Kenya has suffered immensely from health impacts to the extent that the government declared it a national disaster (Mutegi et al. 2018).

References

Autrup, H., Serenici, T., Wakhisi, J. and Wasunna, A. 1987. Aflatoxin exposure measured by urinary excretion of aflatoxin B1-guanine adduct and hepatitis B virus infection in areas with different liver cancer incidence in Kenya. Cancer Research 47: 3430–3433.

Daniel, J.H., Lewis, L.W., Redwood, Y.A., Kieszak, S., Breiman, R.F., Flanders, W.D. et al. 2011. Comprehensive assessment of maize aflatoxin levels in Eastern Kenya, 2005–2007. Environmental Health Perspectives 119: 1794–1799.

Hoffman, V., Jones, K. and Leroy, J.L. 2018. The impact of reducing dietary aflatoxin exposure on child linear growth: A cluster randomised controlled trial in Kenya. BMJ Global Health 3(6): e000983.

Hoffmann, V., Grace, D., Lindahl, J., Mutua, F., Ortega-Beltran, A., Bandyopadhyay, R. et al. 2019. Technologies and strategies for aflatoxin control in Kenya: A synthesis of emerging evidence. Project Note. Washington, D.C.: IFPRI.

International Agency for Research in Cancer (IARC) 2018. Kenya Globocan 2018 report [Online]. Available from: https://gco.iarc.fr/today/data/factsheets/populations/404-kenya-fact-sheets.pdf (Accessed on 03–06–2020).

Kang'ethe, E.K. and Lang'a, K.A. 2009. Aflatoxin B1 and M1 contamination of animal feeds and milk from urban centres of Kenya. African Health Sciences 9: 218–226.

Kenya Bureau of Standards (KEBS). 2018a. Kenya standard KS EAS2:2017. Maize grains specifications. KEBS, Nairobi, Kenya.

Keter, L., Too, R., Mwikwabe, N., Mutai, C., Orwa, J., Mwamburi, L. et al. 2017. Risk of fungi associated with aflatoxin and fumonisin in medicinal herbal products in the Kenyan market. Scientific World Journal 2017: 1–6.

Kiarie, G.M., Dominguez-Salas, P., Kang'ethe, S.K., Grace, D. and Lindhal, J. 2016. Aflatoxin exposure among young children in urban low-income areas of Nairobi and association with child growth. African Journal of Food, Agriculture, Nutrition and Development 16: 10967–10990.

Kirino, Y., Matika, K., Grace, D. and Lindhal, J. 2016. Survey of informal milk retailers in Nairobi, Kenya and prevalence of aflatoxin M1 in marketed milk. African Journal of Food, Agriculture, Nutrition and Development 16: 11022–11038.

Langat, G., Tetsuhiro, M., Gonoi, T., Matiru, V. and Bii, C. 2016. Aflatoxin M1 contamination of milk and its products in Bomet County, Kenya. Advances in Microbiology 6: 528–536.

Leroy, J., Wang, J.S. and Jones, K. 2015. Serum aflatoxin B1-lysine adduct level in adult women from Eastern Province in Kenya depends on household socio-economic status: A cross sectional study. Social Science and Medicine 146: 104–110.

Lewis, L., Onsongo, M., Njapau, H., Schurz-Rogers, H., Luber, G., Kieszak, S. et al. 2005. Aflatoxin contamination of commercial maize products during an outbreak of acute aflatoxicosis in Eastern and Central Kenya. Environmental Health Perspectives 113: 1763–1767.

Menza, N.C., Margaret, M.W. and Lucy, K.M. 2015. Incidence, types and levels of aflatoxin in different peanut varieties produced in Busia and Kisii Central Districts, Kenya. Open Journal of Medical Microbiology 5: 209–221.

Mutegi, C.K., Ngugi, H.K., Hendriks, S.L. and Jones, R.B. 2009. Prevalence and factors associated with aflatoxin contamination of peanuts from Western Kenya. International Journal of Food Microbiology 130: 27–34.

Mutegi, C.K., Wagacha, M., Kimani, J., Otieno, G., Wanyama, R., Hell, K. et al. 2013. Incidence of aflatoxin in peanuts (*Arachis hypogaea* Linnaeus) from markets in Western, Nyanza and Nairobi Provinces of Kenya and related market traits. Journal of Stored Products Research 52: 118–127.

Mutegi, C.K., Cotty, P.J. and Bandyopadhyay, R. 2018. Mycotoxins in Africa prevalence and mitigation of aflatoxins in Kenya (1960-to date). Mycotoxin Journal 3: 341–357. Special issue.

Mutiga, S.K., Were, V., Hoffman, V., Harvey, J.W., Milgroom, M.G. and Nelson, R.J. 2014. Extent and drivers of mycotoxin contamination: Inferences from a survey of Kenyan maize mills. Phytopathology 104: 1221–1231.

Mutiga, S.K., Hoffman, V., Harvey, J.W., Milgroom, M.G. and Nelson, R.J. 2015. Assessment of aflatoxin and fumonisin contamination in western Kenya. Phytopathology 105: 1250–1261.

Mwihia, J.T., Straetemans, M., Ibrahim, A., Njau, J., Muhenje, O., Guracha, A. et al. 2008. Aflatoxin levels in locally grown maize from Makueni District, Kenya. East African Medical Journal 85(7): 311–317. doi: 10.4314/eamj.v85i7.9648

Ndung'u, J.W., Makokha, M.A., Onyango, C.A., Mutegi, C.K., Wagacha,J.M., Christie, M.E. et al. 2013. Prevalence and potential for aflatoxin contamination in groundnuts and peanut butter from farmers and traders in Nairobi and Nyanza provinces of Kenya. Journal of Applied Biosciences 65: 4922–4934.

Nyirahakizimana, H., Mwamburi, L., Wakhisi, J., Mutegi, C.K., Christie, M.E. and Wagacha, J.M. 2013. Occurrence of *Aspergillus* species and aflatoxin contamination in raw and roasted peanuts from formal and informal markets in Eldoret and Kericho towns, Kenya. Advances in Microbiology 3: 333–342.

Oketch, A. 2020. Why cancer should be declared a national disaster in Kenya [Online]. Available from: https://www.nation.co.ke/news/Why-cancer-should-be-declared-national-disaster/1056-5342532-7j2g0i/index.html (Accessed on 03–06–2020).

Okoth, S. and Kola, M. 2012. Market samples as a source of chronic aflatoxin exposure in Kenya. African Journal of Health Sciences 20: 56–61.

Senerwa, D.M., Sirma, A.J., Mtimet, N., Kang'ethe, E.K., Grace, D. and Lindhal, J.F. 2016. Prevalence of aflatoxin in feeds and cow milk from five counties in Kenya. African Journal of Food Agriculture Nutrition and Development 16: 11004–11021.

Sirma, A.J., Senerwa, D.M., Grace, D., Matika, K., Mtimet, N., Kang'ethe, E.K. et al. 2016. Aflatoxin B1 occurrence in millet, sorghum and maize from four agro-ecological zones in Kenya. African Journal of Food, Agriculture, Nutrition and Development 16: 10991–11003.

United Nations Development Programme (UNDP) 2020. Access to affordable quality healthcare through M-Kadi medical services e-card [Online]. Available from: https://www.ke.undp.org/content/kenya/en/home/ourwork/inecgr/successstories/Access-to-Affordable-Quality-Healthcare-through-M-KADI-medical-savings-e-card.html (Accessed on 03–06–2020).

Union for International Cancer Control (UICC). 2019. Addressing Kenya's growing cancer burden [Online]. Available from: https://www.uicc.org/news/addressing-kenya%E2%80%99s-growing-cancer-burden (Accessed on 03–06–2020).

WHO. 2015. Compendium of Food Additive Specifications. Joint FAO/WHO Expert Committee on Food Additives (JECFA), 81st meeting 2015. FAO JECFA Monographs 18.

Yard, E.E., Daniel, J.H., Lewis, L.S., Rybak, M.E., Palikov, E.M., Kim, A.A. et al. 2013. Human aflatoxin exposure in Kenya, 2007: A cross-sectional study. Food Additives and Contaminants Part A 30: 1322–1331.

Benefits/Risks Related to the Consumption of Infant Flours Produced in Burkina Faso

Waré Larissa Yacine[1,4*], Durand Noël[2,3], Barro Nicolas[4] and Montet Didier[2,3]

[1] Centre National de la Recherche Scientifique et Technologique (CNRST)/ Institut de Recherche en Sciences Appliquées et Technologies (IRSAT)/ Département de Technologies Alimentaires (DTA), Kossodo zone industrielle, 03 BP 7047 Ouagadougou 03, Burkina Faso

[2] CIRAD, UMR Qualisud, TA 95/16, 73 Rue J.F. Breton ; 34398 Montpellier, France

[3] Qualisud, Univ Montpellier, CIRAD, Montpellier SupAgro, Univ d'Avignon, Univ de La Réunion, Montpellier, France

[4] Laboratoire de Biologie Moléculaire, d'Epidémiologie et de Surveillance des Bactéries et Virus Transmissibles par les Aliments (LaBESTA), Centre de Recherches en Sciences Biologiques, Alimentaires et Nutritionnelles (CRSBAN), Ecole Doctorale Sciences et Technologies; Université de Ouagadougou 03, BP 7021, Ouagadougou 03, Burkina Faso

1. Introduction

According to current United Nations recommendations, infants should be exclusively breastfed for the first six months of life to achieve optimal growth, development and health (WHO/UNICEF 2009). Thereafter, to meet their evolving nutritional requirements, infants should receive nutritionally adequate and safe complementary foods while breastfeeding continues for up to two years of age or beyond (WHO 2008a). During the first two years of life, adequate nutrition during infancy and early childhood is essential to ensure the growth, health, and development of children to their full potential. Poor nutrition increases the risk of illness, and is responsible, directly or indirectly, for one third of the estimated 9.5 million deaths that occurred in 2006 in children less than five years of age (WHO 2008a, 2013). Thus, the infant flours, also known as infant cereals, are specially designed to meet the nutritional needs of infants and young children because after six months, breast milk

*Corresponding author: ly.ware@gmail.com

is no longer sufficient to provide the infant's nutritional needs in terms of energy and protein (WHO 2008b). Improving the nutrition of young children therefore appears to be a prerequisite to avoid deficient malnutrition (Kouassi et al. 2015).

In Burkina Faso, the prevalence of acute malnutrition, chronic malnutrition and underweight was 7.7%, 27.1% and 16.3% respectively in 2017 according to the WHO 2006 (Ministry of Health 2017). It usually appears during the period corresponding to the introduction of complementary breastmilk food for infants (Dewey and Brown 2003). This hidden hunger is responsible for high rates of infant morbidity and mortality and is a public health problem (Mandjeka 2016).

In order to sustainably improve the nutrition of young children and women of childbearing age and to contribute to reducing the prevalence of malnutrition and micronutrient deficiencies, projects have been initiated in Burkina Faso. These projects consist of support for micro- and small- Burkinabe companies producing or wishing to produce fortified food based on local raw materials. Since the implementation of these projects, several companies have received support and are able to produce infant flour for children aged 6 to 23 months, in accordance with the specifications for the production of infant flour (Gret/Nutrifaso 2007).

In Burkina Faso, most of these infant flours are prepared from local foodstuffs, such as cereals, oilseeds, legumes and tubers (Kouassi et al. 2015).

However, during the processing and storage of designed infant flours, contamination with pathogens or toxins may also occur. These infantile flours are based on raw materials. Work carried out on cereals, legumes and oilseeds has shown the presence of mycotoxins, particularly in maize, millet, sorghum, peanuts, sesame, and soya (Nikièma 1993, Ouattara-Sourabié et al. 2011, Warth et al. 2012, Ezekiel et al. 2014, Afolabi et al. 2015, Waré et al. 2017), and also of pesticides and other micropollutants (Fasonorm 2014). Chronic exposure to toxins has a negative impact on the health of infants and young children. It is therefore very important that caregivers have appropriate recommendations for infant and young child feeding.

2. Infant Flours

2.1 Definition

Infant flour is a cereal-based supplement food that is given as porridge to children from the age of six months as a supplement to breast milk (FAO 2004). It must be specially designed to meet their nutritional needs, taking into account the intake of breast milk and the daily frequency of meals (WHO/ UNICEF 2003). As a result, there are imported infant flours and local infant flours. These infant flours are either cooking flours or instant flours (Gret 1998).

2.2 Industrial Infant Flours

Industrial infant flours are sold in pharmacies, supermarkets and market shops. These flours are stabilized and fortified with vitamins and minerals to meet *Codex alimentarius* standards. But they remain inaccessible to the majority of the African population due to their extremely high price ranging from 2.5 to 4 Euros for 500 g (Hervé et al. 2004).

2.3 Artisanal Infant Flours

Small businesses produce infant flour in Burkina Faso. To play their role in improving infant feeding, entrepreneurs must:

- choose an appropriate formula to cover the nutritional needs of children,
- ppt for appropriate technological processes,
- guaranty the sanitary quality of products, and
- respond to market demand and have good management of their unit.

There are two types of local infant flours: flours to cook, which have to be cooked for at least 15 min and ready-to-use instant flours, which do not require additional cooking.

The ingredients used in the composition of infant flours in Burkina Faso are diverse, but cereals remain the basic raw material: maize, millet, sorghum, rice, and wheat. Other ingredients found in infant flours are peanuts, soybeans, sesame seeds, beans, fruit of *Adansonia digitata*, and fruit of *Parkia biglobosa* or fermentation of *Parkia biglobosa* seeds, dry fish powder, tubers, milk, sugar, salt and a mineral and vitamin supplement (Waré 2018).

3. Benefit of Infant Flours

The cereals (maize, millet, sorghum, rice) that make up infant flour are high in calories, rich in carbohydrates, especially starch but low in protein, fat, minerals and vitamins. It is vital to give other protein-rich foods (FAO 1979) as oilseeds (soy, sesame, peanuts, peas, mahogany nuts) which are rich in protein, in lipids and free of starch. Legumes should therefore be added to the flours. The flours of legumes (lentils, beans, niebe ...) are difficult to assimilate but when roasted, crushed or ground before or after cooking the porridge they are used in anti-diarrhea diets (FAO 1979). Fruit of *Adansonia digitata*, fruit of *Parkia biglobosa* or dry fish powder, and milk constitute the mineral and vitamin intake of the infant flour. Fruits and green leaves can also provide carotene and vitamin C. For example, ripe papayas, guavas, oranges and mangoes are excellent and generally appreciated by children. Each of these ingredients makes the infant flours very rich in macronutrients.

Infant cereals allow infants and young children to be nutritionally satisfied. They can also be introduced very moderately to gently initiate the infant's dietary diversification by introducing flavors (other than milk) and new textures (Doray 2017). In fact, the period from six months to two years is very important from a nutritional point of view in children. But many African children in this age group are not growing as they should and some are suffering from protein-energy malnutrition (FAO 1979).

Research has also shown that infant flour is used in developing countries to reduce the prevalence of malnutrition. In Burkina Faso, Traoré (2005) focused on the development and evaluation of a strategy to improve the complementary feeding of young children. This study consisted in incorporating malted flour into infant flour produced at lower cost in rural areas. The strategy made it possible to improve the energy and nutrient density of the porridges (Traoré 2005). Traoré concluded that the consumption of high energy density and micronutrient fortified porridges (prepared by incorporating germinated sorghum flour and mineral and vitamin supplements) resulted in a very significant improvement in energy intake compared to traditional porridges. It has also been shown to be effective in improving both the growth in size and the iron status of young children (Traoré 2005). Thus, Sanou et al. (2017) showed that infant flours in four production units in Burkina Faso were a significant source of energy and protein for children over six months of age. These infant flours were composed of millet-maize or rice-maize and groundnut soybean or groundnut - bean/niebe in different proportions (Sanou et al. 2017). In 2016, Kayalto improved the production process for local children's flours, incorporating *Moringa oleifera* leaf powder and néré pulp (*Parkia biglobosa*) at various levels into these flours in order to enrich them with vitamin A and iron, and to help reducing the prevalence of malnutrition among children in Tchad (Kayalto 2016).

4. Risks of Infant Flours

Infant flours need good hygiene and proper food handling for their preparation to prevent gastrointestinal diseases. It is thus necessary:

- to ensure that caregivers wash their hands before preparing food,
- to store food safely and serve them immediately after preparation,
- to use clean utensils to prepare and serve food,
- to use clean cups and bowls to feed children, and
- to avoid the use of bottles that are difficult to keep clean.

Indeed, diarrhea is the most frequently observed during the second half of the first year of life (Bern et al. 1992). Microbial contamination of food is the main cause of diarrhea in children and can be prevented through the practices described above.

Infant flours are produced from cereals which often are contaminated with mycotoxins (Bailly and Oswald 2013). To get healthy infant flours, it is thus necessary:

- to sort and dry the raw material (cereals and oilseeds) well, especially for maize and peanuts,

- to store the raw material (cereals and oilseeds) well, especially for maize and peanuts,
- to dry and store the infant flours well,
- cereals must comply with the maximum limits for mycotoxins, and
- infant flours must comply with the maximum limits for mycotoxins.

Indeed, short-term food poisoning and the occurrence of certain chronic illnesses in long-term due to the ingestion of food contaminated with mycotoxins can be reduced if the above measures are applied.

In hot and humid regions such as Burkina Faso, there are species of the genus *Aspergillus* that produce aflatoxins (AFs) and in particular aflatoxin B1 (AFB1). These mycotoxins have genotoxic and carcinogenic properties (IARC 1993). In fact, AFs have been classified as Group 1 human carcinogens with AFB1 being the most toxic followed by AFG1 then AFB2 and AFG2 (EFSA 2007) and then AFM1 that was classified as a group 2B possible human carcinogen (IARC 1993).

Waré et al. (2017) tested 248 of infant flours, cereals flours and raw materials in Burkina Faso and reported the presence of mycotoxins [AFs, ochratoxin A (OTA), fumonisins (F(B1+B2))] in all samples. Only 19 of samples did not exceed EC Regulation limits. In fact, the EC Regulation 1881/2006 limits are:

- for infant formulas: 0.1 µg/kg for AFB1, 0.5 µg/kg for OTA and 200 µg/kg for fumonisins,
- for peanuts, cereals and derived from cereals other than maize and rice: 2 µg/kg for AFB1, 4 µg/kg for AFs, 3 µg/kg for OTA, and 4000 µg/kg for fumonisins,
- for maize and rice: 5 µg/kg for AFB1, 10 µg/kg for AFs, 3 µg/kg for OTA and 4000 µg/kg for fumonisins, and
- for oleaginous grains other than peanuts: 8 µg/kg for AFB1, 15 µg/kg for AFs, and 3 µg/kg for OTA.

5. Mycotoxins in Infant Flours

In the analyzed infant flours these thresholds were 87.4 µg/kg for aflatoxin B1, 3.5 µg/kg for ochratoxin A and 672.9 µg/kg for F(B1+B2) which were respectively 900 times, 6 times and 3 times higher than the EC regulations (1881/2006). They also showed that infant flours composed of dry roasted peanut contained aflatoxin B1 with a mean of 5.8 µg/kg and for AFs 9.9 µg/kg. The maximum content obtained for AFB1 was 84.6 µg/kg of and 135.3 µg/kg for AFs in peanuts dry alone. The maximum content in AFB1 was 87.4 µg/kg for infant flour that was composed of millet and wheat. Waré (2018) showed a significant difference (p <0.05) between the levels of mycotoxin contamination and the nature of the ingredients. Thus, when the infant flours are composed of maize, or rice, or wheat, or peanut there is a high probability of aflatoxin production. In Nigeria Aflatoxin B1 was found in 64.2% of dry roasted groundnuts with a mean of 25.5 ppb (Bankole et al. 2005). In Côte d'Ivoire, peanut paste was analyzed and all the peanut pastes were contaminated by aflatoxin B1 and concentrations reaching up to 4535 µg/kg (Manizan et al. 2018).

Acute high-level aflatoxin exposure has resulted in deaths in some parts of the world, particularly in African countries, and chronic low-level aflatoxin exposure can increase the risk for human hepatocellular carcinoma and result in immune suppression (Zain 2011, IARC 2015). Thus, the presence of mycotoxins in infant flours can lead to the development of serious chronic diseases such as cancer and immunosuppression (Jayaramachandran et al. 2013). Cross-sectional studies have shown a dose-response relationship between aflatoxin exposure and underweight in young West African children less than five years of age (Egal et al. 2005, Raiola et al. 2015).

Kwashiorkor is a disease affecting undernourished children, usually linked to nutritional deficiencies and may also be linked to exposure to aflatoxin (Golden and Ramadath 1987, Gauthier 2016).

Reye's syndrome is a rare but serious disease most often characterized by brain and liver damage that occurs after a flu, respiratory infection or chicken pox. Reye's syndrome mainly affects children, sometimes infants and very rarely adults. Another explanation would be chronic exposure to aflatoxins in South-East Asian countries (Gauthier 2016).

AFM1 is excreted into the milk of dairy animals and human nursing mothers after AFB1 ingestion and bioconversion in the liver (Manizan et al. 2018). Polychronaki et al. (2008) reported that breast milk samples obtained from 388 Egyptian women were found to contain AFM1 with detectable levels in about 36% of samples. This is of interest since a consequence of early exposure to AFs is growth impairment. The daily intakes of AFM1 were in the range of 0.08–0.021 ng/kg body weight (Gong et al. 2002). Adejumo et al. (2013) studied the correlation between AFM1 amounts in breast milk, dietary exposure to AFB1 and socioeconomic status of lactating mothers in Nigeria. AFM1 occurred in 82% of the breast milk samples (3.49–35 ng/L) and 16% exceeded the EU limit of 25 ng/L. Manizan et al. (2018) also showed the presence of aflatoxin M1 in 45% of samples of cereals (maize, rice) and peanut paste from Abidjan (Côte d'Ivoire). AFM1 occurrence in plants, produced by *Aspergillus* spp. through a biosynthetic pattern not involving AFB1 or possibly by insect pests' metabolism from AFB1, was also reported by Giovati et al. (2015). This could explain AFM1 presence in the cereal and peanut paste samples analyzed in Manizan's study (2018). AFM1 was classified in the Group 2B as possible human carcinogen (IARC 1993). Its acute toxicity is nearly equal to that of AFB1, but its potential carcinogenic hazard is about one order of magnitude less than that of AFB1 (Pietri and Piva 2007), but all the group of AFs were in Group 1 recognized carcinogenic for human (IARC 2012). So, AFs including AFM1, seem to be a risk factor that could increase the incidence and prevalence rates of malnutrition of children (Raiola et al. 2015).

Similarly, Ochratoxin A (OTA) was produced by *Aspergillus* spp. and *Penicillium* spp. In hot and humid regions such as Burkina Faso it is mainly *Aspergillus* spp. that produces OTA. An Italian team studied ochratoxin A (OTA) in milk-based infant formulae and showed that 72% of 185 analyzed samples were positive at levels ranging from 35.1 to 689.5 ng/L (Meucci et al. 2010). Baydar et al. (2007) analyzed this mycotoxin in 63 baby foods from Ankara (Turkey) and 40% were positive. A Burkina Faso study showed that 7.5% (15/199) analyzed samples of infant formulas were contaminated by ochratoxin A (Waré et al. 2017) and the maximum content of OTA was 3.5 µg/kg, found in infant flours composed of maize, soybean and peanut. The level ranged from 0.01 to 3.5 µg/kg. These infant formulas are produced in Burkina Faso with a mixture of several cereals (maize, millet, rice, or sorghum), and in Côte d'Ivoire Manizan et al. (2018) showed that 6% of the cereal samples (rice and maize samples) had ochratoxin A content above the EU limit (3 µg/kg). OTA has been classified as a Group 2B possible human carcinogen (IARC, 1993) and was associated with the Balkan Endemic Nephropathy (BEN). It is also known to be teratogenic, hepatotoxic, neurotoxic and immunotoxic (Köszegi and Poor 2016).

The presence of OTA breast milk was found to affect the kidney function and the development of urinary tumors in infants and young children (Skaug 1999). The mechanism of OTA transfer through the human placenta is not fully understood (Malir et al. 2014). Also, both the potential of OTA to cause malformations in humans and its teratogenic mechanism of action are not understood.

Fumonisins (FBs) are a group of mycotoxins produced by *Fusarium* species such as *Fusarium verticillioides* and *Fusarium proliferatum,* which are cereal pathogens. Fumonisins were found in infant formulas from Burkina Faso. In fact, the frequency of fumonisins (F(B1+B2)) contamination was 1.5%, in samples of infant formulas based on maize. The highest concentration of fumonisins (672.9 µg/kg) was recorded in a sample of infant formula based on millet and monkey bread (Waré et al. 2017). D'Arco et al. (2008) analyzed 27 samples of baby food products collected from different markets. One positive sample with a content of 15.9 µg/kg for FB1, 9.2 µg/kg for FB2 and 5.8 µg/kg for FB3 was found. A daily FB intake of 11.3 µg/kg of body weight for FB1, FB2 and FB3 was calculated for child consumers (1–5 years old), based on an average consumption of 200 g of corn meal/day from a study carried out in Argentina (Solovey et al. 1999). Average intake was 0.9 µg/kg of body weight. Recent studies showed that FB1 has been classified as a Group 2B possible human carcinogen (IARC 2002) and associated with esophageal cancer and neural tube defects in humans. FB2 and FB3 toxicity is at the same level of magnitude than that of FB1 (Stockmann-Juvala and Savolainen 2008). Kimanya et al. (2010) investigated the association between FBs exposure from maize and growth retardation among infants in Tanzania.

There is also the possibility of combined effects. The same fungal species can produce different families of mycotoxins and the same mycotoxin can be produced by different species. Manizan et al. (2018) analyzed 238 samples of different foodstuffs collected in the main markets of Côte d'Ivoire (88 rice samples, 79 maize samples and 71 peanut paste samples). They researched aflatoxins, ochratoxine A, ochratoxine B, fumonisins B1, B2, B3, beauvericin, equisetin, aflatoxin M1, cyclopiazonic acid, sterigmatocystin, citrinin and fusaric acid. They found that 91% were contaminated with more than one mycotoxin (about 21% between two and four mycotoxins, 21% between five and seven and 48% with more than eight, 4% with only one mycotoxin (nine rice samples). The largest number of mycotoxins detected in the same sample was found in maize samples (14 in 3 samples) followed by peanut paste (13 in 6 samples) and rice samples (8 in 2 samples). Waré et al. (2017) analyzed 248 samples of infant formulas and raw materials in Burkina Faso. They found that 36.7% (88/240) of the samples were simultaneously contaminated by aflatoxins and ochratoxin A, while 70.8% (170/240) by aflatoxins and fumonisins. Only 31.3% (75/240) of samples had a co-occurrence of total aflatoxins, ochratoxin A and fumonisins (only three mycotoxins were researched in this study). In Tunisia, analysis of 220 samples of pearl millet revealed 57 metabolites. Among major mycotoxins, both aflatoxin B1 (AFB1) and ochratoxin A (OTA) were the most prevalent at rate of 8.6% each and occurring at an average level of 106 and 69.4 μg/kg, respectively. All positive samples were significantly exceeding the European thresholds of 5 and 3 μg/kg, respectively (Houissa et al. 2019).

Each toxin is potentially dangerous. If there are multiple toxins in food, this could have a synergistic effect on consumer health (death or much more frequent occurrence of disease or cancers). All these data raise questions and could help to understand why the malnutrition rate is not decreasing rapidly despite the efforts made by institutions in Burkina Faso.

6. Recommendations

In developing countries, infant nutritionists are faced with the challenge of feeding the children as best as possible to provide them with a satisfying development. However, answer to the difficult question related to the presence of mycotoxins provided mainly by the cereals of the necessary dietary supplements remains. Should we feed the children by poisoning them or leave them with a deficient food? The recommendations we advocate should answer this question.

Diets mainly based on plant-based foods provide insufficient amounts of some 'key' nutrients (including iron, zinc and calcium) to meet recommended nutrient requirements for the ages between six and twenty-four months (Gibson et al. 1998, Dewey and Brown 2003). It is therefore recommended to consume dairy products, fruit, poultry, fish or eggs in addition to infant flours (WHO 2006) or to favor infant flours enriched with micronutrients (Kayalto 2016). We thus proposed:

- raise awareness of the health risks faced by infants and young children following the consumption of infant flours contaminated with germs or mycotoxins, especially aflatoxins,
- buy raw materials that are as fresh (sorted and dried) as possible,
- make sure that infant flour is stored properly, away from insects, in a dry and not too hot place,

- do not keep infant flour for too long before eating it, and
- after opening infant flours, keep them in a dry place.

Moreover, Specifically, in Burkina Faso:

- To sensitize the public authorities to apply the texts on standards for mycotoxins,
- To Burkinabe researchers: to develop good practices as GMP/BPH data sheets for women producers of infant flour in order to reduce the risk of mycotoxin and especially aflatoxin contamination,
- To women producers of infant flours: use the rapid tests (kits) by the production units to detect mycotoxins in raw materials before production, and
- To raise awareness among the various actors in the production chain of processed cereal products, from the field with the raw material to the finished product packaged and stored.

The United States Department of Agriculture (USDA) in partnership with the International Institute of Tropical Agriculture (IITA) has developed a biological control product called "Aflasafe". The Burkinabe's government recently launched the product called "Aflasafe BF01" which aims to reduce mycotoxins in the field, which is a good start in the fight against these toxins. In Burkina Faso, IITA researchers worked with researchers from the Institute for Environment and Agricultural Research (INERA) to develop a country-specific product called "Aflasafe BF01". This product was designed to control aflatoxin contamination in maize and peanuts. This product is natural and is made from local strains of non-toxic fungal strains of *Aspergillus flavus* that coat a sorghum seed in order to progressively displace toxic strains of *A. flavus*. Efficacy tests of the product have been carried out in the regions of Burkina over the last five years and have reduced the rate of aflatoxin contamination by 80 to 99%. "Aflasafe BF01" protects maize and peanuts from the field to storage. Following these conclusive results, the product has been registered in 2017 and is in the process of being commercialized. The above proposals, if properly implemented, could reduce the risks involved by consuming infant flours.

References

Afolabi, C.G., Ezekiel, C.N., Kehinde, I.A., Olaolu1, A.W. and Ogunsanya, O.M. 2015. Contamination of groundnut in South-Western Nigeria by aflatoxigenic fungi and aflatoxins in relation to processing. Journal of Phytopathology 163: 279–286. DOI: 10.1111/jph.12317

Adejumo, O., Atanda, O., Raiola, A., Somorin, Y., Bandyopadhyay, R. and Ritieni, A. 2013. Correlation between aflatoxin M1 content of breast milk, dietary exposure to aflatoxin B1 and socioeconomic status of lactating mothers in Ogun State, Nigeria. Food Chemical and Toxicology 56: 171–177. DOI: 10.1016/j.fct.2013.02.027

Bailly, J.D. and Oswald, I.P. 2013. Toxins: Mycotoxins 1–12. *In:* Paul Worsfold, Colin Poole, Alan Townshend, Manuel Miró, (eds.). Encyclopedia of Analytical Science 3: 129–140. doi.org/10.1016/B978-0-12-409547-2.00547-3.

Bankole, S.A., Ogunsanwo, B.M. and Eseigbe, D.A. 2005. Aflatoxins in Nigerian dry-roasted groundnuts. Food Chemistry 89: 503–506. DOI: 10.1016/j.foodchem.2004.03.004

Baydar, T., Erkekoglu, P., Sipahi, H. and Sahin, G. 2007. Aflatoxin B1, M1 and ochratoxin A levels in infant formulae and baby foods marketed in Ankara, Turkey. Journal of Food and Drug Analysis 15: 89–92.

Bern, C.M.J., de Zoysa, I. and Glass, R.I. 1992. The magnitude of the global problem of diarrhoeal disease: A ten-year update. Bulletin WHO 70: 705–714.

D'Arco, G., Fernández-Franzón, M., Font, G., Damiani, P. and Mañes, J. 2008. Analysis of fumonisins B1, B2 and B3 in corn-based baby food by pressurized liquid extraction and liquid chromatography/tandem mass spectrometry. Journal of Chromatography A 1209: 188–194. DOI: 10.1016/j.chroma.2008.09.032

Dewey, K.G. and Brown, K.H. 2003. Undated on technical issues concerning complementary feeding of young children in developing countries and applications for intervention programs. Food and Nutrition Bulletin 24: 5–28. DOI: 10.1177/156482650302400102

Doray, C.A. 2017. Les céréales pour bébé : l'intérêt nutritionnel des céréales. https://www.passeportsante.net/fr/grossesse/Fiche.aspx?doc=cereales-bebe

European Food Safety Authority (EFSA). 2007. Opinion of the scientific panel on contaminants in the food chain on a request from the commission related to the potential increase of consumer health risk by a possible increase of the existing maximum levels for aflatoxins in almonds, hazelnuts and pistachios and derived products. Question N° EFSA-Q-2006-174. EFSA Journal 446: 1–127.

Egal, S., Hounsa, A., Gong, Y.Y., Turner, P.C., Wild, C.P., Hall, A.J. et al. 2005. Dietary exposure to aflatoxin from maize and groundnut in young children from Benin and Togo, West Africa. Internal Journal of Food Microbiology 104: 215–224. DOI: 10.1016/j.ijfoodmicro.2005.03.004

European Commission. 2006. Commission Regulation (EC) No 1881/2006 of 19 December 2006 setting maximum levels for certain contaminants in foodstuffs (Text with EEA relevance). Official Journal of European Union 364: 5–24.

Ezekiel, C.N., Udom, I.E., Frisvad, J.C., Adetunji, M.C., Houbraken, J., Fapohunda, S.O. et al. 2014. Assessment of aflatoxigenic Aspergillus and other fungi in millet and sesame from Plateau State, Nigeria. Mycology: An International Journal on Fungal Biology 5: 16–22. DOI: 10.1080/21501203.2014.889769

Food and Agriculture Organization (FAO). 1979. Nutrition humaine en Afrique tropicale. Deuxième édition, P-86 ISBN 92-5-200412-2. http://www.fao.org/3/x0081f/X0081F0h.htm consulted on 18/04/2020

Food and Agriculture Organization (FAO). 2004. Undernourishment around the world. *In:* The State of Food Insecurity in the World 2004. The Organization, Rome. 62 pp. http://www.fao.org/3/a-i4646e.pdf consulted on 11/08/2018

Fasonorm. 2014. Farines infantiles spécifications Norme Burkinabè NBF 01-198: 7–9.

Gauthier 2016. Les mycotoxines dans l'alimentation et leur incidence sur la santé. Sciences Pharmaceutiques. Thèse N°43 de l'Université de Bordeaux U.F.R. des sciences pharmaceutiques 132 pp. https://dumas.ccsd.cnrs.fr/dumas-01315198

Gibson, R.S., Ferguson, E.L. and Lehrfeld, J. 1998. Complementary foods for infant feeding in developing countries: Their nutrient adequacy and improvement. European Journal of Clinical Nutrition 52: 764–70. DOI: 10.1038/sj.ejcn.1600645

Giovati, L., Magliani, W., Ciociola, T., Santinoli, C., Conti, S. and Polonelli, L. 2015. AFM1 in milk: Physical, biological, and prophylactic methods to mitigate contamination. Toxins 7: 4330–4349. https://doi.org/10.3390/toxins7104330.

Golden, M.H. and Ramadath, D. 1987. Free radicals in the pathogenesis of Kwashiorkor. Proceedings of the Nutrition Society 46: 53–68. DOI: 10.1079/pns19870008

Gong, Y.Y., Cardwell, K., Hounsa, A., Egal, S., Turner, P.C., Hall, A.J. et al. 2002. Dietary aflatoxin exposure and impaired growth in young children from Benin and Togo: Cross sectional study. British Medical Journal 325: 20–21. DOI: 10.1136/bmj.325.7354.20

Gret. 1998. Dossier "Les farines infantiles". Bulletin du réseau TPA (Technologie et Partenariat en Agroalimentaire) 15: 211–213.

Gret/Nutrifaso. 2007. Cahier des charges d'une farine infantile souhaitant être labellisée Nutrifaso. Projet Nutrifaso - Qualité, 9 p.

Hervé, S., Traoré, T. and Mouquet-Rivier, C. 2004. Etude de marché des farines infantiles et compléments alimentaires en milieu urbain au Burkina Faso. Rapport d'étude du programme Nutrifaso, Ouagadougou, Burkina Faso 56 pp.

Houissa, H., Lasram, S., Sulyok, M., Šarkanj, B., Fontana, A., Strub, C. et al. 2019. Multimycotoxin LC-MS/MS analysis in pearl millet (Pennisetum glaucum) from Tunisia. Food Control 106: 106738. https://doi.org/10.1016/j.foodcont.2019.106738

International Agency for Research on Cancer (IARC). 1993. Monographs on the evaluation of carcinogenic risks to humans. Some naturally occurring substances: Food items and constituents, heterocyclic aromatic amines and mycotoxins. Lyon: France. IARC 56: 599.

International Agency for Research on Cancer (IARC). 2002. IARC monographs on the evaluation of carcinogenic risks to humans. Some traditional herbal medicines, some mycotoxins, naphthalene and styrene. Lyon: France. IARC 82: 601.

International Agency for Research on Cancer (IARC). 2012. Monographs on the evaluation of carcinogenic risks to humans: Chemical agents and related occupations. A review of human carcinogens. Lyon, France: International Agency for Research on Cancer 100 F: 224–248.

International Agency for Research on Cancer (IARC). 2015. Mycotoxin control in low- and middle-income countries. Retrieved from http://www.ncbi.nlm.nih.gov/books/NBK350558/

Jayaramachandran, R., Ghadevaru, S. and Veerapandian, S. 2013. Survey of market samples of food grains and grain flour for Aflatoxin B1 contamination.

International Journal of Current Microbiology and Applied Sciences 2: 184–188.

Kayalto, B. 2016. Caractérisation et amélioration des procédés de fabrication de quelques farines infantiles du Tchad et leur enrichissement en vitamine A et C, en fer et zinc à base de *Moringa oleifera Lam.* et de la pulpe de néré (*Parkia biglobosa* (Jacq.) Benth.). Thèse de Doctorat de l'université Ouaga I Pr Joseph Ki-Zerbo. 153 pp.

Kimanya, M.E., De Meulenaer, B., Roberfroid, D., Lachat, C. and Kolsteren, P. 2010. Fumonisin exposure through maize in complementary foods is inversely associated with linear growth of infants in Tanzania. Molecular Nutrition and Food Reseach 54: 1659–1667. DOI: 10.1002/mnfr.200900483

Köszegi, T. and Poor, M. 2016. Ochratoxin A: Molecular interactions, mechanisms of toxicity and prevention at the molecular level. Toxins 8: 111. https://doi.org/10.3390/toxins8040111

Kouassi, K.A.A.A., Adouko, A.E., Gnahe, D.A., Grodji, G.A., Kouakou, B.D. and Dago, G. 2015. Comparaison des caractéristiques nutritionnelles et rhéologiques des bouillies infantiles préparées par les techniques de germination et de fermentation. International Journal of Biological and Chemical Sciences 9: 944–953. http://dx.doi.org/10.4314/ijbcs.v9i2.31

Malir, F., Ostry, V., Pfohl-Leszkowicz, A. and Novotna, E. 2014. Ochratoxin A: Developmental and reproductive toxicity – An overview. Birth Defects Research (Part B) 00: 1–10. DOI: 10.1002/bdrb.21091

Mandjeka, J.C.A. 2016. Optimisation d'une bouillie infantile et son enrichissement en protéines et vitamines en vue de son utilisation comme aliment de complément chez le nourrisson et le jeune enfant en Centrafrique. Thèse soutenue à Lille 1 en cotutelle avec l'Université de Bangui. 189 pp.

Manizan, A.L., Oplatowska-Stachowiak, M., Piro-Metayer, I., Campbell, K., Koffi-Nevry, R., Elliott, C. et al. 2018. Multi-mycotoxin determination in rice, maize and peanut products most consumed in Côte d'Ivoire by UHPLC-MS/MS. Food Control 87: 22–30. https://doi.org/10.1016/j.foodcont.2017.11.032

Meucci, V., Razzuoli, E., Soldani, G. and Massart, F. 2010. Mycotoxin detection in infant formula milks in Italy. Food Additives and Contaminants 27: 64–71. DOI: 10.1080/02652030903207201

Ministry of Health. 2017. Nutrition Department. National Nutrition Survey 2017. Departmental Report, 90 pp. https://www.humanitarianresponse.info/sites/www.humanitarianresponse.info/files/documents/files/smart_2016.pdf

Nikiema, P.A. 1993. Etude des aflatoxines au Burkina Faso : détermination quantitative et qualitative des aflatoxines de l'arachide par des tests biochimiques et immunologiques. Thèse de doctorat de spécialité : sciences biologiques appliquées, option : biochimie - microbiologie. 118 pp. http://www.beep.ird.fr/collect/uouaga/index/assoc/M07200.dir/M07200.pdf

Ouattara-Sourabié, B.P., Nikièma, A.P. and Traoré, S.A. 2011. Caractérisation de souches d'*Aspergillus* spp isolées des graines d'arachides cultivées au Burkina Faso, Afrique de l'Ouest. International Journal of Biological and Chemical Sciences 5: 1232–1249. DOI: 10.4314/ijbcs.v5i3.72269

Pietri, A. and Piva, G. 2007. Aflatoxins in foods. Italian Journal of Public Health 4. Retrieved from http://ijphjournal.it/article/view/5899.

Polychronaki, N., Wild, C.P., Mykkänen, H., Amra, H., Abdel-Wahhab, M.A., Sylla, A. et al. 2008. Urinary biomarkers of aflatoxin exposure in young children from Egypt and Guinea. Food Chemical and Toxicology 46: 519–526. DOI: 10.1016/j.fct.2007.08.034

Raiola, A., Tenore, G.C., Manyes, L., Meca, G. and Ritieni, A. 2015. Risk analysis of main mycotoxins occurring in food for children: An overview. Food and Chemical Toxicology 80: 0278–6915. DOI: 10.1016/j.fct.2015.08.023

Sanou, A., Tapsoba, F., Zongo, C., Savadogo, A. and Traore Y. 2017. Etude de la qualité nutritionnelle et microbiologique des farines infantiles de quatre unités de production: CMA saint Camille de Nanoro, CSPS Saint Louis de Temnaore, CM saint Camille de Ouagadougou et CHR de Koudougou. Nature & Technology Journal. Vol. B: Agronomic & Biological Sciences, 17: 25–39. http://www.univ-chlef.dz/revuenatec/issue-17/Article_B/Article_421.pdf

Skaug, M.A. 1999. Analysis of Norwegian milk and infant formulas for ochratoxin A. Food Additives and Contaminants 16: 75–78. https://doi.org/10.1080/026520399284235

Solovey, M.M.S., Somoza, C., Cano, G., Pacin, A. and Resnik, S. 1999. A survey of fumonisins, deoxynivalenol, zearalenone and aflatoxins contamination in corn-based food products in Argentina. Food Additives and Contaminants 16: 325–329. DOI: 10.1080/026520399283894

Stockmann-Juvala, H. and Savolainen, K. 2008. A review of the toxic effects and mechanisms of action of fumonisin B1. Human and Experimental Toxicology 27: 799–809. https://doi.org/10.1177/0960327108099525

Traoré, T. 2005. Élaboration et évaluation d'une stratégie d'amélioration de l'alimentation de complément des jeunes enfants au Burkina Faso. Thèse de doctorat de l'université de Ouagadougou, option sciences biologiques appliquées, spécialité nutrition et sciences des aliments. 233 pp. https://horizon.documentation.ird.fr/exl-doc/pleins_textes/divers13-07/010039899.pdf

Waré, L.Y., Durand, N., Nikiema, P.A., Alter, P., Fontana, A., Montet, D. et al. 2017. Occurrence of mycotoxins in commercial infant formulas locally produced in Ouagadougou (Burkina Faso). Food Control 73: 518–523. http://dx.doi.org/10.1016/j.foodcont.2016.08.047

Waré, L.Y. 2018. Évaluation de la qualité sanitaire des farines infantiles produites au Burkina Faso. Thèse de doctorat de l'université Ouaga I Pr Joseph Ki-Zerbo. Option Sciences Biologiques Appliquées, Spécialité Biochimie-Microbiologie 107 pp.

Warth, B., Parich, A., Atehnkeng, J., Bandyopadhyay, R., Schuhmacher, R., Sulyok, M. et al. 2012. Quantitation of mycotoxins in food and feed from Burkina Faso and Mozambique using a modern LC-MS/MS multitoxin method. Journal of Agricultural and Food Chemistry 60: 9352–9363. DOI: 10.1021/jf302003n

WHO. 2006. Guidelines for feeding non-breastfed children aged 6 to 24 months. ISBN 92 4 259343 5, ISBN 978 92 4 259343 3, (NLM classification: WS 120). https://www.who.int/nutrition/publications/infantfeeding/guidingprin_nonbreastfed_child_fr.pdf?ua=1

WHO. 2008a. Indicators for assessing infant and young child feeding practices. Part 1. Definitions. Geneva: World Health Organization, 2008.

WHO. 2008b. The global burden of disease: 2004 update. Geneva: World Health Organization, 2008.

WHO. 2013. Infant and young child feeding: Model Chapter for textbooks for medical students and allied health professionals. Geneva: World Health Organization, 2009. 112 pp.

WHO/UNICEF. 2003. Supplementary feeding for young children in developing countries. Geneva: World Health Organization, 130–131.

WHO/UNICEF. 2009. Baby-friendly Hospital Initiative revised, updated and expanded for integrated care. Geneva: World Health Organization, 2009.

Zain, M.E. 2011. Impact of mycotoxins on humans and animals. Journal of Saudi Chemical Society 15: 129–144. https://doi.org/10.1016/j.jscs.2010.06.006.

Recent Developments in the Analysis of Mycotoxins

Part 1: Sampling, Sample Preparation and Sample Extraction and Clean-up for the Analysis of Mycotoxins

Raymond Coker

Emeritus Professor of Food Safety, Natural Resources Institute of University of Greenwich, Chatham, ME4 4TB, UK
Raymond Coker Consulting Limited, Bromley, BR1 2PJ, UK

1. Introduction

This and the following chapter highlight key developments in methods for the analysis of mycotoxins which occurred, primarily, during 2019.

A previous review has addressed the development of analysis methods from 2017 to 2019 (Tittlemier et al. 2019), and general and advanced methods for the detection and measurement of aflatoxins and their metabolites have been recently reviewed (Mahfuz et al. 2018).

Part 1 describes recent developments in:

- Sampling and Sample Preparation
- Sample Extraction and Clean-up

The analysis of specific mycotoxins is considered under the headings described above.

2. Sampling and Sample Preparation

The importance of employing effective sampling and sample preparation procedures at the beginning of the analytical sequence has previously been comprehensively discussed (Turner et al. 2015). Briefly, it is essential that the method employed for the collection of the aggregate sample, from the

Email: raycokerconsulting@gmail.com

batch of food or feed under evaluation, accommodates the heterogeneous distribution of mycotoxins (especially the aflatoxins) such that the sample is truly representative of the original batch. Similarly, it is equally important that the aggregate sample is comminuted and sub-divided in a manner which maintains the representative nature of the aggregate sample; and, that the final laboratory sample is representative of the sub-sample. These criteria are best met by employing a sub-sampling mill to comminute and sub-divide the aggregate sample, and by converting the resultant sub-sample into a homogeneous aqueous slurry. Typically, 100g representative laboratory samples are then withdrawn from the slurry, prior to their analysis.

In recent developments, a cost effective sampling and analysis method for mycotoxins in cereals has been described (Focker et al. 2019), together with sample preparation and analytical considerations for the US aflatoxin sampling program for shelled groundnuts (Davis et al. 2018). A method for the precise quantitation of aflatoxin, involving the preparation of maize meal slurries has also been described (Kumphanda et al. 2019).

The study performed by Focker et al. (2019) attempted to find the most cost-effective sampling and analysis plan for the determination of deoxynivalenol (DON) in a wheat batch and aflatoxins in a maize batch. Considering the cost of the plan as a major constraint, an optimization model was constructed which maximized the number of correct decisions for accepting or rejecting a batch of cereals. The 'decision variables' were: the number of incremental samples collected from the batch; the number of sample aliquots analysed; the choice of the analytical method (i.e. liquid chromatography combined with mass-spectrometry (LC-MS/MS), enzyme linked-immunosorbent assay (ELISA), or lateral flow devices (LFD). For DON in wheat, the difference between the optimal plans using the three different analytical methods was minimal. However, for aflatoxins in maize, the cost effectiveness of the plan using LC-MS/MS or ELISA were comparable, whereas the plan considering onsite detection with LFDs was least cost effective.

Davis et al. (2018) concluded that sample preparation was a greater source of variation than analytical testing, when determining the aflatoxin content of batches of groundnuts. They recommended that sample preparation should be performed using a vertical cutter mill, which converts a 22 kg aggregate sample into a homogeneous paste. A 1100 g subsample of the latter was prepared by combining randomly collected aliquots of paste before converting the subsample to an aqueous slurry, from which 122.8 g portions (equivalent to 50 g groundnut) were taken for analysis.

Although the authors recognised that disagreement exists within the US groundnut industry regarding the relative performance of a widely used sub-sampling mill and a vertical cutter mill, they observed that a subsampling mill provides significantly less comminution compared to a properly sized vertical cutter mill. Consequently, they argue, the aflatoxin distributions among subsamples produced from a subsampling mill remain more positively skewed than subsample distributions derived from a vertical cutter mill, which are more normally distributed around the sample mean.

However, it was also noted that there is a significant difference between the cost of a sub-sampling mill and an appropriate vertical cutter mill, which are priced at around $5,000 and $40,000, respectively.

The need to maintain the representative nature of the samples, from the original batch to the laboratory sample is especially important when considering so-called 'point-of-need' screening tests (Soares et al. 2018). Here, the volumes of the sample solutions subjected to analysis can be of the order of nanolitres. Needless to say, ensuring that such a small sample volume accurately represents the mycotoxin concentration of, for example, a 20 tonne batch of groundnuts is a very significant challenge! However, it should be remembered that if the representative nature of the samples is not maintained throughout the analytical sequence, then the final analysis result will be meaningless.

3. Sample Extraction and Sample Clean-up

3.1 Introduction

A wide variety of sample extraction and clean-up methods have recently been reported. A selection of these are described below, illustrating how they have been employed in the determination of a variety of mycotoxins.

3.2 Aflatoxins

A variety of sample extraction methods have been recently reported. For example, an *ultrasonic-assisted extraction and dispersive liquid–liquid microextraction* for the analysis of aflatoxin B_1 in egg has been described (Amirkhizi et al. 2017).

Dispersive liquid-liquid-liquid microextraction (DLLME) is a microextraction method based upon a triple component solvent system. Typically, a mixture of extraction and dispersive solvents are rapidly injected into an aqueous sample solution. This generates fine droplets of extraction solvent throughout the aqueous sample solution which, in turn, enhances the extraction process by producing a large surface area between the extraction solvent and sample solution.

In the reported study, a DLLME procedure was developed where the analyte (aflatoxin B_1), in an aqueous dispersive solution mixed with chloroform (the extraction solvent), was rapidly injected into deionised water. After vortexing (where the aflatoxin B_1 was extracted into the fine droplets of chloroform) and centrifugation, the lower chloroform layer was dried under a stream of nitrogen. The residue was reconstituted in methanol and a 20 μL aliquot was analysed by high performance liquid chromatography with UV detection (HPLC-UV).

A recovery of 94–98% was reported with limits of detection and quantitation of 0.12 and 0.32 μg.kg^{-1}, respectively.

The *ultrasonic-assisted extraction process*, which preceded the DLLME, may be briefly described as follows: a homogenised 6 g egg sample was blended with a mixture of acetonitrile/water (80/20) and diatomaceous earth in a magnetic stirrer, followed by sonication at 35°C and filtration. A portion of the filtrate was then subjected to vortex extraction with hexane, followed by centrifugation. The lower layer of aqueous extract was retained for use as the dispersive solution in the DLLME procedure.

The sample clean-up performance of the ultrasonic-assisted extraction and dispersive liquid–liquid microextraction procedure was comparable to that of a traditional immunoaffinity column.

Although the method involved multiple steps, it was also considered to be efficient, rapid, inexpensive, simple and environmentally friendly.

An *ultrasonic-assisted extraction* procedure, used in combination with traditional *immunoaffinity cartridge (IAC) clean-up* and high performance liquid chromatography with fluorescence detection (HPLC-FLD), has been developed for the determination of aflatoxins in *Hibiscus sabdariffa* and other highly acidic traditional medicines (Liu et al. 2018).

Five grams of *H. sabdariffa* was mixed in a 50 mL centrifuge tube with 1.00 g sodium chloride and 25 mL of methanol/water (70/30, v/v) solution. The mixture was homogenized by swirling for 2 min., and was then extracted by ultrasonication in an ultrasonic bath for 20 min. The resultant slurry was separated by centrifuging at 10,000 rpm for 5 min at 20°C. Five mL of the upper layer was immediately transferred into a new centrifuge tube and 40 mL of 0.1 M phosphate buffer (pH 8.0) was added, in order to dilute the sample 1:8. Finally, the diluent was filtered through 0.45 µm glass fibre filter paper, prior to the IAC clean-up procedure and HPLC-FD.

Trace levels of aflatoxins have been extracted from soy milk using a sensitive, simple and green graphene oxide (GO)-based *stir bar sorptive extraction* (SBSE) method, prior to analysis by high performance liquid chromatography-laser-induced fluorescence detection (HPLC-LIF) (Ma et al. 2018). Under optimal conditions, the quantitative method had low limits of detection ranging from 2.4 to 8.0 pg.mL^{-1}, and recoveries of 80.5 to 102.3%.

A commercially available, *water-based*, 'solvent free' extraction procedure has been successfully applied to the extraction of aflatoxins in maize (Lattanzio et al. 2018). Analysis of the extract was performed by lateral flow immunoassay (LFIA), and the complete method was evaluated according to Commission Regulation (EU) No. 519/2014, encompassing EU maximum permitted aflatoxin B1 levels up to 8 µg.kg^{-1}. The evaluation provided information on the method precision profile, cut off values, and false positive and false negative rates.

The total precision expressed as relative standard deviation, varied from 14 to 29% for contaminated samples. The 'cut off' values, calculated assuming that 2 µg.kg^{-1} or 4 µg.kg^{-1} were the screening target concentrations, were 1.24 and 2.18 µg.kg^{-1}, respectively. The false 'suspect' rates for blank samples were 42% and 8%, respectively. The false negative rate for samples containing the analyte at higher concentrations was found to be lower than 1%. (The 'cut off' value is defined as 'the response, signal, or concentration obtained with the screening method, above which the sample is described as 'suspect', with a false negative rate of 5%.)

When naturally contaminated maize samples were analysed using the LFIA method and the Official AOAC Method 991.31, there was a good correlation between the methods (linear regression, r = 0.97, slope = 0.96). There was also a satisfactory agreement between the recorded and analytical results when reference samples of maize were analysed using the LFIA method.

Finally, a QC protocol was developed for evaluating the performance of screening kits, based upon guidelines in EU Regulation 519/2014/EU.

A *deep eutectic solvent-based matrix solid phase dispersion* (DES-MSPD) method has been applied to the determination of aflatoxins in crops (e.g. millet, groundnuts and hempseed), in conjunction with HPLC-FLD (Wu et al. 2019).

The main challenge was the identification of a simple, cost-effective and environment friendly MSPD method.

A normal MSPD process involves: grinding a mixture of the sample with a suitable 'functional dispersant' in a pestle and mortar; packing the homogenised mixture into a SPE cartridge; and, the elution of the packed cartridge with a suitable solvent.

A DES-MSPD procedure involves the addition of a 'deep eutectic solvent' (DES) to the mixture in the pestle and mortar, during the homogenisation process.

A DES is a mixture of two or more compounds with a freezing point well below the melting point for any of the original mixture components. The ingredients of naturally occurring deep eutectic solvents (NADES) are common cellular ingredients, such as polyalcohols, sugars, organic acids and bases, and amino acids. Combinations of such compounds, in specific molar ratios, are liquids that solubilize a wide range of natural products that are poorly water-soluble.

A mixture of tetrabutylammonium chloride and hexyl alcohol was used as the DES in the developed method, together with diatomaceous earth and silica gel as dispersants, for the analysis of groundnuts/hempseeds and millet, respectively.

Acetonitrile was used as the elution solvent, once the homogenised mixture of sample, DES and dispersant had been packed into the SPE cartridge. The eluate was evaporated to dryness under nitrogen at 50°C, and redissolved in acetonitrile. The aflatoxin content of the resultant solution was then determined by HPLC-FLD.

The intra-day and inter-day variability of the aflatoxins in all crops studied was less than 7.5%; and, the limits of detection and quantitation were 0.03–0.1 and 0.1–0.33 µg.kg^{-1}, respectively.

A procedure for the *selection of optimal aflatoxin-specific monoclonal antibodies,* and the subsequent production of an immunoaffinity column (IAC), has been developed (Ertekin et al. 2019).

The procedure may be summarised as follows:

1. Determination of antibody specificity with aflatoxins B_1, B_2, G_1, G_2, and M_1.
2. Determination of the tolerance of antibodies to the solvent employed in the extraction of aflatoxins from food samples.
3. Evaluation of the impact of food matrix on the performance of the antibodies with relevant food extracts, and on the determination of low concentrations of aflatoxins.
4. Production of an immunoaffinity column, and evaluation of the resin-binding efficacy and overall column performance.

Using these criteria, nine monoclonal antibodies were selected for further evaluation, and one was found to be superior to the others.

IACs were produced on a pilot scale, using the selected antibody, and their performance was subjected to a validation study, involving 15 independent, certified analysis laboratories.

The IACs demonstrated more than 90% recovery of 5 ng of aflatoxins B_1, B_2, G_1, and more than 80% recovery of 5 ng of aflatoxin G_1 in wheat flour and hazelnut extracts, with an inter-laboratory precision of less than 6% relative standard deviation for all aflatoxins.

The selective extraction and enrichment of aflatoxin from food samples, involving the employment of *molecularly imprinted polymers* (MIPs) has been reported (Rui et al. 2019).

A highly selective surface molecular imprinted polymer (FDU-12@ MIPs) was prepared using structural analogues of aflatoxins as the template molecule, and mesoporous silica FDU-12 as the carrier.

A highly ordered mesoporous nanostructure was observed in the fabricated polymer when it was examined by scanning electron microscopy, energy dispersive X-ray spectroscopy, X-ray diffraction and attenuated total reflection Fourier transform infrared spectrometry. The adsorption capacity of FDU-12@MIPs for aflatoxins were higher than that of non-imprinted mesoporous silica polymer (FDU-12@NIPs).

FDU-12@MIPs were successfully used as a solid-phase extraction sorbent, for the determination of aflatoxins in food samples, when coupled with HPLC.

Recoveries from spiked cereals were 82.6–116.7%, with relative standard deviations ranging from 2.73 to 4.21%.

The detection limits of aflatoxin G_2, G_1, B_2 and B_1 were estimated as 0.05, 0.06, 0.06 and 0.05 $\mu g.kg^{-1}$, respectively; and. limits of quantification were estimated to be 0.15, 0.2, 0.2 and 0.15 $\mu g.kg^{-1}$, respectively.

The method exhibited a linear response toward aflatoxin G_2, G_1, B_2 and B_1 in the range of 0.1–50.0 $\mu g.kg^{-1}$ (R^2 = 0.9992–0.9996).

These results demonstrated that FDU-12@MIPs could be used as an efficient adsorbent for the solid phase extraction and enrichment of aflatoxins in real samples.

The quantitation of aflatoxin B_1 (AFB1) in vegetable oils has been successfully performed, using a *low temperature clean-up (LTC)* followed by *immuno-magnetic solid phase extraction (IMSPE)* and simple fluorimetry (Yu et al. 2019).

LTC (20 h in a freezer, at –20°C) removed the interference from the oil matrix, whilst the analyte was preconcentrated by IMSPE.

The analysis sequence for a 5 mL sample of vegetable oil may be summarised as follows:

1. Add acetonitrile to sample and vortex and freeze, producing a solid bottom layer of fat.
2. Remove and dry the upper organic layer and reconstitute in a mixture of acetonitrile and deionised water.
3. Add the IMSPE to the resultant solution and vortex blend for 5 min., in order to bind the AFB1 to the adsorbent.
4. Magnetically separate the IMSPE/bound AFB1 and elute the AFB1 using methanol.
5. Dry the methanolic solution and reconstitute in 120 µL methanol, and add bromine water to enhance the fluorescence of the AFB1.
6. Measure the fluorescence of the resultant solution using a fluorimeter, with an excitation wavelength of 365–380 nm, and an emission wavelength of 450–550 nm.

The proposed method showed satisfactory efficiency and reproducibility with recovery rates being within the range of 79.6–117.9%, with a relative standard deviation not greater than 11.48%.

The sensitivity of the method was good with the limits of detection and quantification being as low as 0.0048 and 0.0126 ng.g^{-1}, respectively.

The analysis of real samples was successfully performed for five different types of vegetable oils.

Given the 20 h freezing period included in the method, it is clearly best applied to the analysis of a batch of accumulated samples, where a quick result is not required.

Furthermore, it is important that the necessary steps are taken to ensure that the 5 mL sample of vegetable oil is representative of the original batch from which it was taken.

Ultra-trace amounts of aflatoxin M_1 have been effectively extracted from milk and dairy products, by employing *aptamer functionalized magnetic nanoparticles* (AMNPs) (Khodadadi et al. 2018).

When this extraction and pre-concentration procedure was used in conjunction with HPLC, the limit of detection of the method was 0.2 ng.L^{-1}.

A porous *monolithic column, based on covalent cross-linked polymer gels*, has been reported for the *online extraction* and analysis of trace aflatoxins in food samples, with complex matrices, in combination with HPLC-UV (Wei et al. 2018).

The low detection limits were in the range of 0.08–0.2 µg.kg^{-1}; and, the recoveries for spiked samples fell between 76.1 and 113% with RSDs of 1.1–9.6%.

A *nano-graphene adsorbent-based SPE clean-up* method has been developed, and used in conjunction with HPLC-FLD, for the determination of aflatoxins in food (Feizy et al. 2019).

The developed clean-up method was used for the extraction and pre-concentration of the aflatoxins (AFs) in rice and wheat samples.

The limits of detection were 0.63, 0.47, 0.62 and 0.83 ng.g^{-1}, and limits of quantification were 1.92, 2.65, 1.88 and 2.83 ng.g^{-1} for aflatoxins B_1, B_2, G_1 and G_2, respectively.

Acceptable levels of accuracy were achieved—the relative recovery values (RR%) of 75.88–113.30% and 70.61–110.75% were obtained in cereal samples, at the spiked level of 2 and 5 ng.g^{-1}, respectively. Within-laboratory relative standard deviations for repeatability (n = 6) were in the range of 2.14–3.17%.

3.3 Ochratoxin A

A simple and high-throughput sample extraction and clean-up procedure, based on the use of a *'supramolecular solvent with restricted access properties'* (SUPRAS-RAM), was developed for the quantification of ochratoxin A (OTA) in spices subjected to EU regulation. (Caballero-Casero et al. 2018).

The aim of the method development program was the identification of a cost-effective and high throughput procedure, which obviated the use of a lengthy and expensive clean-up step, for the routine surveillance of a wide variety of spices.

SUPRAS are nanostructured liquids obtained from colloidal suspensions of amphiphiles (possess both hydrophilic and lipophilic properties) by spontaneous processes of self-assembly and coacervation (the formation of organic-rich droplets) (Caballo et al. 2017).

SUPRAS have been developed which behave as *'restricted access liquids'*, thus facilitating the removal of macromolecules by chemical and physical mechanisms, such that pigments, and essential oils do not influence the extraction of OTA.

Sample extraction and clean-up took 10 min and required minute amounts of spice (0.2 g) and volume of SUPRAS (0.4 mL). After centrifugation, the crude extracts were directly analyzed by HPLC-FLD, using external calibration. Method validation proved that the full analysis met the performance criteria set by the EU guidelines in terms of sensitivity, recoveries, precision, selectivity and trueness.

Although spices are classified as difficult and unique commodities (e.g. *Capsicum, Piper,* nutmeg, ginger and turmeric) by the EU, recoveries for the determination of OTA in these commodities ranged from 81 to 101%; and, the limits of detection and quantification were 1.2 and 3.0 µg.kg^{-1}, respectively.

The analytical procedure was considered to meet the requirements for employment in enforcement and surveillance programs related to OTA in spices.

However, it should be considered whether a method requiring only 0.2 g sample will produce results which represent the batch (lot) of the commodity under evaluation.

The determination of ochratoxins A, B and C has been performed by HPLC, after sample clean-up using *polydopamine-based molecularly imprinted polymers on magnetic nanoparticles* (Fe_3O_4@PDA MIPs) (Hu et al. 2018).

The sorbent was produced by depositing a polydopamine-based molecularly imprinted polymer (MIP) on the surface of magnetite (ferric oxide) nanoparticles, which were characterized by IR spectroscopy and transmission electron microscopy.

Ochratoxins A, B and C were quantified by HPLC with fluorometric detection (HPLC-FLD), following desorption from the MIP with acetonitrile.

The LODs were between 1.8 and 18 pg.mL^{-1}, and the recoveries from spiked samples were 71.0–88.5%, with relative standard deviations (RSDs) of 2.3–3.8%, when rice and wine samples were analysed.

The preparation and evaluation of a *highly hydrophilic, aptamer-based organic-silica hybrid affinity monolith* for on-column specific discrimination of ochratoxin A (OTA) has been reported (Chen et al. 2019).

Polyhedral oligomeric silsesquioxanes (POSS) are a family of molecules that consist of a silica-like core surrounded by a shell of organic groups.

The hybrid affinity monolith was produced using a one-pot approach, by vortexing a mixture of POSS components, hydrophilic monomers and aptamer.

When applied to beer samples, the hybrid monolith showed a very high specificity towards OTA, with minimal recognition of ochratoxin B. The reported recovery of OTA was between 94.9 and 99.8%.

The performance of the highly hydrophilic monolith was far superior to that of a hydrophobic POSS-based affinity monolith.

Levels of ochratoxin A in grape juice and wine have been determined using a *β-cyclodextrin, polyurethane nanosponge solid phase extraction clean-up* together with HPLC-FLD (Appell et al. 2018).

The procedure was capable of detecting ochratoxin A levels below regulatory levels in grape juice and wine; and, the recoveries of spiked ochratoxin A (0.5–20 ng.mL^{-1}) were between 77.0–89% in wine and 69.1–86.5% in grape juice.

A procedure for the determination of ochratoxin A in wine has been reported, employing a *rapid and automated on-line solid phase extraction step*, followed by HPLC–MS/MS with peak focusing (Campone et al. 2018).

This study reports a fast and automated analytical procedure based on an on-line SPE-HPLC-MS/MS method for the automatic pre-concentration, clean up and sensitive determination of OTA in wine.

Direct injection of adjusted pH wine samples was performed without any other sample pre-treatment. OTA contained in a 100 µL sample (pH ~5.5) was selectively retained and concentrated on a mixed mode ion exchange chromatography SPE cartridge. After a washing step to remove matrix interferents, the analyte was eluted in back-flush mode from the SPE cartridge, and the eluent was diluted through a mixing Tee, using an aqueous solution.

Peak-focusing was used to concentrate the OTA on top of a monolithic chromatographic column prior to chromatographic separation.

The developed method was validated according to EC Commission Decision 657/2002 and applied to the analysis of 41 red and 17 white wines.

The linearity range was 0.5–5.0 ng.mL^{-1}, and the recovery values and relative standard deviation (<7%) were in agreement with Commission Regulation EC 519/2014.

The developed method features minimal sample handling, low solvent consumption, high sample throughput, low analysis cost and provides accurate and highly selective results.

Importantly, since the method involves the analysis of a 100 μL sample of untreated wine, it is imperative that the sample is representative of the batch under test.

The determination of ochratoxin A in green coffee has been performed using *an aptamer assisted ultrafiltration clean-up* in combination with HPLC fluorescence detection (El Saadani et al. 2020). A limit of detection (LOD) of 0.05 ng/mL was reported, with a recovery of up to 97.7%.

3.4   Citrinin

A *bull-frog immunoaffinity clean-up column* for the determination of citrinin (CIT) in red yeast rice has been developed and validated (Gu et al. 2019).

Anti-citrinin antibodies were readily obtained when bullfrogs (*Rana catesbeiana*) were directly immunized by thigh intramuscular injection with citrinin hapten, thus indicating that bullfrogs could respond to CIT hapten and produce anti-CIT immunoglobulin.

The performance of bullfrog-immunoaffinity columns, produced by covalently coupling bullfrog serum against CIT onto membrane, was evaluated by HPLC-FLD. Recoveries ranged from 75.5% to 80.1% for CIT-spiked samples, demonstrating that bullfrog serum proteins were specific to CIT with a high affinity.

Immunofluorescence images indicated that the antibodies exhibited stronger binding capacities for CIT in serum supernatant, which was identified as an immunoglobulin M (IgM) pentamer.

3.5   Zearalenone

Sample clean-up has been performed using *dummy molecularly imprinted polymers*, prior to the determination of zearalenone in grain samples using HPLC-FLD (Huang et al. 2019).

The MIPs were synthesized using coumarin-3-carboxylic acid as a dummy template of zearalenone, methacrylic acid as functional monomer, and ethylene glycol dimethacrylate hydroxyethyl methacrylate as cross-linker. The polymers were characterized by scanning electron microscopy, transmission electron microscopy, energy-dispersive X-ray spectroscopy, X-ray diffraction, and particle-size distribution analyses.

A self-made cartridge was used in the purification and enrichment of zearalenone extracted from real samples, and each cartridge could be re-used on, at least, seven occasions.

Limits of detection and quantitation varied from 2.09–4.16 ng.kg^{-1}, and from 6.25–12.50 ng.kg^{-1}, respectively. The spiking recoveries of zearalenone varied from 81.70–90.10%, with relative standard deviations lower than 5.56%.

3.6 Trichothecenes

'Sub-critical' water extraction, followed by *strongly hydrophilic reversed-phase solid-phase clean-up* and UPLC-MS/MS, has been successfully applied to the determination of six trichothecenes (deoxynivalenol, deoxynivalenol-3-glucoside, 3-acetyl-deoxynivalenol, 15-acetyl-deoxynivalenol, HT-2 toxin and T-2 toxin) in cereal matrices (Miró-Abella et al. 2017).

The sub-critical water extraction (pressurised liquid extraction) was performed using a commercially available 11 mL stainless steel extraction shell, packed with 1 g sample and diatomaceous earth, in combination with an accelerated solvent extractor.

Using water containing 1% formic acid as the extraction solvent, method detection limits between 0.05 and 4.0 µg.kg^{-1}, and method quantitation limits between 0.4 and 20.0 µg.kg^{-1} were achieved.

However, the challenge of ensuring that a 1 g laboratory sample is sufficiently representative of the original batch, when measuring very low concentrations of mycotoxins is not addressed. Needless to say, methods requiring very small quantities of sample will only be acceptable if all the necessary sampling and sample preparation criteria are met.

The same authors have also described optimised extraction methods for the determination of trichothecenes in rat faeces (Miró-Abella et al., 2019), for use in combination with HPLC-MS/MS. The aim of the study was the development of a method for the determination of low levels of trichothecenes, and their derivatives, in biological samples.

A *QuEChERS* (Quick, Easy, Cheap, Effective, Rugged and Safe) clean-up, followed by *dispersive solid phase extraction* (dSPE) with activated carbon, was the superior approach when compared with a Pressurised Liquid Extraction (PLE) procedure.

When the full quantitation method was applied to the determination of deoxynivalenol, 3-acetyl-deoxynivalenol, 15-acetyl-deoxynivalenol, deoxynivalenol-3-glucoside and de-epoxy deoxynivalenol, in rat faecal samples, recovery results ranged from 78 to 83% (except for deoxynivalenol-3-glucoside). Low quantitation limits were reported as 0.2 to 3.4 µg.kg^{-1}.

3.7 Patulin

The determination of patulin in apple juice by *single-drop liquid-liquid-liquid microextraction,* coupled with liquid chromatography-mass spectrometry, has also been reported (Li et al. 2018).

Ten mL of patulin-contaminated apple juice was mixed with 1.5 mL of ethyl acetate, first by manual shaking and then by heated magnetic stirring, in order to transfer the analyte to the ethyl acetate phase. A 5 µL microdrop of water, suspended from a blunt-end 25 µL Hamilton syringe, was then lowered into the upper ethyl acetate later for 20 min., in order to allow the transfer of the patulin from the ethyl acetate to the microdrop. The DLLME procedure was completed in 20 min. and was considered to hold significant potential for the successful analysis of high-sugar samples.

Finally, the microdrop was retracted into the syringe and immediately transferred into HPLC sample vials, followed by analysis by high performance liquid chromatography-tandem mass spectrometry (HPLC-MS/MS).

The limits of quantitation and detection for patulin in apple juice were 2.0 and 0.5 µg.L^{-1}, respectively. The linear range covered three orders of magnitude from 2.0 to 2000 µg.L^{-1}.

3.8 Alternaria Toxins

The level of alternariol (AOH) and alternariol monomethyl ether (AME) in foodstuffs has been determined by *molecularly imprinted solid-phase extraction (MISPSE)* and ultra-high-performance liquid chromatography tandem mass spectrometry (Rico-Yuste et al. 2018).

MIP microparticles were prepared using 4-vinylpyridine (VIPY) and methacrylamide (MAM) as functional monomers, ethylene glycol dimethacrylate (EDMA) as cross-linker and 3,8,9-trihydroxy-6 H-dibenzo [b,d]pyran-6-one (S2) as a surrogate template for alternariol.

The MISPE method was validated, as stipulated within the European Commission Decision 2002/657/EC, for the determination of AOH and AME in tomato juice and sesame oil, by UPLC–MS/MS.

The method performance was satisfactory with recoveries varying between 92.5 and 106.2%, and limits of quantification within the 1.1–2.8 µg.kg^{-1} range in both products. The method also performed favorably when compared with previously validated methods employing a QuEChERS-based clean-up (Walravens et al. 2016).

3.9 Multiple Mycotoxins

The optimization of an effective, rapid, simple, inexpensive, safe and rugged method for the extraction of mycotoxins and veterinary drugs from eggs and milk by 'response surface technology' has been reported (Zhou et al. 2018).

Response Surface Methodology (RSM) is a collection of statistical and mathematical techniques which are useful for developing, improving, and optimizing processes. The most extensive applications of RSM are in those situations where several input variables potentially influence specific performance measures of the process. The performance measure is called the response, and the input variables are subject to the control of the analyst.

A multiclass method was developed for the simultaneous determination of various classes of veterinary drugs, mycotoxins and their metabolites in egg and milk by ultra-high performance liquid chromatography-tandem mass spectrometry (UPLC-MS/MS).

The analytes were extracted by a *QuEChERS*-based procedure, which included *salt-out partitioning and dispersive solid-phase extraction*, to afford additional clean-up.

With the aim of maximizing throughput and extraction efficiency, a Plackett-Burman design was initially employed to screen significant variables. A response surface methodology, based on central composite design, was conducted in order to achieve optimal experimental conditions. The matrix effects were evaluated, and compensated for, using matrix-matched calibration curves ($R^2 > 0.987$).

Satisfactory performance characteristics were reported, and the occurrence of sterigmatocystin in eggs was reported and confirmed for the first time.

An analysis method involving extraction with acetonitrile-1% formic acid and sequential vortex together with ultrasonic extraction, centrifugation and single-step *strongly hydrophilic reverse phase solid phase extraction*, has been developed, in conjunction with UHPLC-MS/MS, for the analysis of seven mycotoxins in fruit and vegetables (especially during storage) (Dong et al. 2019).

The method was applicable to the screening of alternariol (AOH), alternariol monomethyl ether (AME), tenuazonic (TeA), tentoxin (TEN), ochratoxin A, patulin and deoxynivalenol, without using standard addition or matrix-matched calibration.

Limits of detection and quantitation varied between 0.05–3.0 µg.kg^{-1}, and 0.2–10.0 µg.kg^{-1}, respectively.

None of the mycotoxins were detected in fresh samples of strawberries and tomatoes. However, AOH and AME were detected in strawberries and TeA in tomatoes after long-term storage, increasing in concentration with storage time.

A method employing sample pre-concentration, a *QuEChERS based and dispersive solid-phase clean-up*, combined with UPLC-MS/MS, has been developed for the determination of 23 mycotoxins in beer – including emerging and masked toxins (González-Jartín et al. 2019).

For most of the mycotoxins evaluated, recoveries ranged from 70 to 110% with LOQs (from 0.038 to 30.43 µg.L^{-1}).

It is proposed that the developed method can be used for the routine analysis of beer, involving the simultaneous determination of the aflatoxins, ochratoxin A, deoxynivalenol, 3-acetyldeoxynivalenol, deoxynivalenol-3-glucoside, 15-acetyldeoxynivalenol, zearalenone, zearalenols, fumonisins, T-2 toxin, HT-2 toxin, neosolaniol, moniliformin, enniatins and beauvericin.

A similar approach has been applied to the determination of the enniatins and beauvericin in animal feeds (see below, Tolosa et al. 2019).

Mycotoxin analysis methods that involve sample extraction with *carbonaceous nanosorbents, multi-walled carbon nanotubes (MWCNTs)* and graphene oxide (GO) have been reviewed from 2011 onwards (Reinholds et al. 2019).

Recent studies have highlighted the advantages of magnetically modified MWCNTs and GO in mycotoxin analysis.

The developments covered in this review point to promising applications of functionalised carbonaceous nanosorbents in mycotoxin analysis.

GO based sorbents can be effective for the adsorption of relatively polar aflatoxins whereas MWCNTs, with high specific surface area and reduced agglomeration achieved through modification with silica and magnetic particles, are preferred for the extraction of less polar mycotoxins.

Unmodified or magnetically modified MWCNTs and GO adsorbents have been applied to a variety of mycotoxins (including aflatoxins, ochratoxin A, deoxynivalenol, nivalenol, zearalenone [and their metabolites], T-2 toxin, HT-2 toxin, neosolaniol and diacetoxyscirpenol).

It was concluded that unmodified or magnetic MWCNTs may show good adsorption capacity of low polarity mycoestrogens (ZEA, ZAN and their metabolites).

Functionalisation of MWCNTs, and optimisation of acidity, may enhance their application in dSPE or SPE methods for the concentration of relatively polar trichothecenes.

In order to achieve multi-mycotoxin analysis, it was reported that future studies on extraction procedures should be based on functionalised GO adsorbents.

A procedure involving a single-step clean-up method, using a *multiwalled carbon nanotube (MWCNT)*, in conjunction with UPLC-MS/MS, has been developed for the simultaneous analysis of 21 mycotoxins in corn and wheat (Jiang et al. 2018).

The mycotoxins determined included nine trichothecenes, zearalenone and its derivatives, four aflatoxins, and two ochratoxins.

The rapid, sensitive, reliable and inexpensive optimal clean-up step involved a combination of carboxylic MWCNT and C18 adsorbents.

The optimal clean-up procedure in combination with UPLC-MS/MS analysis afforded a satisfactory linearity ($r^2 \geq 0.9910$), high sensitivity, good recovery (75.6–110%), and acceptable precision (relative standard deviation (RSD), 0.3–10%) across the range of mycotoxins determined.

A method has been developed and validated for the simultaneous determination of pesticides, mycotoxins, tropane alkaloids, growth regulators, and pyrrolizidine alkaloids in oats and whole wheat grains, after online clean-up via two-dimensional liquid chromatography tandem mass spectrometry (Urban et al. 2019).

The samples were extracted with a mixture of acetonitrile and water and were injected into the two-dimensional LC-MS/MS system without any further clean-up or sample preparation.

Even though co-extracted matrix components were minimized by the online clean-up, the use of a matrix matched calibration was still necessary and advisable. Matrix effects were evaluated by comparing analyte signals in the respective matrix matched standard with the neat solvent standards.

The limits of detection for the mycotoxins studied (aflatoxin B1, B2, G1 and G2, ochratoxin A, deoxynivalenol, HT2-toxin, T-2 toxin and zearalenone) were below $8\,\mu\text{g.kg}^{-1}$ in both matrices.

The final method was validated according to the current Eurachem validation guide and SANTE document. The number of successfully validated analytes throughout all three validation levels in oats and wheat, respectively, were as follows: 330 and 316 out of 370 pesticides, 6 and 13 out of 18 pyrrolizidine alkaloids, and seven out of nine regulated mycotoxins, respectively.

The developed procedure was considered to be the first 2D-LC-MS/MS method for the determination of multiple analyte classes in one analytical run, and of value to a routine laboratory wishing to effectively save time and costs, and to further improve the monitoring of contaminants in foods.

The determination of mycotoxins (aflatoxins and ochratoxin A, OTA) using *hollow fibre dispersive liquid–liquid–microextraction (HF-DLLME)*, prior to high-performance liquid chromatography–tandem mass spectrometry (HPLC-MS/MS) has been discussed (Alsharif et al. 2019).

The linearity of the method was 0.1 to 30 μg.L^{-1} for the aflatoxins, and 0.1 to 20 μg.L^{-1} for ochratoxin A, with regression coefficients (R^2) exceeding 0.9990.

The precision was satisfactory with relative standard deviation values less than 11%, and the recovery was between 83 and 101%.

The limits of detection and quantitation were in the range from 0.04 to 0.06 μg.L^{-1} and 0.08 to 0.13 μg.L^{-1}, respectively, for the aflatoxins; and, 0.02 to 0.04 μg.L^{-1} and 0.08 to 0.10 μg.L^{-1}, respectively, for OTA.

The developed method was successfully applied for the determination of the aflatoxins and OTA in food samples.

A method has been developed and validated, based on *Captiva EMR-lipid clean-up* and LC-MS/MS, for the simultaneous determination of 19 mycotoxins in human plasma (Arce-López et al. 2020).

The studied mycotoxins were deepoxy-deoxynivalenol, aflatoxins (B$_1$, B$_2$, G$_1$, G$_2$ and M$_1$), T-2 and HT-2, ochratoxins A and B, zearalenone, sterigmatocystin, nivalenol, deoxynivalenol, 3-acetyldeoxynivalenol, 15-acetyldeoxynivalenol, neosolaniol, diacetoxyscirpenol and fusarenon-X.

Sample deproteinization and clean-up were easily and rapidly performed in one step using Captiva EMR-lipid (3 mL) cartridges and acetonitrile (with 1% formic acid), prior to LC-MS/MS analysis.

Validation was based on the evaluation of limits of detection (LOD) and quantitation, linearity, precision, recovery, matrix effect, and stability.

Limit of detection values ranged from 0.04 ng.mL^{-1} for aflatoxin B1 to 2.7 ng.mL^{-1} for HT-2 (excepting for nivalenol, where the LOD was 9.1 ng.mL^{-1}).

Recovery was obtained in intermediate precision conditions and at three concentration levels. Mean recovery values ranged from 68.8% for sterigmatocystin to 97.6% for diacetoxyscirpenol (relative standard deviation, RDS ≤ 15% for all the mycotoxins).

Although quantitation of the 19 mycotoxins is performed in two groups, involving two runs, the analyses of all 19 mycotoxins was completed in approximately 1 h.

Acknowledgements

I should like to express my gratitude to the editors for the opportunity to contribute to this important new book, and I am grateful to Dr Didier Montet, Prof. Sabine Galindo, Dr Catherine Brabet and Ramesh Chandra Ray for kindly reviewing my draft submission, and for their very helpful comments.

References

See end of Part 2 from the same author.

Recent Developments in the Analysis of Mycotoxins

Part 2: Quantitation Methods for the Analysis of Mycotoxins

Raymond Coker

Emeritus Professor of Food Safety, Natural Resources Institute of University of Greenwich, Chatham, ME4 4TB, UK
Raymond Coker Consulting Limited, Bromley, BR1 2PJ, UK

1. Introduction

Recent developments, in the many different types of recently reported quantitation methods, are considered under the following headings:

- Chromatography
- Immunochemical methods
- Sensors
- Biomonitoring

2. Chromatography

Most of the recently reported developments have involved liquid chromatography-tandem mass spectrometry (LC-MS/MS) and ultra performance chromatography-tandem mass spectrometry (UPLC-MS/MS).

A significantly smaller number of developments have also recently been reported in: liquid chromatography-mass spectrometry (LC-MS); high performance liquid chromatography with fluorescence, ultra-violet, or diode array detection; UPLC with fluorescence, or high resolution mass spectrometry (HRMS) detection; ultra-fast liquid chromatography (UFLC); and, gas liquid chromatography-mass spectrometry (GLC-MS). A variety of extraction and clean-up methods have also been employed prior to the chromatographic determination of the mycotoxins.

Email: raycokerconsulting@gmail.com

2.1 HPLC-UV Analysis with Immunoaffinity Clean-up

Trichothecenes and their derivatives. A simple method involving immunoaffinity clean-up followed by HPLC-UV analysis has been developed for the determination of deoxynivalenol, nivalenol and their 3-β-D-glucosides (DON-3-glucoside and NIV-3-glucoside) in baby formula and Korean rice wine (Lee et al. 2019). The method was validated in-house, demonstrating: very good linearity ($R^2 > 0.99$) within the range of 20–1000 µg.kg^{-1}; accuracy ranging from 78.7 to 106.5% for all the analytes; good intermediate precision (relative standard deviation < 12%); and adequate detection and quantitation limits (< 4.4 and < 13.3 µg. kg^{-1}), respectively. The validated method was successfully applied to the analysis of 31 baby formulas and Korean rice wines marketed in Korea.

2.2 Liquid Chromatography-Tandem Mass Spectrometry (LC-MS/MS)

Developments have been reported for the analysis of patulin, ochratoxin A and citrinin combined, moniliformin, *Alternaria* mycotoxins, enniatins, beauvericin, and multiple mycotoxins.

Patulin. A method has been developed and validated for the determination of patulin in apple and apple juice, by liquid chromatography-tandem mass spectrometry; and, the method has been used for the surveillance of products in Brazilian supermarkets (Dias et al. 2019). Samples were extracted with ethyl acetate acidified with acetic acid, followed by the addition of sodium sulphate. The extracts were submitted to dispersive solid phase clean-up employing sodium sulphate and silica gel, prior to the determination of patulin. The method was validated by assessing linearity and linearity range, limits of detection and quantitation, matrix effect, precision (RSD%) and accuracy (recovery from samples spiked at 25, 50 and 100 µg.kg^{-1}). The recovery ranged from 76 to 84% for apple and from 102 to 108% for apple juice. In both cases the RSD was lower than 15%.

The matrix effect was also evaluated and was shown to be less intense for apple juice (-22%) than for apples (-42%). The limit of detection was 50 µg.L^{-1} and 12.5 µg.kg^{-1}, for apple juice and apples, respectively; whereas, the limit of quantitation was 42 µg.L^{-1} and 42 µg.kg^{-1}, for apple juice and apples, respectively. The validated method was applied to 62 samples of apple and apple juice collected from supermarkets of southern Brazil, during the years 2016 and 2017, and the patulin level in all samples was above the established Brazilian permitted limit.

Moniliformin. A rapid LC-MS/MS method has been reported for the determination of moniliformin (MON) in maize products from the Bavarian market (Barthel et al. 2017). The homogenized samples were extracted with acetonitrile/water (1/1; v/v). The centrifuged extracts were diluted with acetonitrile and subjected to LC-MS/MS without any further clean-up, using a hydrophobic interaction liquid chromatography column (HILIC). The LOD and LOQ achieved by this procedure were 2.6 and 8.8 µg.kg^{-1}, respectively.

A small number (39) of samples (popcorn, maize meal, and semolina) were collected in 2014 and 2015 at mills, cinemas, wholesale, and retail from the Bavarian market (Germany). The rate of contamination with MON was very high (97%) with levels ranging between the limit of detection and 847 $\mu g.kg^{-1}$.

Multiple Mycotoxins. During 2019, a variety of methods have been reported for the simultaneous determination of mycotoxins (single- or multi-class), in various commodities, by LC-MS/MS.

Recent examples are summarised in Table 1, together with methods involving LC-HRMS, LC-MS/MS and LC-HRMS in combination, LC-LC-MS/MS, UHPLC/UHPLC-MS/MS, UHPLC-Q-Orbitrap HRMS, UHPLC-ToF-MS, UFLC-MS/MS, and HPLC-ESI-TQ-MS/MS.

Some of these methods are further discussed below.

Alternaria toxins. An LC-MS/MS method has been developed for monitoring free and modified *Alternaria* toxins mycotoxins in food (Puntscher et al. 2018). The developed method included 17 toxins and their occurrence in tomato sauce, sunflower seed oil, and wheat flour.

The target analytes included *alternariol (AOH), AOH-3-glucoside, AOH-9-glucoside, AOH-3-sulfate, alternariol monomethyl ether (AME), AME-3-glucoside, AME-3-sulfate, altenuene, isoaltenuene, tenuazonic acid (TeA), tentoxin (TEN), altertoxin I and II, alterperylenol, stemphyltoxin III, altenusin, and altenuic acid III.*

Sample extracts were centrifuged and diluted by a factor of two resulting in an overall dilution of 1/10 (w/v). Importantly, since the LC-MS/MS method was thoroughly optimized and allowed for highly sensitive and selective quantitation, no derivatization or solid-phase extraction steps were required.

However, due to fine particles suspended in the wheat flour extracts, an additional filtration was performed prior to analysis, which ensured reproducible pressure conditions of the LC system, even after a high number of injections. Furthermore, the addition of n-hexane to the sunflower seed oil samples simplified their handling and led to enhanced extraction efficiency. The optimized method reportedly provided a time- and cost-effective sample preparation protocol and a chromatographic baseline separation. Since certified reference materials for the analysis of *Alternaria* toxins were not commercially available, the development and validation of the method was based on the fortification of blank matrix samples with reference standards.

Limits of detection (0.03–9 $ng.g^{-1}$) and quantitation (0.6–18 $ng.g^{-1}$) were reported, together with an intermediate (within laboratory) precision of 9 to 44%, and relative recovery values (calculated as the ratio of analyte concentration quantified using the matrix-matched calibration curve and the known spiking level in the fortified blank samples) of 75 to 100%.

However, stemphyltoxin III, AOH-3-sulfate, AME-3-sulfate, altenusin, and altenuic acid III showed recoveries in wheat flour below 70%, although the results were stable and reproducible.

A pilot study with samples from the Austrian retail market demonstrated that tomato sauces (n = 12) contained AOH, AME, TeA, and TEN in

Table 1: Chromatographic quantitation methods recently employed for the simultaneous determination of multiple mycotoxins

Toxins	Matrix	Quantitation Method	Reference
Aflatoxins B_1, B_2, G_1, G_2 and ochratoxin A; zearalenone, deoxynivalenol, fumonisins B1, B2, B3 and T-2 toxin	Functional and medicinal herbs	LC-MS/MS	Cho et al. 2019
Aflatoxin B_1, deoxynivalenol, fumonisins B1 and B2, ochratoxin A, T-2 and HT-2 toxin and zearalenone	Animal feed	LC-MS/MS	Jedziniak et al. 2019
Aflatoxins B_1, B_2, G_1, G_2, DON and 15-acetyl-DON, fumonisins B1, B2, B3; OTA, T2-toxin, HT-2-toxin and ZON	Maize and wheat	LC-MS/MS	Meyer et al. 2019
Ochratoxin A, aflatoxins B1, B2, G1, G2; deoxynivalenol, fumonisins B1, B2, and B3; T-2 toxin, HT-2 toxin, zearalenone	Edible oils	LC-MS/MS	Zhang and Xu 2019
Alternaria toxins	Food	LC-MS/MS	Puntscher et al. 2018
Aflatoxin B_1, B_2, G_1, G_2; ochratoxin A, fumonisins B1, B2, B3, deoxynivalenol; HT-2 toxin and T-2 toxin; zearalenone	Corn flour, peanut powder, brown rice flour, and corn with wheat flour	LC-MS/MS	Li et al. 2019
Ochratoxin A, NIV, DON, DON-3-glucoside, 3-acetyl-DON, 15-acetyl-DON; T-2 toxin, HT-2 toxin, ZON, enniatins A, A1, B, B1; ergotamine, ergosine, ergocornine, α-ergocryptine and ergocrystine	Plant based fish feed and fish	LC-HRMS	Johny et al. 2019
Aflatoxins, ochratoxin A, Fusarium mycotoxins, Alternaria mycotoxins and enniatins	Biological matrices from swine and broiler chickens	LC-MS/MS and LC-HRMS	Lauwers et al. 2019

Aflatoxins, ochratoxin A, deoxynivalenol and zearalenone, DON-3-glucoside, ZON-3-sulphate, tropane alkaloids (atropine/scopolamine), ergot alkaloids	Cereals	LC-LC-MS/MS	Kresse et al. 2019
Emerging fusariotoxins, enniatins and beauvericin	Animal feeds	LC-QTRAP/MS/MS	Tolosa et al. 2019
Aflatoxins, OTA, fumonisins, DON, 3-acetyl-DON, 15-acetyl-DON, T2-toxin, HT2-toxin, nivalenol, diacetoxyscirpenol, fusarenone X, beauvericin, enniatins, alternariol, alternariol monomethyl ether	Pet food	[a]UHPLC-Q-Orbitrap HRMS	Castaldo et al. 2019
Aflatoxins, zearalenone, α-zearalenol	Vegetable oil	UHPLC-MS/MS	Hidalgo-Ruiz et al. 2019
Aflatoxin B_1, aflatoxin B_2, aflatoxin G_1, aflatoxin G_2, aflatoxin M_1, aflatoxin M_2, ochratoxin A, ochratoxin B, zearalenone, zearalanone, α-zeralanol, β-zeralanol, α-zeralenol, β-zeralenol	Raw milk	UHPLC-MS/MS	Mao et al. 2018
Aflatoxins, ochratoxin A, zearalenone, T2 toxin, fumonisins	Maize	UHPLC-ToF-MS	Silva et al. 2019
Twelve parent mycotoxins and 14 mycotoxin metabolites	Paired plasma and urine	UPLC-MS/MS	Fan et al. 2019
Twenty-three mycotoxins with differing chemical characteristic including regulated, emerging and masked compounds	Beer	UPLC-MS/MS	González-Jartín et al. 2019
Pesticides and mycotoxins	Red wine	UPLC-MS/MS	Dias et al. 2019
Zearalenone and its metabolites	Human serum	UPLC-MS/MS	Sun et al. 2019

(Contd.)

Table 1: (*Contd.*)

Toxins	Matrix	Quantitation Method	Reference
Ochratoxin A and citrinin	Chicken and swine feed; food (cereal-based products, fruit, vegetable juices, nuts, seeds, herbs, spices, vegetarian and soy products, alcoholic beverages, baby food products and food supplements)	UPLC-MS/MS	Meerpoel et al. 2018
Emerging mycotoxins (beauvericin, enniatin A, enniatin A1, enniatin B, enniatin B1, alternariol, alternariol monomethyl ether, altenuene, tentoxin, and tenuazonic acid)	Cereals, legumes, potatoes, meats, eggs, aquatic foods, dairy products, vegetables, fruits, sugars, beverages, and alcoholic beverages	UPLC-MS/MS	Sun et al. 2019
Aflatoxins B1, B2, G1, and G2,; ochratoxin A, T-2 toxin, and HT-2 toxin	*Polygonum multiflorum*	UFLC-MS/MS	Huang et al. 2018
Ochratoxin A, ochratoxin B, fumonisins B2, B3, plus zearalenone, zearalenol, citrinin, nivalenol, alternariol, patulin, alternariol methyl ether, T-2 and HT-2 toxins, deoxynivalenol, tentoxin and aflatoxins B1, B2, G1 and G2	Lotus seeds	UFLC-MS/MS	Wei et al. 2019

[a]UHPLC is equivalent to UPLC, and recognises that UPLC is a registered trademark

concentrations up to 20, 4, 322, and 0.6 ng.g^{-1}, respectively, while sunflower seed oil (n = 7) and wheat flour samples (n = 9) were contaminated at comparatively lower levels. Interestingly and of relevance for risk assessment, AOH-9-glucoside, discovered for the first time in naturally contaminated food items, and AME-3-sulfate were found in concentrations similar to their parent toxins.

Importantly, the role of modified/masked toxins and the toxicological effects of combinations of these and parent toxins must be carefully considered. However, as acknowledged by the authors, it is important that given the very small number of samples involved in the study, the relevance of the reported concentrations of toxins reported in this study should be confirmed by a well-designed surveillance programme.

The identification and quantitation of the *emerging fusariotoxins, enniatins and beauvericin*, in animal feeds and their ingredients, has been performed using LC-QTRAP/MS/MS (Tolosa et al. 2019).

An analytical procedure for the simultaneous determination of emerging *Fusarium* mycotoxins was validated, involving an acetonitrile-based extraction followed by a QuEChERS based and dispersive solid-phase clean-up, prior to LC-MS/MS.

The analytical method was validated and the key performance characteristics fulfilled the criteria set out by the Commission Decision 2002/657/EC, and guidance document on the identification of mycotoxins in food and feed (SANTE/12089/2016).

All the studied mycotoxins (enniatins A, A1, B and B1; beauvericin) showed correlation coefficients (R^2) greater than 0.990 over the working range (0.1–200 µg.kg^{-1}), when matrix-matched calibration curves were employed. LODs and LOQs were in a range from 0.2 to 1.0 µg.kg,$^{-1}$ and from 1.0 to 5.0 µg.kg^{-1} for all toxins, respectively. Recoveries range from 86 to 98%, from 112 to 136%, and from 89 to 117%, at low, intermediate and high toxin spiking levels, respectively. Repeatability studies showed a relative standard deviation (RSD) lower than 5% at two spiking levels, whereas RSDs lower than 15% were obtained in the reproducibility studies. The validated method was successfully applied to raw materials (n = 39) and to complete feedstuffs (n = 48).

Both raw materials and complete feedstuffs showed mycotoxin contamination at incidences of 18% and 92%, respectively. Enniatin B was the most commonly found toxin, occurring at concentrations up to several thousand µg.kg^{-1}.

The co-occurrence of mycotoxins was also observed in 47% of samples, enniatin B and beauvericin being the most common combination. Although these results highlight the occurrence of emerging fusariotoxins in feedstuffs, the samples tested do not appear to have been collected from well-defined populations, or to have been representative of those populations.

Stable isotope dilution assay (SIDA) and matrix matched calibration methods are both employed to combat matrix interference during the determination of mycotoxins and other analytes.

The reported study compared the effectiveness of SIDA and matrix-matched calibration methods for the quantitation of mycotoxins by LC–MS/MS, in commercially available commodities and two reference materials (Li et al. 2019).

The selected mycotoxins were: *aflatoxin B_1, B_2, G_1, and G_2; ochratoxin A; fumonisins B1, B2, and B3; deoxynivalenol; HT-2 toxin; T-2 toxin and zearalenone;* and, the matrices were corn flour, peanut powder, brown rice flour, and corn with wheat flour (purchased from commercially available sources), together with two maize reference materials.

All samples were extracted with water–acetonitrile, followed by filtration and LC–MS/MS analysis.

Matrix-matched calibration has been widely used to compensate for matrix effects, and it may eliminate the need for costly internal standards (ISs). The preparation of matrix extracts for matrix-matched calibration standards conveniently uses the same process used to prepare the sample for analysis. However, it is cumbersome when many different matrices are being studied, as each one needs its own matrix-matched calibration curve.

SIDA is a technique that measures the relative response of a ^{13}C-labeled IS that has been added to the test samples prior to the extraction process. The labelled IS is able to compensate for issues in extraction efficiency as well as any ionization effects, caused by matrix components. It is a very effective procedure, but labelled ISs are expensive and are not available for a significant number of mycotoxins.

In the reported study, the SIDA calibration standards were produced by adding the ^{13}C-IS spiking solution to the mycotoxin calibration standards (although labelled ISs were not available for every mycotoxin used in the study). Using matrix-matched standards, a calibration curve with a 100-fold dynamic range and excellent linearity ($r^2 \geq 0.9949$) was constructed.

The one exception was for HT-2 toxin in peanut powder. A less-abundant transition was used for quantitation to avoid isobaric interferences with matrix components and, as a result, the LOQ for HT-2 was reported as 10.0 ng.g^{-1}, as opposed to 5.00 ng.g^{-1} for the other toxins.

Furthermore, the presence of incurred mycotoxins negatively affects the recoveries of matrix-matched calibration analyses. In the reported study, mycotoxins were observed in commercially purchased commodities, which made it impossible to accurately quantify those particular mycotoxins using a matrix-matched calibration technique. A wide dynamic range (1000-fold) and excellent linearity ($r^2 \geq 0.9996$) were observed using the SIDA calibration technique.

The accuracy and precision for the six mycotoxins that had corresponding ^{13}C-ISs was good across all four matrices. The accuracies ranged from 78.6 to 112% with RSDs $\leq 16\%$ across all four matrices. It was concluded that the employment of ^{13}C-IS is the preferred manner to compensate for matrix effects, notwithstanding the high cost involved and the unavailability of labelled ISs for a significant number of mycotoxins. Matrix matching was considered as providing a viable alternative of compensating for matrix effects, although it could not match the efficacy of labelled ISs. The difficulty of obtaining mycotoxin-free samples of specific commodities was also recognised.

A comprehensive analysis of multi-class mycotoxins (*aflatoxins B_1, B_2, G_1 and G_2, ochratoxin A, zearalenone, deoxynivalenol, fumonisins B1, B2, B3 and T-2 toxin*) in twenty different species of functional and medicinal herbs, using liquid chromatography–tandem mass spectrometry, has been described (Cho et al. 2019).

The complexity and variability of the matrices was best accommodated by IAC clean-up, supported by QuEChERS as required. The developed method was initially validated by evaluating specificity, linearity, precision, recovery, limit of quantification and uncertainty, and was then applied to the analysis of 100 real samples. Finally, the method was further verified by comparing the analysis results with those obtained by LC-Orbitrap HRMS. The method's performance was superior to other previously developed LC-MS/MS procedures for the analysis of herbal materials.

Mycotoxin derivatives in thermally processed food. The challenges faced during the determination of mycotoxin derivatives produced during thermal food processing has been discussed (Stadler et al. 2019). The authors conclude that an issue of paramount importance, for both targeted and untargeted approaches, is the availability of reference standards in order to produce high quality data on the identity, quantity and toxicity of degradation products.

An untargeted analysis of mycotoxin derivatives is reported, where LC-MS based ^{13}C labelling provides a full mass balance of deoxynivalenol, and its degradation products, formed during baking of crackers, biscuits and bread. Untargeted stable isotope assisted LC-HRMS was used to determine all extractable degradation products (Stadler et al. 2019).

2.3 Ultra-performance Liquid Chromatography-Tandem Mass Spectrometry (UPLC-MS/MS)

Recently reported developments in UPLC-MS/MS include the determination of *zearalenone*, a mixture of *ochratoxin A and citrinin, and emerging toxins*, in human serum, in a variety of foods and feeds, and in foods, respectively.

2.3.1 Zearalenone and its Metabolites

This study described an improved method for the high-throughput and sensitive determination of zearalenone and its five metabolites (*zearalanone, α-zearalenol, β-zearalenol, α-zearalanol and β-zearalanol*) in human serum (Sun et al. 2019).

Serum samples were analysed both before and after enzyme hydrolysis to assess the free and total amount of each compound by UPLC-MS/MS, in multi reaction monitoring (MRM) mode, following off-line 96-well µElution solid-phase extraction (SPE).

All the analytes were fully resolved on a C18 column within 6 min. It enabled multiple sample preparation at the same time, eliminating tedious evaporation and reconstitution steps. Ninety six samples (one 96-well plate) could be processed and analyzed within 24 h. using an isotope labelled internal standard (^{13}C-ZEN). High recoveries were achieved for all the compounds in the range 91.6 to 119.5%, with intra-day and inter-day relative standard deviations (RSDs) of less than 8%.

The LODs and LOQs were 0.02–0.06 ng.mL^{-1} (0.6–2 fmol) and 0.1–0.2 ng.mL^{-1} (3–6 fmol), respectively, demonstrating a significant enhancement in sensitivity compared to the existing methods.

The validated method was applied to the analysis of paired urine and serum samples collected from 125 healthy individuals in Henan Province, China. ZEN metabolites in human serum were significantly lower than those in urine. Only one serum sample was positive for ZEN after enzyme digestion, whereas at least one of the ZEN metabolites was detected in 75.2% of the paired urine samples.

Ochratoxin A and Citrinin. The development and validation of an UPLC–MS/MS method, for the simultaneous determination of citrinin (CIT) and ochratoxin A (OTA) in a variety of feed (chicken and swine) and foodstuffs (cereal-based products, fruit, vegetable juices, nuts, seeds, herbs, spices, vegetarian and soy products, alcoholic beverages, baby food products and food supplement) has been reported (Meerpoel et al. 2018).

A QuEChERS-based extraction method was employed, without any further clean-up step. The extracts were then 5-fold concentrated and the mycotoxin concentrations determined by UPLC–MS/MS system.

The developed method was validated according to the criteria described in Commission Regulation No. 401/2006/EC and Commission Decision No. 2002/657/EC. The limits of concentration of CIT and OTA were 3.9 and 5.6 µg.kg^{-1} respectively. The simultaneous occurrence of CIT and OTA was reported in more than 50% of 90 Belgian chicken and pig feed samples.

Emerging mycotoxins. Sensitive and reliable UPLC-MS/MS methods have been developed for the determination of emerging mycotoxins in China (Sun et al. 2019).

In this study, a sensitive and reliable method for the determination of 10 emerging mycotoxins (*beauvericin, enniatin A, enniatin A1, enniatin B, enniatin B1, alternariol, alternariol monomethyl ether, altenuene, tentoxin (TEN), and tenuazonic acid (TeA)*) in 12 different food matrices (cereals, legumes, potatoes, meats, eggs, aquatic foods, dairy products, vegetables, fruits, sugars, beverages, and alcohol beverages) was developed and validated.

A simple extraction, followed by a one-step sample clean-up using a HLB solid phase extraction (SPE) column was sufficient for all 12 food matrices, prior to analysis with ultra-high performance liquid chromatography coupled to tandem mass spectrometry (UHPLC-MS/MS). Multi reaction monitoring (MRM) mode was used in mass detection including soft ionization, trapping precursor ions, fragmentation to product ions, and quantitation. Isotope internal standards ^{13}C-TeA, TEN-d$_3$, and ^{13}C-AFB$_2$ were used for accurate quantitation. ^{13}C-AFB1 was employed as the reference IS for the quantitation of the enniatins.

The developed method was validated for the 10 mycotoxins in all selected matrices. The limits of detection and quantitation were 0.0004 to 0.3 ng.mL^{-1} and 0.002 to 0.9 ng.mL^{-1}, respectively.

The recoveries of 10 mycotoxins in fortified samples were 60.6 to 164%, including very low spiking levels, in all 12 food matrices. The relative standard deviations (RSDs) were less than 12%.

The developed method was applied to the analysis of 60 samples collected during the 6th China Total Diet Study, and enabled the inclusion of emerging mycotoxins for the first time.

The simultaneous determination of pesticides and mycotoxins (*aflatoxins B$_1$, B$_2$, G$_1$, G$_2$; ochratoxin A, deoxynivalenol, diacetoxyscirpenol, T-2 toxin, HT2-toxin, zearalenone, fumonisins B1 and B2*) in red wine by direct injection and UPLC-MS/MS analysis has been successfully performed (Dias et al. 2019).

Direct injection involved: (a) centrifugation of the wine sample, (b) dilution of an aliquot of the supernatant with methanol:water (1:1) and (c) syringe filtration of the diluted sample, and injection of a portion into the UPLC-MS/MS system. Satisfactory results were reported for the accuracy and repeatability for the developed method, together with a low limit of quantitation for all the mycotoxins, apart from aflatoxins B$_2$ and G$_2$.

2.4 Ultra-fast Liquid Chromatography (UFLC)

Multiple mycotoxins. An economic analytical procedure has been developed, for the determination of seven mycotoxins of different chemical classes (aflatoxin B$_1$, B$_2$, G$_1$, and G$_2$, ochratoxin A, T-2 toxin, and HT-2 toxin) in dried roots of *Polygonum multiflorum* (Huang et al. 2018).

The procedure employed a simple clean-up step together with ultra-fast liquid chromatography (UFLC), with a mobile phase composed of acetonitrile and water containing 0.1% formic acid, in combination with hybrid triple quadrupole/linear ion trap mass spectrometry.

The mycotoxins were completely extracted using a modified QuEChERs method without additional clean-up steps. The types of extraction solvents and adsorbents for the extraction procedure were optimized to achieve high recoveries and reduce co-extractives in the final extracts.

Given the high level of matrix interference observed, matrix-matched calibration curves were developed to ensure accurate mass spectrometric quantitation. Excellent linearity was reported for the seven mycotoxins, together with limits of detection and quantitation in the range of 0.031–2.5 µg.kg^{-1} and 0.078–6.25 µg.kg^{-1}, respectively. Recoveries were between 74.3 and 119.8%, with relative standard deviations below 7.43%.

The developed method was successfully applied to 24 batches of *P. multiflorum* roots; and, six samples were contaminated with aflatoxins B$_1$, B$_2$, G$_1$, or ochratoxin A.

A similar approach has been successfully developed and used for the determination of 19 mycotoxins mycotoxins (ochratoxin A, ochratoxin B, fumonisins B2, B3, plus zearalenone, zearalenol, citrinin, nivalenol, alternariol, patulin, alternariol methyl ether, T-2 and HT-2 toxins, deoxynivalenol, tentoxin and aflatoxins B_1, B_2, G_1 and G_2) in starch-rich lotus seeds (Wei et al. 2019). A modified ultrasonication-assisted extraction with QuEChERS purification, was employed in combination with ultrafast liquid chromatography tandem mass spectrometry (UFLC-MS/MS). The validated method exhibited satisfactory linearity, together with acceptable limits of detection (LODs of 0.1–15.0 µg.kg^{-1}), good precision (RSDs <13.0%) and satisfactory recoveries (79.4-131.6%).

A matrix effect was particularly associated with aflatoxins B_1 and B_2, deoxynivalenol and T-2 toxin. Matrix-matched curve-based quantitation showed that 26 (57.8%) out of 45 lotus seed samples were contaminated with one or more mycotoxins, ochratoxin A, aflatoxins B_1, B_2 and citrinin being the most prevalent.

2.5 UPLC Time of Flight Mass Spectrometry (UPLC-ToF-MS)

A UHPLC-ToF-MS method has been developed for the determination of nine mycotoxins (aflatoxins B_1, B_2, G_1, G_2), ochratoxin A, zearalenone, T2-toxin and fumonisins B1 and B2) in maize (Silva et al. 2019). The method included a two-step extraction with aqueous acetonitrile. Good results were reported for repeatability (RSDr ≤ 15.4%), reproducibility (RSDR ≤ 15.9%) and recovery (77.8–110.4%), except for aflatoxin G_2 at 2 µg/kg where a recovery of 73.4% was achieved. The reported performance criteria met those imposed by Commission Regulation (EC) no. 401/2006.

3. Immunochemical Methods

A wide variety of immunochemical methods have been reported during 2019, some of which are shown in Table 2. Selected methods are further described below.

3.1 Aflatoxins

A novel nanobody and mimotope based immunoassay has been reported for the rapid analysis of aflatoxin B_1. (Where a mimotope is a macromolecule, often a short-chain peptide, which can take on the structural appearance of the epitope of a harmful biological substance.) (Zhao et al. 2019).

This publication is believed to be the first reported development of an immunoassay which employs both an aflatoxin recombinant antibody and its mimotope peptide.

A rapid magnetic beads-based directed competitive ELISA (MB-dcELISA) was developed utilizing Nb28 and its mimotope ME17. Using aflatoxin B_1 as a model system, mimotopes of an aflatoxin nanobody Nb28 were screened by phage display.

A nanobody with a known amino acid sequence is easier and less expensive to prepare than a monoclonal antibody. Similarly, mimotopes

Table 2: Immunochemical methods

Toxins	Matrix	Immunochemical Method	Reference
Aflatoxins	Foods, including maize, chili powder, chocolate, green coffee beans, and roasted coffee beans	IAC plus ELISA	Hojo et al. 2019
	Spiked corn, rice, peanut, feedstuff, corn germ oil and peanut oil samples	Rapid magnetic beads-based directed competitive ELISA (MB-dcELISA)	Zhao et al. 2019
Sterigmatocystin	Wheat and corn flours	Indirect competitive ELISA	Singh et al. 2019
Ochratoxin A	Coffee	Fluorescence Resonance Energy Transfer on Lateral Flow Immunoassay (FRET-LFI)	Oh et al. 2019
	Cereals	Anti-Idiotypic VHH antibody and toxin-free enzyme Immunoassay	Zhang et al. 2019
Fumonisin B1	Wheat	Comparison of mimotope-based immunoassays. (Magnetic bead-based ELISA, MB-ELISA)	Peltomaa et al. 2019
Cyclopiazonic acid	Maize and cheese	Imaging surface plasmon resonance (iSPR)	Hossain et al. 2019
Aflatoxin B_1 and fumonisins	Wheat and wheat products	Colour encoded lateral flow immunoassay	Di Nardo et al. 2019
Multiple toxins (including including aflatoxin M_1, ochratoxin A, deoxynivalenol, fumonisin B1)	Milk	Quadplex flow cytometric immunoassay (FCIA)	Qu et al. 2019
Aflatoxins and pesticides	Maize	Multi-TRFICA (time-resolved fluorescence) paper sensor	Tang et al. 2019

should be cost-effectively produced by commercial peptide synthesis, and obviate the production of toxic aflatoxin conjugates. The detection limit of the MB-dcELISA was 0.13 ng.mL[1], with a linear range of 0.24–2.21 ng.mL[-1]. Further validation study indicated good recovery (84.2–116.2%) with low coefficient of variation (2.2–15.9%) in spiked corn, rice, peanut, feedstuff, corn germ oil and peanut oil samples. The developed immunoassay based on a nanobody and mimotope provides a new strategy for the monitoring of aflatoxin B_1 and other toxic small molecular weight compounds.

3.2 Ochratoxin A

A rapid and simple method for the determination of ochratoxin has been reported, using fluorescence resonance energy transfer on a lateral flow immunoassay (FRET-LFI) (Oh et al. 2019).

The authors posit that a label-free, direct, and non-competitive FRET-LFI system detects mycotoxins more rapidly and easily than other currently available methods, and with a high level of selectivity.

Importantly, the developed method is also reported as effectively counteracting any significant matrix effect in the sample (e.g. as in coffee) by utilising the capillary action occurring in a nitrocellulose membrane. The combination of antibody and antigen produced emission bands at both 330 nm and 440 nm upon excitation at 280 nm, because of the FRET between the two moieties. The determination of ochratoxin A in coffee was successfully completed in 30 min. with a limit of detection of 0.88 ng.mL^{-1}.

The development of an anti-idiotypic VHH antibody and toxin-free, indirect competitive ELISA has also been reported for the determination of ochratoxin A (OTA) in cereals (Zhang et al. 2019).

Here, the anti-idiotypic nanobody VHH 2-24 was successfully produced by immunizing alpaca with the anti-OTA mAb 1H2 antibody, and the resultant nanobody was then used as a surrogate OTA standard, in a toxin-free ELISA. When applied to corn, rice and wheat samples the average recoveries of the immunoassay ranged from 81.8 to 105.0%.

Excellent agreement was observed when the performance of the developed ELISA was compared with that of HPLC. A good linear relationship was demonstrated (y = 1.0383x – 0.3787, R2 = 0.994).

The toxin-free assay developed in this study clearly demonstrated the potential use of anti-idiotypic VHH as a surrogate calibration standard for mycotoxins, and other highly toxic small molecules, in the determination of low levels of toxins in agricultural products.

3.3 Fumonisin B1

A dodecapeptide (A2), which functioned as a mimotope for fumonisin B1 (FB1), has previously been selected by phage display. The development and comparison of immunoassays for FB1 have now been reported which employed the mimotope (A2), a fluorescent recombinant fusion protein and a synthetic peptide with a biotin linker (Peltomaa et al. 2019). The immunoassays employed magnetic beads and enzymatic detection. The highest sensitivity was obtained with a magnetic bead–based assay using the synthetic peptide and enzymatic detection, which provided a detection limit of 0.029 ng.mL^{-1}.

3.4 Cyclopiazonic Acid

A sensitive and rapid immunoassay utilizing imaging surface plasmon resonance (iSPCR) for the determination of cyclopiazonic acid (CPA) in maize and Camembert cheese has been reported. The method combined an indirect competitive immunoassay, and signal amplification based upon a secondary antibody (Ab2) conjugated with gold nanoparticles (Hossain et al. 2019).

Matrix-matched calibration curves were used to optimise the performance of the method. Recoveries, at two spiking levels in maize and cheese, were 89 to 126%, with relative standard deviations of repeatability (RSDr) of less than 16%. The limits of detection were 17 and 6 µg.kg^{-1} in maize and cheese, respectively. To separate the CPA-contaminated samples from uncontaminated samples, a cut-off validation level of 40 µg.kg^{-1} was introduced. The assay was applied to samples of naturally contaminated maize and was compared with competitive inhibition enzyme-linked immunosorbent assay (CI-ELISA). This is the first report of the determination of CPA using an immuno-biosensor iSPR format.

3.5 Multiple Mycotoxins

Multiplex flow cytometric immunoassays (MFCI) have been developed for the high-throughput screening of multiple mycotoxin residues in milk (Qu et al. 2019).

Microsphere-based immunoassays involving flow cytometry have recently gained popularity for use in protein detection and infectious disease diagnosis due to their simple assay format and capacity for multiplexed analysis.

In FCIs, although the procedure followed is similar to a conventional ELISA method, the microspheres are used as immune reaction carriers, as opposed to the surface of the microtitre plate in conventional ELISAs. In the reported study, mycotoxin ovalbumin conjugates were produced by a variety of methods, and the resultant antigens were immobilised onto superparamagnetic microspheres via carbodiimide bonding.

MFCIs utilize a flow-based dual layer system which employs encoded microspheres, which are analysed in a flow-stream by two lasers. The first laser (635 nm) identifies each microsphere-associated analyte by fluorescence, and the second green laser (532 nm) measures the reporter molecules attached to the analytes.

In this study, four mycotoxins, including aflatoxin M$_1$ (AFM1), ochratoxin A (OTA), deoxynivalenol (DON), and fumonisin B1 (FB1), were selected as model analytes, and a quadplex flow cytometric immunoassay (FCIA) was developed for detecting multiple mycotoxin residues in milk.

The performance of the optimized assay was characterised, by limits of detection of 0.045 µg.L^{-1} for AFM1, 0.94 µg.L^{-1} for OTA, 7.48 µg.L^{-1} for DON, and 2.45 µg.L^{-1} for FB1. Furthermore, the recoveries of the target mycotoxins from spiked milk were 76–95%, with a relative standard deviation of less than 13.4%. The developed FCIA did not require multiple washing steps, and exhibited lower detection limits for multiple mycotoxin determination than traditional ELISA formats.

4. Sensors

Many countries that are adversely affected by the mycotoxin contamination of their crops, do not have resources for the purchase and implementation of expensive instrumentation, including LC-MS/MS equipment.

Consequently, there is an urgent need for robust, user-friendly, affordable and accurate instrumentation (usually referred to as 'sensors') that may be used by producers, processors and traders in emerging economies.

Existing sensors have been compared using the so-called 'Mycotoxin Testing Paradigm', which addresses the integration of three parameters associated with sensor technology (method performance, speed of analysis, and cost) (Renaud et al. 2019).

It is especially important that those countries wishing to exploit the opportunities offered by lucrative but highly regulated markets in the developed world, have access to analytical methods that will ensure that their exports do not exceed the mycotoxin regulatory levels of their importers, and that they meet their customers' requirements.

Work has continued on the further development of a wide variety of sensors (Table 3), including:

- Aptamer-based sensors
- Immunosensors
- Biosensors
- Biomimetic sensors
- Surface Plasmon Resonance sensors
  ○ Spectroscopy-based sensors: Fluorescence, Near Infrared, Raman
- Other sensors

Current trends in rapid tests for mycotoxins have been reviewed (Nolan et al. 2019). Currently available testing kits are described, including enzyme-linked immunosorbent assays, membrane-based immunoassays, fluorescence polarisation immunoassays and fluorometric assays, together with recent advances in biosensor technology. The biosensors discussed include surface plasmon resonance, mass sensitive sensors, quartz crystal microbalance and film bulk acoustic resonators. The importance of multiplexing and rapid on-site analysis capabilities are highlighted.

Specific antigen-based and emerging detection technologies for mycotoxins has also been reviewed (Rahman et al. 2019), together with the use of thin-film devices for mycotoxin detection in foods (Santos et al. 2019).

Table 3: Sensors employed for the determination of mycotoxins

Toxins	Type of Sensor	Reference
Reviews	Specific antigen-based and emerging detection technologies of mycotoxins	Rahman et al. 2019
	Thin Films Sensor Devices for Mycotoxins Detection in Foods: Applications and Challenges	Santos et al. 2019
	Specific antigen-based and emerging detection technologies of mycotoxins	Rahman et al. 2019
	Thin Films Sensor Devices for Mycotoxins Detection in Foods: Applications and Challenges	Santos et al. 2019
Aptamer-based		
Aflatoxin B$_1$	A rapid fluorometric method using a thioflavin T-based aptasensor	Li et al. 2019
	GO-amplified fluorescence polarization assay with a low dosage aptamer probe	Ye et al. 2019
	Electrochemical sensor based on smart host-guest recognition of β-cyclodextrin polymer.	Wu et al. 2019
Ochratoxin A	Electrochemiluminescence aptasensor based on resonance energy transfer system between CdTe quantum dots and cyanine dyes	Gao et al. 2019
	Electrochemiluminescence aptasensor based on a nicking endonuclease-powered DNA walking machine	Wei et al. 2019
	Fluorometric assay based on photoinduced electron transfer	Zhao et al. 2019
	Label-free sensor based on Aptamer/NH2 Janus particles for ultrasensitive electrochemical detection	Yang et al. 2019
Patulin	Impedimetric sensor for Label Free Detection	Khan et al. 2019
Multiple toxins	Label-free fluorescent sensing of via aggregation-induced emission dye	Zhu et al. 2019
	Systematic truncating of aptamers to create high-performance graphene oxide (GO)-based aptasensors for the multiplex detection of mycotoxins.	Wang et al. 2019
Immunosensors		
Ochratoxin A	Ultrasensitive and green electrochemical sensor based on phage displayed mimotope peptide	Hou et al. 2019

(Contd.)

Table 3: *(Contd.)*

Toxins	Type of Sensor	Reference
	Development of a new format of competitive immunochromatographic assay using secondary antibody–europium nanoparticle conjugates	Majdinasab et al., 2019
Deoxynivalenol	Electrochemical nanoprobe-based sensor for deoxynivalenol mycotoxin residues analysis in wheat samples.	Valera et al. 2019
Zearalenone	A smartphone-based quantitative detection device integrated with latex microsphere immunochromatography for on-site detection	Li et al. 2019
Biosensors		
Sterigmatocystin	Development of a third-generation biosensor	Díaz Nieto et al. 2019
Mycotoxins (Review)	Novel optical biosensing technologies	Nabok et al. 2019
Biometic sensors		
Aflatoxin B$_1$	Smartphone-based biomimetic sensor using molecularly imprinted polymer membranes	Sergeyeva et al. 2019
Surface Plasmon Resonance		
Fusarium mycotoxins	Gold nanoparticle-enhanced multiplexed imaging surface plasmon resonance (iSPR) detection	Hossain and Maragos 2018
Fluorescence Spectroscopy		
Aflatoxin B$_1$	Application of multiplexing fibre optic laser induced fluorescence spectroscopy for detection of aflatoxin B1 contaminated pistachio kernels	Wu and Xu 2019
Multiple mycotoxins	Evaluation of an alternative spectroscopic approach (ToxiMet System) for aflatoxin analysis: Comparative analysis of food and feed samples with UPLC–MS/MS.	Campbell et al. 2017
Near-Infrared Spectroscopy		
Review	Updated overview of infrared spectroscopy methods for detecting mycotoxins on cereals (corn, wheat, and barley).	Levasseur-Garcia 2018

Aflatoxin B$_1$	Shortwave Near-Infrared Spectroscopy for rapid detection of Aflatoxin B1 contamination in polished rice.	Putthang et al. 2019
Aflatoxigenic moulds and aflatoxin	Rapid and non-destructive method for simultaneous determination of aflatoxigenic moulds and aflatoxin contamination on corn kernels using visible–near-infrared spectroscopy	Tao et al. 2019
Ochratoxin A	Rapid screening of ochratoxin A in wheat by infrared spectroscopy	De Girolamo et al. 2019
Deoxynivalenol	Fourier transform near-infrared and mid-infrared spectroscopy as efficient tools for rapid screening of deoxynivalenol contamination in wheat bran	De Girolamo et al. 2018
Raman Spectroscopy		
Zearalenone	Quantitative assessment of zearalenone in maize using multivariate algorithms coupled to Raman spectroscopy	Guo et al. 2019
Alternariol	Simple approach for the rapid detection of alternariol in pear fruit by surface-enhanced Raman scattering with pyridine-modified silver nanoparticles	Pan et al. 2018
Multiple mycotoxins	Cauliflower-Inspired 3D SERS Substrate for Multiple Mycotoxins Detection	Li et al. 2019
Other Sensors		
Aflatoxin B$_1$	Rapid detection of aflatoxin B$_1$ by dummy template molecularly imprinted polymer capped CdTe quantum dots.	Guo et al. 2019
Ochratoxin A	DNA template-mediated click chemistry-based portable signal-on sensor	Qiu et al. 2019
	An iridium (III) complex/G-quadruplex based on long-lifetime luminescent.	Zhang et al. 2019
Multiple mycotoxins (Fumonisin B1 and ochratoxin A)	Magnetic Reduced Graphene Oxide/Nickel/Platinum Nanoparticles Micromotors (Molinero-Fernández, Jodra, Moreno-Guzmán, López, and Escarpa, 2018)	Molinero-Fernández et al. 2018
DON	Miniature mass spectrometer	Renaud et al. 2019

The former discusses a variety of immunoassays including ELISA, flow injection immunoassay, chemiluminescence immunoassay, lateral flow immunoassay, and flow-through immunoassay.

Emerging biosensor technologies are also addressed including fibre-optic immunosensor, surface plasmon resonance biosensor, near-IR spectroscopy, fluorescence imaging spectroscopy, fluorescence polarization and bendable paper-based lab-on-a-chip.

The latter review describes how thin film biosensor devices are currently one of the most active research areas. A thin film generally refers to a layer of material that ranges from a few nm to several μm in thickness, which can be employed as the analyte immobilization surface in biosensors.

Their main reported benefits, compared to traditional analytical methods used for mycotoxins analysis, are fast analysis time and rapid detection, high sensitivity, easy sample preparation, reusability and low cost. The capability of nanomaterials (e.g. gold, silver, metal oxides and quantum dots) to enhance the detection capability of biosensors is also discussed.

A variety of thin film biosensors are described, based upon the type of transducer employed, including: optical (colorimetry, fluorescence, luminescence, interferometry, spectroscopy, surface plasmon resonance, and total internal reflection ellipsometry, TIRE); electrochemical (amperometry, conductimetry, potentiometry, and voltammetry); and piezoelectric (quartz crystal microbalance, (QCM). Selected examples of the ongoing development, and employment of sensors for the determination of specific toxins, are summarised below:

4.1 Aptamer-based Sensors

An aptamer is a specific nucleotide sequence which has a high affinity and strong selectivity towards specific analytes. Aptamer-based sensors have been developed for the determination of aflatoxin B_1, ochratoxin A, patulin and multiple mycotoxins.

A *graphene oxide, GO-amplified fluorescence polarization (FP) assay* has been reported for the low cost, high-sensitivity determination of aflatoxin B_1, employing a low dosage aptamer probe (Ye et al. 2019).

Aptamers labelled with fluorescein amidite (FAM) were adsorbed on the surface of GO through π–π stacking and electrostatic interaction, thus forming aptamer/GO macromolecular complexes. Under these conditions, the local rotation of fluorophores was limited and the system had a high FP value.

When the biosensor was presented with an aflatoxin contaminated sample, the aptamers formed an aptamer/AFB1 complex and were dissociated from the GO surface. The resultant changes in the molecular weights of aptamers produced a change in the FP value.

The results showed that when only 10 nM of aptamer was used, the changes in FP and the AFB1 concentration showed a good linear relationship within the range of 0.05 to 5 nM of AFB1, and the lowest detection limit (LOD) was 0.05 nM. The recoveries of rice sample extract ranged from 89.2% to 112%.

A simple and sensitive *electrochemiluminescence (ECL)* aptasensor has been developed for the determination of ochratoxin A (OTA), based on electrochemiluminescence resonance energy transfer (ECL-RET) and a nicking endonuclease-powered DNA walking machine (Wei et al. 2019).

ECL is a highly sensitive method which combines the benefits offered by electrochemistry and chemiluminescence; and, cadmium sulphide semiconductor quantum dots (CdS QDs) demonstrate excellent quantum yield and good stability.

Cy5 is a bright, far-red-fluorescent synthetic Cyanine dye, whilst a DNA "walking machine" is comprised of a so-called DNA "walker strand" which facilitates signal amplification.

The developed sensor was based on the ECL-RET between Cy5 and CdS QDS, and the apparent ability of a DNA walking machine to improve the sensitivity of the aptasensor.

The aptasensor produced results for the determination of OTA which were linear over the range 0.05 nM to 5 nM, with a detection limit of 0.012 nM (S/N = 3). It also showed an excellent selectivity for OTA over other mycotoxins. The aptasensor has been successfully applied to the determination of OTA in wine and beer.

4.2 Immunosensors

A smartphone, integrated with latex microsphere immunochromatography (SIAP), produced using cost-effective 3D printing technology, has been reported for the on-site determination of zearalenone (ZEN) in cereals and feed (Li et al. 2019). The cut-off values of latex microsphere immunochromatography for ZEN in cereals and feed were 2.5 and 3.0 µg.kg^{-1}, respectively. The SIAP detection system could quantitatively detect ZEN in cereals and feed with detection limits of 0.08 and 0.18 µg.kg^{-1} respectively.

The reported recovery rates for corn, and feed samples were between 92.0-105.0, and 86.0-107.5%, respectively, with the coefficient of variation ranging from 2.7 to 8.9, and 7.8 to 10.9%, respectively.

4.3 Biosensors

Progress made on the recent development of *novel optical biosensing techniques* for the determination of mycotoxins has been reported (Nabok et al. 2019), involving direct assays with either specific antibodies or aptamers.

The main development employed *total internal reflection ellipsometry (TIRE) combined with localized surface plasmon resonance (LSPR) transducers,* based on gold nano-structures.

The combination of TIRE and LSPR offers superior refractive index sensitivity as compared to traditional UV–vis absorption spectroscopy. The short evanescent field decay period associated with LSPR was overcome using small-size bio-receptors, such as half-antibodies and aptamers.

A limit of detection of mycotoxins (aflatoxins B$_1$ and M$_1$, OTA and ZEN) at a concentration 0.01 µg.kg^{-1} was reportedly achieved, when this method was applied to the analysis of agriculture products, food and feed. A LOD at the sub-ppt level was reported for another recently developed optical biosensor, where an optical planar waveguide operated as a polarization interferometer (PI). The reported methods are portable, highly sensitive, and simple to use biosensors, which are suitable for the point-of-need detection of mycotoxins.

5. Spectroscopy Sensors

5.1 Fluorescence Spectroscopy

The performance of the *ToxiMet System* has been evaluated (Campbell et al. 2017), which included its comparison with UPLC-MS/MS. The analytical sequence for the ToxiMet System is composed of four main steps: sample extraction; sample clean-up (using the ToxiSep cartridge); analyte immobilisation (on a ToxiTrace cartridge); followed by simultaneous determination of the individual, immobilised aflatoxins B_1, B_2, G_1 and G_2 by the ToxiQuant instrument, without the need for derivatisation or chromatographic separation.

Chemometric analysis of the fluorescence spectrum, produced by the UV irradiation of the ToxiTrace cartridge, was automatically performed by the ToxiQuant instrument, and afforded the concentration levels of each of the four aflatoxins. A major advantage of the ToxiMet System is that the handling of toxic analytical standards is not required, since the ToxiMet System is supplied with in-built calibrations for a variety of applications.

During the described evaluation study, it appears that the sample extract was not sufficiently diluted when a high concentration (>30 µg.kg^{-1}) of aflatoxin was initially recorded for a sample of peanut butter (the correct procedure was confirmed by the supplier of the ToxiMet System after the completion of the evaluation study); and, that the ToxiMet System was employed for the aflatoxin analysis of dried distillers grains, for which the instrument had not been calibrated.

The comparison between the ToxiMet System and UPLC-MS/MS was potentially compromised by the nature of the sample preparation methods employed, where the laboratory samples were 25g and 1g, respectively. However, when the ToxiMet System and UPLC-MS/MS were applied to the determination of the aflatoxins in rice and maize, promising results were achieved.

The R^2 correlations of the concentrations determined for B_1, B_2, G_1, G_2 and total aflatoxins were 0.90, 0.45, 0.90, 0.54 and 0.92. For B_2 and G_2 the correlation was poor as the number of positive samples were low and bordering on the limit of quantification for both methods.

Similarly, when both methods were used to determine the aflatoxin content of samples of maize, the R^2 correlations of the concentrations determined for B_1, B_2, G_1, G_2 and total aflatoxins were 0.93, 0.51, 0.97, 0.77 and 0.94.

As was observed for the analysis of the rice samples, for B_2 and G_2 the correlation was poor as the number of positive samples were low and bordering on the limit of quantification for both methods.

The reported LOQs for the ToxiMet System were 0.67 (B$_1$), 0.4 (B$_2$), 0.72 (G$_1$) and 0.14 (G$_2$) µg.kg^{-1} and for UPLC-MS/MS, 0.25 (B$_1$), 0.25 (B$_2$), 0.25 (G$_1$) and 0.25 (G$_2$) µg.kg^{-1}.

It was concluded that the ToxiMet system provides an alternative, highly accurate analysis method for the determination of the aflatoxins, which is significantly more affordable, faster and user-friendly than UPLC-MS/MS analysis.

5.2 Infra-red Spectroscopy

Fourier transform near-infrared and mid-infrared spectroscopy have been studied, as potentially efficient tools for the rapid screening of deoxynivalenol in wheat bran (De Girolamo et al. 2018).

Although bran fractions have numerous reported health benefits, cereal bran is usually the part of the grain with the highest concentration of DON. The applicability of (a) Fourier transform near-infrared (FTNIR), or (b) mid-infrared (FTMIR) spectroscopy, or (c) their combination, to the rapid screening for DON in wheat bran was reported, with reference to the maximum permitted level recognised by the EU (750 µg.kg^{-1}). Partial least squares-discriminant analysis (PLS-DA) and principal component-linear discriminant analysis (PC-LDA) were employed as classification techniques, using a cut-off value of 400 µg.kg^{-1} DON.

The overall discrimination rates, for compliant and non-compliant samples, were from 87 to 91% for FTNIR, and from 86 to 87% for the FTMIR spectral range. FTNIR spectroscopy, in combination with the PC-LDA classification model, performed best with no false compliant samples and 18% false noncompliant samples. The combination of the two spectral ranges did not provide a substantial improvement in classification accuracy.

5.3 Raman Spectroscopy

The quantitative determination of zearalenone in maize has been performed using *multivariate algorithms coupled to Raman spectroscopy* (Guo et al. 2019).

Maize samples were finely ground using an ultra-centrifugal mill, fitted with a 0.5 mm trapezoidal sieve. 40 g samples of the ground material were taken for Raman spectroscopy analysis.

A representative Raman spectrum was obtained, using a SmartRaman spectrometer, by collecting four 40 g samples from the ground material and producing an average spectrum, in an attempt to avoid the adverse effect of a heterogeneous mixture. Regression models were built using a selection of variable selection algorithms, and the synergy interval PLS (SiPLS)-ant colony optimisation (ACO) algorithm was superior to others in terms of predictive power.

The best model based on SiPLS-ACO achieved coefficients of correlation (Rp) of 0.9260 and Root Mean Square Error of Prediction (RMSEP) of 87.9132 µg.kg^{-1} in the prediction set, respectively.

Raman spectroscopy combined with multivariate calibration showed promising results for the rapid screening of large numbers of zearalenone maize contaminations in bulk quantities, without the need for sample-extraction steps.

When employing the type of sample preparation procedure described above, it is essential that the 40 g laboratory sample selected for analysis is representative of the original aggregate sample collected from the batch under investigation. It is not clear from the publication how this criterion was met. Furthermore, it is equally important that the 40 g subsample is homogenised as thoroughly as possible before Raman analysis.

5.4   Other Sensors

A novel and sensitive fluorescence sensor for the rapid and highly selective determination of aflatoxin B_1 (AFB1) has been produced, *by combining molecular imprinting techniques with quantum dot technology* (Guo et al. 2019).

Molecularly imprinted polymers coated CdTe quantum dots (MIP@CdTe QDs) were prepared using 5,7-dimethoxycoumarin as a dummy template. 3-Aminopropyltriethoxysilane was employed as the functional monomer, and tetraethyl orthosilicate was used as the cross-linking agent. The MIP@CdTe QD composites were characterized by Fourier transform infrared spectroscopy, transmission electron microscopy, and fluorescence spectroscopy.

When applied to the determination of AFB1 in spiked samples of coix and wheat, the relative fluorescence intensity of the MIP@CdTe QDs showed adequate linearity with AFB1 concentration over the range from 80 to 400 ng.g^{-1}, and a limit of detection of 4 ng.g^{-1}. The method was successfully applied to the quantitative determination of AFB1 in real samples. The recoveries at different spiking levels ranged from 99.20 to 101.78%, and compared favourably with those measured by UHPLC-MS/MS.

The reported LOD was significantly lower than the regulatory level of AFB1 in wheat in all countries apart from the EU, where the permitted levels are ≥ 20 µg.kg^{-1} (total aflatoxins) and 2.0 µg.kg^{-1}, respectively.

A low cost, simple, portable *G-quadruplex-based platform* has been developed for the time-resolved monitoring of ochratoxin A (OTA) (Zhang et al. 2019). The developed method exploits the structural transformation of an OTA aptamer and a G-quadruplex-selective Ir(III) probe.

In aqueous solution, in the absence of OTA, the aptamer exists in a random coil conformation. The probe does not bind to the aptamer, leading to a weak orange luminescent emission. The addition of OTA induces the aptamer to undergo a structural transition to a G-quadruplex, which binds strongly to the Ir(III) probe, producing an intense yellow luminescent signal.

The platform demonstrated a high sensitivity and selectivity for OTA, with a detection limit of 40 nM via steady-state emission spectroscopy. Importantly, the sensitivity of the platform was enhanced by the employment of time-resolved emission spectroscopy (TRES), which afforded a detection limit of 10.8 nM. OTA was successfully determined in herbal plant extracts.

Preliminary studies employing a *miniature mass spectrometer, fitted with a quadrupole or ion trap* for mass detection, have been reported (Renaud et al. 2019).

DON was determined in wheat without any sample preparation apart from grinding. The ground material was directly introduced by a sample loop and thermally desorbed, a rugged ion trap allowing for the detection of the $[M+H]^+$ ion at m/z 297. An analysis time of <1 min. is reported.

However, as with other procedures which involve the use of very small analytical samples, the need to maintain the representative nature of the sample, from the original batch to the final laboratory sample, needs to be addressed.

6. Biomonitoring

The ability to measure the exposure of Man and animal to mycotoxins is still compromised by an incomplete understanding of the metabolites of mycotoxins in differing species, and a lack of analytical methods with the capability of detecting key metabolic markers of exposure.

The challenges associated with the identification and validation of biomarkers of exposure to mycotoxins have been *reviewed* (Vidal et al. 2018).

Recent advances in *high-resolution mass spectrometry*, along with new analytical techniques (post-acquisition data-mining) are expected to improve the quality and output of the biomarker identification process. The review provides a comprehensive overview of reported *in vitro* and *in vivo* mycotoxin metabolism studies in relation to biomarkers of exposure for deoxynivalenol, nivalenol, fusarenon-X, T-2 toxin, diacetoxyscirpenol, ochratoxin A, citrinin, fumonisins, zearalenone, aflatoxins, and sterigmatocystin.

Recent progress has been made in addressing the challenges (Table 4) described in the review, and some of these developments are described below.

A highly sensitive urinary assay, for the most relevant regulated and emerging mycotoxins, has been developed using *SIDA-based UHPLC-MS/MS* and ^{13}C-labelled or deuterated internal standards (Šarkanj et al. 2018). Challenging matrix effects were effectively addressed using the SIDA method.

The combination of enzymatic pre-treatment, solid phase extraction (HLB) and UHPLC-MS/MS provided a highly sensitive procedure. The method was validated in-house and used to re-assess mycotoxin exposure in urine samples obtained from Nigerian children, adolescents and adults, naturally exposed through their regular diet. The high sensitivity of the developed method enabled the detection of biomarkers in all samples.

Zearalenone was the most frequently detected contaminant (82%) together with ochratoxin A (76%), aflatoxin M_1 (73%) and fumonisin B1 (71%) which were determined in a large share of urine samples.

Overall, 57% of 120 urine samples were contaminated with both aflatoxin M_1 and fumonisin B1, and other co-exposures were frequently observed.

These results clearly demonstrated the advanced performance of the developed method to assess lowest background exposures (pg.mL^{-1} level)

Table 4: Biomonitoring of human and animal exposure to mycotoxins

Biomonitoring		
Review	Mycotoxin biomarkers of exposure: A comprehensive review	Vidal et al. 2018
	Biomonitoring of mycotoxin exposure using urinary biomarker approaches: a review	Tuanny et al. 2019
Deoxynivalenol and derivatives	Identification and determination of deoxynivalenol and deepoxy-deoxynivalenol in pig colostrum and serum using liquid chromatography in combination with high resolution mass spectrometry	Stastny et al. 2019
	Humans significantly metabolize and excrete the mycotoxin deoxynivalenol and its modified form deoxynivalenol-3-glucoside within 24 hours	Vidal et al. 2018
Zearalenone and derivatives	Biomonitoring of zearalenone and its main metabolites in urines of Bangladeshi adults	Ali and Degen 2019
Nephrotoxic mycotoxins	Analyses of biomarkers of exposure to nephrotoxic mycotoxins in a cohort of patients with renal tumours	Malir et al. 2019
Multiple mycotoxins	Ultra-sensitive, stable isotope assisted quantification of multiple urinary mycotoxin exposure biomarkers	Šarkanj et al. 2018
Efficacy of mycotoxin detoxification	Biomarkers for Exposure as a Tool for Efficacy Testing of a Mycotoxin Detoxifier in Broiler Chickens and Pigs.	Lauwers et al. 2019

using a single, highly robust assay that enable the systematic investigation of low concentrations of urinary mycotoxins.

7. Conclusions

Needless to say, sampling continues to be the most important step within the whole analytical sequence, since a sample which fails to represent the batch (lot) from which it was collected will, ultimately, lead to a meaningless result.

A relatively limited amount of progress on the development of sampling plans has been reported during the period under review.

Interestingly, an optimization model was developed which enabled the design of cost-effective sampling and analysis methods for the determination of mycotoxins in cereals.

A significant amount of recent work was reported on developments in sample extraction and sample clean-up procedures.

Ultrasonic extraction methods were featured, together with dispersive liquid-liquid microextraction, stir bar sorptive extraction and automated on-line extraction, together with a variety of other methods.

Apart from reports on the continued employment of immunochemical, MIP and QuEChERs-based clean-up procedures, developments on the employment of other techniques were discussed, including strongly hydrophilic reverse phase solid phase extraction, dispersive solid phase clean-up and multi-walled carbon nanotubes.

A very considerable body of work on new developments in the analysis of mycotoxins has been generated during 2019.

Work has continued on the employment of LC-MS/MS procedures, including UPLC-MS/MS and LC-HRMS.

There has also been considerable progress with immunochemical methods above and beyond ELISA and lateral flow devices.

However, most interest appears to have centred around the development of a wide variety of sensors, in order to address the need for simpler, robust, affordable and portable analysis methods which are accessible to those with limited resources and/or technical skills throughout the food and feed supply chains, including within the emerging economies.

Interest has continued in the development of methods for the biomonitoring of human and animal exposure to mycotoxins, using a variety of biomarkers in urine, plasma and other materials.

Clearly, the continued introduction and employment of innovative techniques, throughout the food and feed supply chain, will lead to further significant advances in mycotoxin analysis, especially in the development of effective procedures which are accessible to those with a limited supply of skills and resources.

References

Ali, N. and Degen, G.H. 2019. Biomonitoring of zearalenone and its main metabolites in urines of Bangladeshi adults. Food and Chemical Toxicology 130: 276–283. DOI: 10.1016/j.fct.2019.05.036

Alsharif, A.M.A., Choo, Y.M., Tan, G.H. and Abdulráuf, L.B. 2019. Determination of Mycotoxins Using Hollow Fiber Dispersive Liquid–Liquid–Microextraction (HF-DLLME) Prior to High-Performance Liquid Chromatography – Tandem Mass Spectrometry (HPLC-MS/MS). Analytical Letters 52(12): 1976–1990. DOI: 10.1080/00032719.2019.1587766

Amirkhizi, B., Nemati, M., Arefhosseini, S.R. and Shahraki, S.H. 2017. Application of the Ultrasonic-Assisted Extraction and Dispersive Liquid–Liquid Microextraction for the Analysis of AFB1 in Egg. Food Analytical Methods 11(3): 913–920. DOI: 10.1007/s12161-017-1052-6

Appell, M., Evans, K.O., Jackson, M.A. and Compton, D.L. 2018. Determination of ochratoxin A in grape juice and wine using nanosponge solid phase extraction clean-up and liquid chromatography with fluorescence detection. Journal of Liquid Chromatography and Related Technologies 41(15-16): 949–954. DOI: 10.1080/10826076.2018.1544148

Arce-López, B., Lizarraga, E., Flores-Flores, M., Irigoyen, Á. and González-Peñas, E. 2020. Development and validation of a methodology based on Captiva EMR-lipid clean-up and LC-MS/MS analysis for the simultaneous determination of mycotoxins in human plasma. Talanta 206: 120193. DOI: 10.1016/j.talanta.2019.120193

Barthel, J., Rapp, M., Holtmannspötter, H. and Gottschalk, C. 2017. A rapid LC-MS/MS method for the determination of moniliformin and occurrence of this mycotoxin in maize products from the Bavarian market. Mycotoxin Research 34(1): 9–13. DOI: 10.1007/s12550-017-0293-y

Caballero-Casero, N., García-Fonseca, S. and Rubio, S. 2018. Restricted access supramolecular solvents for the simultaneous extraction and cleanup of ochratoxin A in spices subjected to EU regulation. Food Control 88: 33–39. DOI: 10.1016/j.foodcont.2018.01.003

Caballo, C., Sicilia, M.D. and Rubio, S. 2017. Supramolecular solvents for green chemistry. (pp. 111–137) *In:* The Application of Green Solvents in Separation Processes. Elsevier.

Campbell, K., Ferreira Cavalcante, A.L., Galvin-King, P., Oplatowska-Stachowiak, M., Brabet, C., Metayer, I. et al. 2017. Evaluation of an alternative spectroscopic approach for aflatoxin analysis: Comparative analysis of food and feed samples with UPLC–MS/MS. Sensors and Actuators B: Chemical 239: 1087–1097. DOI: 10.1016/j.snb.2016.08.115

Campone, L., Piccinelli, A.L., Celano, R., Pagano, I., Russo, M. and Rastrelli, L. 2018. Rapid and automated on-line solid phase extraction HPLC-MS/MS with peak focusing for the determination of ochratoxin A in wine samples. Food Chemistry 244: 128–135. DOI: 10.1016/j.foodchem.2017.10.023

Castaldo, L., Graziani, G., Gaspari, A., Izzo, L., Tolosa, J., Rodríguez-Carrasco, Y. et al. 2019. Target analysis and retrospective screening of multiple mycotoxins in pet food using UHPLC-Q-Orbitrap HRMS. Toxins 11(8): 434. DOI: 10.3390/toxins11080434

Chen, Y., Ding, X., Zhu, D., Lin, X. and Xie, Z. 2019. Preparation and evaluation of highly hydrophilic aptamer-based hybrid affinity monolith for on-column specific discrimination of ochratoxin A. Talanta 200: 193–202. DOI: 10.1016/j.talanta.2019.03.053

Cho, H.D., Suh, J.H., Feng, S., Eom, T., Kim, Junghyun, Hyun, S.M. et al. 2019. Comprehensive analysis of multi-class mycotoxins in twenty different species of functional and medicinal herbs using liquid chromatography-tandem mass spectrometry. Food Control 96: 517–526. DOI: 10.1016/j.foodcont.2018.10.007

Davis, J.P., Jackson, M.D., Leek, J.M. and Samadpour, M. 2018. Sample preparation and analytical considerations for the US aflatoxin sampling program for shelled peanuts. Peanut Science 45(1): 19-31. DOI: 10.3146/ps17–12.1

De Girolamo, A., Cervellieri, S., Cortese, M., Porricelli, A.C.R., Pascale, M., Longobardi, F. et al. 2018. Fourier transform near-infrared and mid-infrared spectroscopy as efficient tools for rapid screening of deoxynivalenol contamination in wheat bran. Journal of the Science of Food and Agriculture 99(4): 1946–1953. DOI: 10.1002/jsfa.9392

De Girolamo, A., Cervellieri, S., Cortese, M., Porricelli, A.C.R., Pascale, M., Longobardi, F. et al. 2019. Rapid screening of ochratoxin A in wheat by infrared spectroscopy. Food Chemistry 282: 95–100. DOI: 10.1016/j.foodchem.2019.01.008

Di Nardo, F., Alladio, E., Baggiani, C., Cavalera, S., Giovannoli, C., Spano, G. et al. 2019. Colour-encoded lateral flow immunoassay for the simultaneous detection of aflatoxin B1 and type-B fumonisins in a single test line. Talanta 192: 288–294. DOI: 10.1016/j.talanta.2018.09.037

Dias, J.V., da Silva, R.C., Pizzutti, I.R., dos Santos, I.D., Dassi, M. and Cardoso, C.D. 2019. Patulin in apple and apple juice: Method development, validation by liquid chromatography-tandem mass spectrometry and survey in Brazilian south supermarkets. Journal of Food Composition and Analysis 82: 103242. DOI: 10.1016/j.jfca.2019.103242

Dias, J., Nunes, M.D.G.P., Pizzutti, I.R., Reichert, B., Jung, A.A. and Cardoso, C.D. 2019. Simultaneous determination of pesticides and mycotoxins in wine by direct injection and liquid chromatography-tandem mass spectrometry analysis. Food Chemistry 293: 83–91. DOI: 10.1016/j.foodchem.2019.04.088

Díaz Nieto, C.H., Granero, A.M., Garcia, D., Nesci, A., Barros, G., Zon, M. A. et al. 2019. Development of a third-generation biosensor to determine sterigmatocystin mycotoxin: An early warning system to detect aflatoxin B1. Talanta 194: 253–258. DOI: 10.1016/j.talanta.2018.10.032

Dong, H., Xian, Y., Xiao, K., Wu, Y., Zhu, L. and He, J. 2019. Development and comparison of single-step solid phase extraction and QuEChERS clean-up for the analysis of 7 mycotoxins in fruits and vegetables during storage by UHPLC-MS/MS. Food Chemistry 274: 471–479. DOI: 10.1016/j.foodchem.2018.09.035

El Saadani, M., Durand, N., Sorli, B., Guibert, B., Alter, P. and Montet D. 2020. Aptamer assisted ultrafiltration cleanup with high performance liquid chromatography-fluorescence detector for the determination of OTA in green coffee. Food Chemistry 310: 125851. DOI: 10.1016/j.foodchem.2019.125851

Ertekin, Ö., Kaymak, T., Pirinçci, Ş.Ş., Akçael, E. and Öztürk, S. 2019. Aflatoxin-specific monoclonal antibody selection for immunoaffinity column development. BioTechniques 66(6): 261–268. DOI: 10.2144/btn-2018-0143

Fan, K., Xu, J., Jiang, K., Liu, X., Meng, J., Di Mavungu, J.D. et al. 2019. Determination of multiple mycotoxins in paired plasma and urine samples to assess human exposure in Nanjing, China. Environmental Pollution 248: 865–873. DOI: 10.1016/j.envpol.2019.02.091

Feizy, J., Jahani, M. and Beigbabaei, A. 2019. Graphene adsorbent-based solid-phase extraction for aflatoxins clean-up in food samples. Chromatographia 82(6): 917–926. DOI: 10.1007/s10337-019-03725-w

Focker, M., Fels-Klerx, V.D.H.J. and Oude Lansink, A.G.J.M. 2019. Cost-effective sampling and analysis for mycotoxins in a cereal batch. Risk Analysis 39(4): 926–939. DOI: urn:nbn:nl:ui:32-542151

Gao, J., Chen, Z., Mao, L., Zhang, W., Wen, W., Zhang, X. et al. 2019. Electrochemiluminescent aptasensor based on resonance energy transfer system between CdTe quantum dots and cyanine dyes for the sensitive detection of Ochratoxin A. Talanta 199: 178–183. DOI: 10.1016/j.talanta.2019.02.044

González-Jartín, J.M., Alfonso, A., Rodríguez, I., Sainz, M.J., Vieytes, M.R. and Botana, L.M. 2019. A QuEChERS based extraction procedure coupled to UPLC-MS/MS detection for mycotoxins analysis in beer. Food Chemistry 275: 703–710. DOI: 10.1016/j.foodchem.2018.09.162

Gu, S., Wang, X., Yang, L. and Chen, J. 2019. Development and validation of a bullfrog-immunoaffinity column clean-up for citrinin determination in red yeast rice. Process Biochemistry 78: 200–206. DOI: 10.1016/j.procbio.2019.01.021

Guo, P., Yang, W., Hu, H., Wang, Y. and Li, P. 2019. Rapid detection of aflatoxin B1 by dummy template molecularly imprinted polymer capped CdTe quantum dots. Analytical and Bioanalytical Chemistry 411(12): 2607–2617. DOI: 10.1007/s00216-019-01708-2

Guo, Z., Wang, M., Wu, J., Tao, F., Chen, Q., Wang, Q. et al. 2019. Quantitative assessment of zearalenone in maize using multivariate algorithms coupled to Raman spectroscopy. Food Chemistry 286: 282–288. DOI: 10.1016/j.foodchem.2019.02.020

Hidalgo-Ruiz, J.L., Romero-González, R., Martínez Vidal, J.L. and Garrido Frenich, A. 2019. A rapid method for the determination of mycotoxins in edible vegetable oils by ultra-high performance liquid chromatography-tandem mass spectrometry. Food Chemistry 288: 22–28. DOI: 10.1016/j.foodchem.2019.03.003

Hojo, E.R.I., Matsuura, N., Kamiya, K., Yonekita, T., Morishita, N., Murakami, H. et al. 2019. Development of a rapid and versatile method of enzyme-linked immunoassay combined with immunoaffinity column for aflatoxin analysis. Journal of Food Protection 82(9): 1472–1478. DOI: 10.4315/0362-028x.jfp-19-036

Hossain, M.Z. and Maragos, C.M. 2018. Gold nanoparticle-enhanced multiplexed imaging surface plasmon resonance (iSPR) detection of Fusarium mycotoxins in wheat. Biosensors and Bioelectronics 101: 245–252. DOI: 10.1016/j.bios.2017.10.033

Hossain, Z., Busman, M. and Maragos, C.M. 2019. Immunoassay utilizing imaging surface plasmon resonance for the detection of cyclopiazonic acid (CPA) in maize and cheese. Analytical and Bioanalytical Chemistry 411(16): 3543–3552. DOI: 10.1007/s00216-019-01835-w

Hou, S.L., Ma, Z.E., Meng, H., Xu, Y. and He, Q.H. 2019. Ultrasensitive and green electrochemical immunosensor for mycotoxin ochratoxin A based on phage displayed mimotope peptide. Talanta 194: 919–924. DOI: 10.1016/j.talanta.2018.10.081

Hu, M., Huang, P., Suo, L. and Wu, F. 2018. Polydopamine-based molecularly imprinting polymers on magnetic nanoparticles for recognition and enrichment of ochratoxins prior to their determination by HPLC. Microchimica Acta 185(6). DOI: 10.1007/s00604-018-2826-2

Huang, P., Liu, Q., Wang, J., Ma, Z., Lu, J. and Kong, W. 2018. Development of an economic ultrafast liquid chromatography with tandem mass spectrometry method for trace analysis of multiclass mycotoxins in Polygonum multiflorum. Journal of Separation Science. DOI: 10.1002/jssc.201800602

Huang, Z., He, J., Li, Y., Wu, C., You, L., Wei, H. et al. 2019. Preparation of dummy molecularly imprinted polymers for extraction of Zearalenone in grain samples. Journal of Chromatography A 1602: 11–18. DOI: 10.1016/j.chroma.2019.05.022

Jedziniak, P., Panasiuk, Ł., Pietruszka, K. and Posyniak, A. 2019. Multiple mycotoxins analysis in animal feed with LC-MS/MS: Comparison of extract dilution and immunoaffinity clean-up. Journal of Separation Science 42(6): 1240–1247. DOI: 10.1002/jssc.201801113

Jiang, D., Wei, D., Wang, L., Ma, S., Du, Y. and Wang, M. 2018. Multiwalled carbon nanotube for one-step cleanup of 21 mycotoxins in corn and wheat prior to ultraperformance liquid chromatography-tandem mass spectrometry analysis. Toxins 10(10): 409. DOI: 10.3390/toxins10100409

Johny, A., Kruse Fæste, C., Bogevik, A.S., Marit Berge, G., Fernandes, J.M.O. and Ivanova L. 2019. Development and validation of a liquid chromatography high-resolution mass spectrometry method for the simultaneous determination of mycotoxins and phytoestrogens in plant-based fish feed and exposed fish. Toxins 11(4): 222. DOI: 10.3390/toxins11040222

Khan, R., Ben Aissa, S., Sherazi, T., Catanante, G., Hayat, A. and Marty, J. 2019. Development of an impedimetric aptasensor for label free detection of patulin in apple juice. Molecules 24(6): 1017. DOI: 10.3390/molecules24061017

Khodadadi, M., Malekpour, A. and Mehrgardi, M.A. 2018. Aptamer functionalized magnetic nanoparticles for effective extraction of ultratrace amounts of aflatoxin M1 prior its determination by HPLC. Journal of Chromatography A 1564: 85–93. DOI: 10.1016/j.chroma.2018.06.022

Kresse, M., Drinda, H., Romanotto, A. and Speer, K. 2019. Simultaneous determination of pesticides, mycotoxins, and metabolites as well as other contaminants in cereals by LC-LC-MS/MS. Journal of Chromatography B 1117: 86–102. DOI: 10.1016/j.jchromb.2019.04.013

Kumphanda, J., Matumba, L., Whitaker, T.B., Kasapila, W. and Sandahl, J. 2019. Maize meal slurry mixing: an economical recipe for precise aflatoxin quantitation. World Mycotoxin Journal 12(3): 203–212. DOI: 10.3920/wmj2018.2415

Lattanzio, V.M.T., Guarducci, N., Powers, S., Ciasca, B., Pascale, M. and von Holst, C. 2018. Validation of a lateral flow immunoassay for the rapid determination of aflatoxins in maize by solvent free extraction. Analytical Methods 10(1): 123–130. DOI: 10.1039/c7ay02249b

Lauwers, M., Croubels, S., Letor, B., Gougoulias, C. and Devreese, M. 2019. Biomarkers for Exposure as a tool for efficacy testing of a mycotoxin detoxifier in broiler chickens and pigs. Toxins 11(4): 187. DOI: 10.3390/toxins11040187

Lauwers, M., De Baere, S., Letor, B., Rychlik, M., Croubels, S. and Devreese, M. 2019. Multi LC-MS/MS and LC-HRMS methods for determination of 24 mycotoxins including major phase I and II biomarker metabolites in biological matrices from pigs and broiler chickens. Toxins 11(3): 171. DOI: 10.3390/toxins11030171

Levasseur-Garcia, C. 2018. Updated overview of infrared spectroscopy methods for detecting mycotoxins on cereals (corn, wheat, and barley). Toxins 10(1): 38. DOI: 10.3390/toxins10010038

Li, D., Steimling, J.A., Konschnik, J.D., Grossman, S.L. and Kahler, T.W. 2019. Quantitation of mycotoxins in four food matrices comparing stable isotope dilution assay (SIDA) with matrix-matched calibration methods by LC–MS/MS. Journal of AOAC International 102(6): 1673–1680. DOI: 10.5740/jaoacint.19-0028

Li, J., Yan, H., Tan, X., Lu, Z. and Han, H. 2019. Cauliflower-inspired 3D SERS substrate for multiple mycotoxins detection. Analytical Chemistry 91(6): 3885–3892. DOI: 10.1021/acs.analchem.8b04622

Li, X., Li, H., Ma, W., Guo, Z. and Zhang, Q. 2018. Determination of patulin in apple juice by single-drop liquid-liquid-liquid microextraction coupled with liquid chromatography-mass spectrometry. Food Chem 257: 1–6. DOI: 10.1016/j.foodchem.2018.02.077

Li, X., Wang, J., Yi, C., Jiang, L., Wu, J., Chen, X. et al. 2019. A smartphone-based quantitative detection device integrated with latex microsphere immunochromatography for on-site detection of zearalenone in cereals and feed. Sensors and Actuators B: Chemical 290: 170–179. DOI: 10.1016/j.snb.2019.03.108

Li, Y., Wang, J., Zhang, B., He, Y., Wang, J. and Wang, S. 2019. A rapid fluorometric method for determination of aflatoxin B1 in plant-derived food by using a thioflavin T-based aptasensor. Microchimica Acta 186(4). DOI: 10.1007/s00604-019-3325-9

Liu, X., Ying, G., Sun, C., Yang, M., Zhang, L., Zhang, S. et al. 2018. Development of an Ultrasonication-assisted extraction based HPLC with a fluorescence method for sensitive determination of aflatoxins in highly acidic Hibiscus sabdariffa. Frontiers in Pharmacology 9. DOI: 10.3389/fphar.2018.00284

Ma, H., Ran, C., Li, M., Gao, J., Wang, X., Zhang, L. et al. 2018. Graphene oxide-coated stir bar sorptive extraction of trace aflatoxins from soy milk followed by high performance liquid chromatography-laser-induced fluorescence detection. Food Additives and Contaminants, Part A 35(4): 773–782. DOI: 10.1080/19440049.2017.1416182

Mahfuz, M., Gazi, M.A., Hossain, M., Islam, M.R., Fahim, S.M. and Ahmed, T. 2018. General and advanced methods for the detection and measurement of aflatoxins and aflatoxin metabolites: A review. Toxin Reviews 39(2): 123–137. DOI: 10.1080/15569543.2018.1514638

Majdinasab, M., Zareian, M., Zhang, Q. and Li, P. 2019. Development of a new format of competitive immunochromatographic assay using secondary antibody–europium nanoparticle conjugates for ultrasensitive and quantitative determination of ochratoxin A. Food Chemistry 275: 721–729. DOI: 10.1016/j.foodchem.2018.09.112

Malir, F., Louda, M., Ostry, V., Toman, J., Ali, N., Grosse, Y. et al. 2019. Analyses of biomarkers of exposure to nephrotoxic mycotoxins in a cohort of patients with renal tumours. Mycotoxin Research 35(4): 391–403. DOI: 10.1007/s12550-019-00365-9

Mao, J., Zheng, N., Wen, F., Guo, L., Fu, C., Ouyang, H. et al. 2018. Multi-mycotoxins analysis in raw milk by ultra high performance liquid chromatography coupled to quadrupole orbitrap mass spectrometry. Food Control 84: 305–311. DOI: 10.1016/j.foodcont.2017.08.009

Meerpoel, C., Vidal, A., di Mavungu, J.D., Huybrechts, B., Tangni, E.K., Devreese, M. et al. 2018. Development and validation of an LC-MS/MS method for the simultaneous determination of citrinin and ochratoxin a in a variety of feed and foodstuffs. Journal of Chromatography A 1580: 100–109. DOI: 10.1016/j.chroma.2018.10.039

Meyer, H., Skhosana, Z.D., Motlanthe, M., Louw, W. and Rohwer, E. 2019. Long term monitoring (2014–2018) of multi-mycotoxins in South African commercial maize and wheat with a locally developed and validated LC-MS/MS method. Toxins 11(5): 271. DOI: 10.3390/toxins11050271

Miró-Abella, E., Herrero, P., Canela, N., Arola, L., Ras, R., Borrull, F. et al. 2019. Optimised extraction methods for the determination of trichothecenes in rat faeces followed by liquid chromatography-tandem mass spectrometry. Journal of Chromatography B 1105: 47–53. DOI: 10.1016/j.jchromb.2018.12.013

Miró-Abella, E., Herrero, P., Canela, N., Arola, L., Ras, R., Fontanals, N. et al. 2017. Determination of trichothecenes in cereal matrices using subcritical water extraction followed by solid-phase extraction and liquid chromatography-tandem mass spectrometry. Food Analytical Methods 11(4): 1113–1121. DOI: 10.1007/s12161-017-1089-6

Molinero-Fernández, Á., Jodra, A., Moreno-Guzmán, M., López, M.Á. and Escarpa, A. 2018. Magnetic reduced graphene oxide/nickel/platinum nanoparticles micromotors for mycotoxin analysis. Chemistry – A European Journal 24(28): 7172–7176. DOI: 10.1002/chem.201706095

Nabok, A., Al-Rubaye, A.G., Al-Jawdah, A.M., Tsargorodska, A., Marty, J.L., Catanante, G. et al. 2019. Novel optical biosensing technologies for detection of mycotoxins. Optics and Laser Technology 109: 212–221. DOI: 10.1016/j.optlastec.2018.07.076

Nolan, P., Auer, S., Spehar, A., Elliott, C.T. and Campbell, K. 2019. Current trends in rapid tests for mycotoxins. Food Additives and Contaminants Part A 36(5): 800–814. DOI: 10.1080/19440049.2019.1595171

Oh, H.K., Joung, H.A., Jung, M., Lee, H. and Kim, M.G. 2019. Rapid and simple detection of ochratoxin A using fluorescence resonance energy transfer on lateral flow immunoassay (FRET-LFI). Toxins 11(5): 292. DOI: 10.3390/toxins11050292

Pan, T.T., Sun, D.W., Pu, H. and Wei, Q. 2018. Simple approach for the rapid detection of alternariol in pear fruit by surface-enhanced raman scattering with pyridine-modified silver nanoparticles. Journal of Agricultural and Food Chemistry 66(9): 2180–2187. DOI: 10.1021/acs.jafc.7b05664

Peltomaa, R., Agudo-Maestro, I., Más, V., Barderas, R., Benito-Peña, E. and Moreno-Bondi, M.C. 2019. Development and comparison of mimotope-based immunoassays for the analysis of fumonisin B1. Analytical and Bioanalytical Chemistry 411(26): 6801–6811. DOI: 10.1007/s00216-019-02068-7

Puntscher, H., Kütt, M.-L., Skrinjar, P., Mikula, H., Podlech, J., Fröhlich, J. et al. 2018. Tracking emerging mycotoxins in food: Development of an LC-MS/MS method for free and modified alternaria toxins. Analytical and Bioanalytical Chemistry 410(18): 4481–4494. DOI: 10.1007/s00216-018-1105-8

Putthang, R., Sirisomboon, P. and Sirisomboon, C.D. 2019. Shortwave near-infrared spectroscopy for rapid detection of aflatoxin B1 contamination in polished rice. Journal of Food Protection 82(5): 796–803. DOI: 10.4315/0362-028x.jfp-18-318

Qiu, S., Yuan, L., Wei, Y., Zhang, D., Chen, Q., Lin, Z. et al. 2019. DNA template-mediated click chemistry-based portable signal-on sensor for ochratoxin A detection. Food Chemistry 297: 124929. DOI: 10.1016/j.foodchem.2019.05.203

Qu, J., Xie, H., Zhang, S., Luo, P., Guo, P., Chen, X. et al. 2019. Multiplex flow cytometric immunoassays for high-throughput screening of multiple mycotoxin residues in milk. Food Analytical Methods 12(4): 877–886. DOI: 10.1007/s12161-018-01412-4

Rahman, H.U., Yue, X., Yu, Q., Xie, H., Zhang, W., Zhang, Q. et al. Specific antigen-based and emerging detection technologies of mycotoxins. Journal of the Science of Food and Agriculture 99(11): 4869–4877. DOI: 10.1002/jsfa.9686

Reinholds, I., Pugajeva, I., Bogdanova, E., Jaunbergs, J. and Bartkevics, V. 2019. Recent applications of carbonaceous nanosorbents for the analysis of mycotoxins in food by liquid chromatography: A short review. World Mycotoxin Journal 12(1): 31–43. DOI: 10.3920/wmj2018.2339

Renaud, J.B., Miller, J.D. and Sumarah, M.W. 2019. Mycotoxin testing paradigm: Challenges and opportunities for the future. Journal of AOAC International 102(6): 1681–1688. DOI: 10.5740/jaoacint.19-0046

Rico-Yuste, A., Walravens, J., Urraca, J.L., Abou-Hany, R.A.G., Descalzo, A.B., Orellana, G. et al. 2018. Analysis of alternariol and alternariol monomethyl ether in foodstuffs by molecularly imprinted solid-phase extraction and ultra-high-performance liquid chromatography tandem mass spectrometry. Food Chemistry 243: 357–364. DOI: 10.1016/j.foodchem.2017.09.125

Rui, C., He, J., Li, Y., Liang, Y., You, L., He, L. et al. 2019. Selective extraction and enrichment of aflatoxins from food samples by mesoporous silica FDU-12 supported aflatoxins imprinted polymers based on surface molecularly imprinting technique. Talanta 201: 342–349. DOI: 10.1016/j.talanta.2019.04.019

Santos, A.O., Vaz, A., Rodrigues, P., Veloso, A.C.A., Venâncio, A. and Peres, A.M. 2019. Thin films sensor devices for mycotoxins detection in foods: Applications and challenges. Chemosensors 7(1): 3. DOI: 10.3390/chemosensors7010003

Šarkanj, B., Ezekiel, C.N., Turner, P.C., Abia, W.A., Rychlik, M., Krska, R. et al. 2018. Ultra-sensitive, stable isotope assisted quantification of multiple urinary mycotoxin exposure biomarkers. Analytica Chimica Acta 1019: 84–92. DOI: 10.1016/j.aca.2018.02.036

Sergeyeva, T., Yarynka, D., Piletska, E., Linnik, R., Zaporozhets, O., Brovko, O. et al. 2019. Development of a smartphone-based biomimetic sensor for aflatoxin B1 detection using molecularly imprinted polymer membranes. Talanta 201: 204–210. DOI: 10.1016/j.talanta.2019.04.016

Silva, A.S., Brites, C., Pouca, A.V., Barbosa, J. and Freitas, A. 2019. UHPLC-ToF-MS method for determination of multi-mycotoxins in maize: Development and validation. Current Research in Food Science 1: 1–7. DOI: 10.1016/j.crfs.2019.07.001

Singh, G., Velasquez, L., Huet, A.C., Delahaut, P., Gillard, N. and Koerner, T. 2019. Development of a polyclonal antibody-based indirect competitive ELISA for determination of sterigmatocystin in wheat and corn flours. Food Additives and Contaminants Part A 36(2): 327–335. DOI: 10.1080/19440049.2019.1567943

Soares, R.R.G., Ricelli, A., Fanelli, C., Caputo, D., de Cesare, G., Chu, V. et al. 2018. Advances, challenges and opportunities for point-of-need screening of mycotoxins in foods and feeds. The Analyst 143(5): 1015–1035. DOI: 10.1039/c7an01762f

Stadler, D., Berthiller, F., Suman, M., Schuhmacher, R. and Krska, R. 2019. Novel analytical methods to study the fate of mycotoxins during thermal food processing. Analytical and Bioanalytical Chemistry 412(1): 9–16. DOI: 10.1007/s00216-019-02101-9

Stadler, D., Lambertini, F., Bueschl, C., Wiesenberger, G., Hametner, C., Schwartz-Zimmermann, H. et al. 2019. Untargeted LC-MS based 13C labelling provides a full mass balance of deoxynivalenol and its degradation products formed during baking of crackers, biscuits and bread. Food Chemistry 279: 303–311.

Stastny, K., Stepanova, H., Hlavova, K. and Faldyna, M. 2019. Identification and determination of deoxynivalenol (DON) and deepoxy-deoxynivalenol (DOM-

1) in pig colostrum and serum using liquid chromatography in combination with high resolution mass spectrometry (LC-MS/MS (HR)). Journal of Chromatography B 121735. DOI: 10.1016/j.jchromb.2019.121735

Sun, D., Li, C., Zhou, S., Zhao, Y., Gong, Y., Gong, Z. et al. 2019. Determination of trace zearalenone and its metabolites in human serum by a high-throughput UPLC-MS/MS analysis. Applied Sciences 9(4): 741. DOI: 10.3390/app9040741

Sun, D., Qiu, N., Zhou, S., Lyu, B., Zhang, S., Li, J. et al. 2019. Development of sensitive and reliable UPLC-MS/MS methods for food analysis of emerging mycotoxins in China total diet study. Toxins 11(3): 166. DOI: 10.3390/toxins11030166

Tang, X., Zhang, Q., Zhang, Z., Ding, X., Jiang, J., Zhang, W. et al. 2019. Rapid, on-site and quantitative paper-based immunoassay platform for concurrent determination of pesticide residues and mycotoxins. Analytica Chimica Acta 1078: 142–150. DOI: 10.1016/j.aca.2019.06.015

Tao, F., Yao, H., Zhu, F., Hruska, Z., Liu, Y., Rajasekaran, K. et al. 2019. A rapid and nondestructive method for simultaneous determination of aflatoxigenic fungus and aflatoxin contamination on corn kernels. Journal of Agricultural and Food Chemistry 67(18): 5230–5239. DOI: 10.1021/acs.jafc.9b01044

Tittlemier, S.A., Cramer, B., Dall'Asta, C., Iha, M.H., Lattanzio, V.M.T., Malone, R.J. et al. 2019. Developments in mycotoxin analysis: An update for 2017–2018. World Mycotoxin Journal 12(1): 3–29. DOI: 10.3920/wmj2018.2398

Tolosa, J., Rodríguez-Carrasco, Y., Ferrer, E. and Mañes, J. 2019. Identification and quantification of enniatins and beauvericin in animal feeds and their ingredients by LC-QTRAP/MS/MS. Metabolites 9(2): 33. DOI: 10.3390/metabo9020033

Tuanny, F.L., Mousavi, K.A., In Lee, S.H. and Fernandes Oliveira, C.A. 2019. Biomonitoring of mycotoxin exposure using urinary biomarker approaches: A review. Toxin Reviews 1–21. DOI: 10.1080/15569543.2019.1619086

Turner, N.W., Bramhmbhatt, H., Szabo-Vezse, M., Poma, A., Coker, R. and Piletsky, S.A. 2015. Analytical methods for determination of mycotoxins: An update (2009–2014). Analytica Chimica Acta 12: 11.

Urban, M., Hann, S. and Rost, H. 2019. Simultaneous determination of pesticides, mycotoxins, tropane alkaloids, growth regulators, and pyrrolizidine alkaloids in oats and whole wheat grains after online clean-up via two-dimensional liquid chromatography tandem mass spectrometry. Journal of Environmental Science and Health Part B 54(2): 98–111. DOI: 10.1080/03601234.2018.1531662

Valera, E., García-Febrero, R., Elliott, C.T., Sánchez-Baeza, F. and Marco, M.P. 2019. Electrochemical nanoprobe-based immunosensor for deoxynivalenol mycotoxin residues analysis in wheat samples. Analytical and Bioanalytical Chemistry 411(9): 1915–1926. DOI: 10.1007/s00216-018-1538-0

Vidal, A., Claeys, L., Mengelers, M., Vanhoorne, V., Vervaet, C., Huybrechts, B. et al. 2018. Humans significantly metabolize and excrete the mycotoxin deoxynivalenol and its modified form deoxynivalenol-3-glucoside within 24 hours. Scientific Reports 8(1). DOI: 10.1038/s41598-018-23526-9

Vidal, A., Mengelers, M., Yang, S., De Saeger, S. and De Boevre, M. 2018. Mycotoxin biomarkers of exposure: A comprehensive review. Comprehensive Reviews in Food Science and Food Safety 17(5): 1127–1155. DOI: 10.1111/1541-4337.12367

Walravens, J., Mikula, H., Rychlik, M., Asam, S., Devos, T., Njumbe Ediage, E. et al. 2016. Validated UPLC-MS/MS methods to quantitate free and conjugated alternaria toxins in commercially available tomato products and fruit and vegetable juices in Belgium. Journal of Agricultural and Food Chemistry 64(24): 5101–5109. DOI: 10.1021/acs.jafc.6b01029

Wang, X., Gao, X., He, J., Hu, X., Li, Y., Li, X. et al. 2019. Systematic truncating of aptamers to create high-performance graphene oxide (GO)-based aptasensors for the multiplex detection of mycotoxins. The Analyst 144(12): 3826–3835. DOI: 10.1039/c9an00624a

Wei, F., Liu, X., Liao, X., Shi, L., Zhang, S., Lu, J. et al. 2019. Simultaneous determination of 19 mycotoxins in lotus seed using a multimycotoxin UFLC-MS/MS method. Journal of Pharmacy and Pharmacology 71(7): 1172–1183. DOI: 10.1111/jphp.13101

Wei, M., Wang, C., Xu, E., Chen, J., Xu, X., Wei, W. et al. 2019. A simple and sensitive electrochemiluminescence aptasensor for determination of ochratoxin A based on a nicking endonuclease-powered DNA walking machine. Food Chemistry 282: 141–146. DOI: 10.1016/j.foodchem.2019.01.011

Wei, T., Chen, Z., Li, G. and Zhang, Z. 2018. A monolithic column based on covalent cross-linked polymer gels for online extraction and analysis of trace aflatoxins in food sample. Journal of Chromatography A 1548: 27–36. DOI: 10.1016/j.chroma.2018.03.015

Wu, Q. and Xu, H. 2019. Application of multiplexing fiber optic laser induced fluorescence spectroscopy for detection of aflatoxin B1 contaminated pistachio kernels. Food Chemistry 290: 24–31. DOI: 10.1016/j.foodchem.2019.03.079

Wu, S.S., Wei, M., Wei, W., Liu, Y. and Liu, S. 2019. Electrochemical aptasensor for aflatoxin B1 based on smart host-guest recognition of β-cyclodextrin polymer. Biosensors and Bioelectronics 129: 58–63. DOI: 10.1016/j.bios.2019.01.022

Wu, X., Zhang, X., Yang, Y., Liu, Y. and Chen, X. 2019. Development of a deep eutectic solvent-based matrix solid phase dispersion methodology for the determination of aflatoxins in crops. Food Chemistry 291: 239–244. DOI: 10.1016/j.foodchem.2019.04.030

Yang, Y.J., Zhou, Y., Xing, Y., Zhang, G.M., Zhang, Y., Zhang, C.H. et al. 2019. A label-free aptasensor based on Aptamer/NH2 Janus particles for ultrasensitive electrochemical detection of ochratoxin A. Talanta 199: 310–316. DOI: 10.1016/j.talanta.2019.02.015

Ye, H., Lu, Q., Duan, N. and Wang, Z. 2019. GO-amplified fluorescence polarization assay for high-sensitivity detection of aflatoxin B1 with low dosage aptamer probe. Analytical and Bioanalytical Chemistry 411(5): 1107–1115. DOI: 10.1007/s00216-018-1540-6

Yu, X., Li, Z., Zhao, M., Lau, S.C.S., Ru Tan, H., Teh, W.J. et al. 2019. Quantification of aflatoxin B1 in vegetable oils using low temperature clean-up followed by immuno-magnetic solid phase extraction. Food Chemistry 275: 390–396. DOI: 10.1016/j.foodchem.2018.09.132

Zhang, C., Zhang, Q., Tang, X., Zhang, W. and Li, P. 2019. Development of an anti-idiotypic VHH antibody and toxin-free enzyme immunoassay for ochratoxin A in cereals. Toxins 11(5): 280. DOI: 10.3390/toxins11050280

Zhang, J.T., Kang, T.S., Wong, S.Y., Pei, R.J., Ma, D.L. and Leung, C.H. 2019. An iridium(III) complex/G-quadruplex ensemble for detection of ochratoxin A based on long-lifetime luminescent. Analytical Biochemistry 580: 49–55. DOI: 10.1016/j.ab.2019.06.005

Zhang, K. and Xu, D. (2019). Application of stable isotope dilution and liquid chromatography tandem mass spectrometry for multi-mycotoxin analysis in edible oils. J AOAC Int 102(6): 1651–1656. DOI: 10.5740/jaoacint.18-0252

Zhao, F., Tian, Y., Shen, Q., Liu, R., Shi, R., Wang, H. et al. 2019. A novel nanobody and mimotope based immunoassay for rapid analysis of aflatoxin B1. Talanta 195: 55–61. DOI: 10.1016/j.talanta.2018.11.013

Zhao, H., Xiang, X., Chen, M. and Ma, C. 2019. Aptamer-based fluorometric ochratoxin A assay based on photoinduced electron transfer. Toxins 11(2): 65. DOI: 10.3390/toxins11020065

Zhou, J., Xu, J.J., Cong, J.M., Cai, Z.X., Zhang, J.S., Wang, J.L. et al. 2018. Optimization for quick, easy, cheap, effective, rugged and safe extraction of mycotoxins and veterinary drugs by response surface methodology for application to egg and milk. Journal of Chromatography A 1532: 20–29. DOI: 10.1016/j.chroma.2017.11.050

Zhu, Y., Xia, X., Deng, S., Yan, B., Dong, Y., Zhang, K. et al. 2019. Label-free fluorescent aptasensing of mycotoxins via aggregation-induced emission dye. Dyes and Pigments 170: 107572. DOI: 10.1016/j.dyepig.2019.107572

Biochip Array Technology: Innovative Multi-analytical Methodology for the Simultaneous Screening of a Broad Range of Mycotoxins from a Single Food or Feed Cereal Based Sample

María Luz Rodríguez[*], Monika Plotan, Jonathan Porter, R. Ivan McConnell and S. Peter FitzGerald

Randox Food Diagnostics, 55 Diamond Road, Crumlin, Co Antrim BT29 4QY, United Kingdom

1. Introduction

Mycotoxins, a group of naturally occurring toxins produced by certain fungi such as those belonging to the genus *Alternaria, Aspergillus, Fusarium* and *Penicillum,* are harmful to humans, domestic animals and livestock. Aflatoxin B1 is the most potent natural carcinogen known and is usually the major aflatoxin produced by toxigenic strains (Squire 1981, World Health Organisation and International Agency for research on Cancer 2002). Deoxynivalenol is associated with gastrointestinal adverse effects to both humans and animals and long-term dietary exposure of animals causes weight gain suppression, anorexia and altered nutritional efficiency (EFSA 2013). The presence of deoxynivalenol in the diet exacerbates the intestinal damage induced by the colibactin-producing *Escherichia coli* strains, which raises questions about the synergism between food contaminants and gut microbiota in intestinal carcinogenesis (Payros et al. 2017). Experimentally, fumonisin has been shown to cause liver damage in multiple species (pigs, horses, cattle and primates) as well as species-specific target organ toxicity, such as lung in pigs, brain in horses, kidney in rats, rabbits and sheep and oesophagus in rats and pigs. Fumonisins (B1 and B2), usually found in corn, have been implicated in field cases of porcine pulmonary oedema and equine leukoencephalomacia. Zearalenone has major effects on reproduction, swine being the most sensitive species (Tala and Kebede 2016).

*Corresponding author: scientific.publications@randox.com

Mycotoxins are found in a wide range of foods and feeds, particularly in areas with climates of high temperature and humidity. Mycotoxins can enter the food or feed chain through contaminated crops, in particular cereals, but also nuts, beans, spices, dried fruits, oilseeds, coffee and cocoa (Milićević et al. 2008, Tala and Kebede, 2016). They also occur in animal derived products by carryover when animals are fed with contaminated feedstuffs (Dänicke and Winkler 2015). Milk and milk products from livestock that have ingested contaminated feed are the main source of aflatoxin contamination for humans (Boudra et al. 2007, Gizachew et al. 2016). Contamination may also occur post-harvest during storage, transport, and processing stages of the food or feed supply chain. The potential impact of climate change-factors on the mycotoxin contamination of food crops pre- and post-harvest by mycotoxigenic fungi has been reviewed (Magan et al. 2011, Paterson and Lima 2010, 2017). The predicted increase in emission of greenhouse gases and subsequently rise of global temperature with a potential increase of rainfall could make better conditions for fungus to grow and could subsequently cause significant increase in the mycotoxin's presence. Mycotoxin occurrence is widespread throughout the world, mycotoxin producing fungi favour different climate conditions. Current mycotoxin occurrence above the EU and Codex limits appears to confirm the Food and Agriculture Organisation (FAO) 25% estimate, though it has been recently estimated to be up to 60-80% for detected mycotoxins above detectable levels (Eskola et al. 2019). Mycotoxins represent a major foodborne risk susceptible to climate change. Favourable climatic conditions to the growth of *Aspergillus flavus* are likely to extend the aflatoxin contamination risk in maize in South and Central Europe in the next 30 years. The mycotoxigenic *Fusarium* species profile on wheat in Europe is in continuous change, in particular a worrisome growing contamination of *Fusarium graminearum* in the Central and Northern Europe (Moretti et al. 2019).

In addition, most fungi are able to produce several mycotoxins concurrently, food commodities can be contaminated by several fungi simultaneously or in quick succession, which can explain the co-occurrence of mycotoxins in food, including cereal grains (Smith et al. 2016).

Mycotoxins of interest for food safety and regulatory purposes are aflatoxins (B1, B2, G1, G2 and M1), citrinin, deoxynivalenol, ergot alkaloids, fumonisins (FB1, FB2 and FB3), HT-2 toxin, patulin, ochratoxin A, T-2 toxin and zearalenone. Mycotoxins constitute a significant problem for the animal feed industry and ongoing risk to feed supply security (Bryden 2012). Due to health and welfare concerns, regulations and recommendations have been established in many countries for some mycotoxins. Globally, maximum levels have been established for aflatoxins (total), aflatoxin M1, ochratoxin A, and patulin in the Codex standard 193–1995 (*Codex Alimentarius* 1995). In the European Union (EU), aflatoxins, deoxynivalenol, fumonisins, ochratoxin A, patulin and zearalenone are currently regulated (EC No 1881/2006, EC No 165/2013). Maximum permitted limits for aflatoxin B1 (Commission Directive 2003/100/EC) were set and guidance levels were recommended for deoxynivalenol, zearalenone, ochratoxin A, T-2, HT-2 toxins and fumonisins (Commission Recommendation 2006/576/EC and 2013/165/EU) in products intended for animal feed. National legislations are applied in other countries worldwide, which establish regulatory maximum limits and guidance levels in food crops.

Mycotoxins, which are in contact with highly metabolically active plants in the field, are specifically prone to be metabolised, the formed substances are often referred to as "masked mycotoxins." In the scientific opinion on the risks for human and animal health related to the presence of modified forms of certain mycotoxins in food and feed (European Food Safety Authority Panel (EFSA) on Contaminants in the Food Chain (CONTAM) 2014a), the term modified mycotoxins was applied to masked and bound mycotoxins and mycotoxins' metabolites. The mycotoxins in their unchanged forms were referred to as "parent compounds". Mycotoxins can also be modified by bacteria, mammals and by processing of the edible plants.

Within this context, analytical screening methods are needed to detect the presence of mycotoxins and modified forms in different matrices. This chapter refers to an innovative multi-mycotoxin multiplex screening approach: Biochip Array technology (BAT). BAT is an immunoassay based technology that maximises the screening capacity by enabling the simultaneous measurement of multiple mycotoxins from a single food or feed cereal based sample.

2. Mycotoxins: Analytical Detection

In the interest of animal health and to protect consumer safety by verifying the compliance of feed and food with acceptable safety standards, analytical tools are essential to identify sample compliance. The available risk assessment together with recent scientific reports underline a significance of predominant mycotoxins detection not only at established legislation but also at lower levels, which may have a potential health risk to humans and animals due to their natural co-occurrence and subsequently, potential of additive or synergistic effect.

Chromatographic, spectrometric and immunoassay based techniques are used for the detection of mycotoxins. The progression of the instrument intensive techniques, such as mass spectrometry, and screening techniques such as immunoassays with time has led to the proliferation of techniques capable of detecting groups of mycotoxins.

The use of a broad range of analytical techniques for the single or multiplex detection of mycotoxins, including thin-layer chromatography (Scott et al. 1970, Schaafsma et al. 1998), gas chromatography-mass spectrometry with an on-column injector (Hossain and Goto 2015), liquid chromatography high resolution mass spectrometry (Lattanzio et al. 2012a), reverse phase high performance liquid chromatography with a photodiode array and fluorescence detector (Soleimany et al. 2011), liquid chromatography-tandem mass spectrometry (Schuhmacher et al. 2005, Malachová et al. 2014, 2018, Zhao et al. 2015, Habler and Rychlik 2016, Habler et al. 2017, Adekoya et al. 2019), near-infrared reflectance spectroscopy (Berardo et al. 2005), enzyme-linked immunosorbent assays (ELISAs) (Pleadin et al. 2012), dipsticks (Lattanzio et al. 2012b, 2013), and biosensors (Al-Rubaye et al. 2018) has been reported. Using liquid chromatography-tandem mass spectrometry, simultaneous determination of target and modified "masked mycotoxins" was attained in different matrices such as, cereals (Schuhmacher et al. 2005, Habler and Rychlik 2016), beer (Habler et al. 2017), four model food matrices -apple puree for infants, hazelnuts, maize, green pepper (Malachová et al. 2014).

There is a significant difference in the research and development achievements in mycotoxin analysis implementing previously described specialised laboratory technology in comparison with industrial 'on field' mycotoxin testing. Published peer reviewed articles are capable of quantifying over hundred mycotoxins simultaneously including conjugated, masked forms and metabolites of the main legislated mycotoxins. In majority, routine analysis performed in the food and feed supply chain is covered by immunoassays including lateral flows and ELISAs to assure the final product security following the global jurisdictions. Only for confirmatory analysis more sophisticated technology is utilised due to the cost, maintenance and time.

An innovative multi-analytical screening approach, using a simple easy to use immunological strategy consolidated in one platform, biochip array technology (BAT), allows the simultaneous determination of regulated and modified mycotoxins in cereals and their milling products as well as cereal based feed. With multiplex Myco arrays, the time to result is within 3 hours. It is comparable to the time needed to perform analysis using a single ELISA kit and much shorter than chromatography-based methods. The methodology is cost-effective, easy to implement and use and no highly skilled personnel or specialized equipment is required. The flexibility of the technology allows extension of analytical profile and implementation of new assays for a wider range of mycotoxins, once corresponding analytical standards will be commercially available to follow up the most recent developments in mycotoxins analysis.

Another challenge for development of detection technologies for various analytes including mycotoxins is sample extraction protocol. Mycotoxins distinct chemical structure and the diverse matrices in which they occur pose challenges for analytical chemistry. Each group of mycotoxins and each matrix have different chemical and physical properties, so the methods for the separation of toxins for multiplex analysis must be considered on an individual basis before generic sample preparation is being developed.

3. Biochip Array Technology (BAT): Overview of the Biochip Platform

The concept of assays based on ligand binding on microarrays was introduced by Ekins et al. (1990a, 1990b, 1998). Panels of microarrays can increase the capacity of detection as multiple analytes can be assessed at the same time from a single sample. The development of stable, reproducible microarray technology in a fully automated Evidence series analyser has been pronounced (FitzGerald et al. 2005). BAT is based upon a standard immunoassay format but provides multiple results from a single sample. Applications of this technology to drug residues in food related products screening have been reported (O'Mahony et al. 2011, Porter et al. 2012, Oruc, et al. 2012, 2013, Popa I. et al. 2012a, 2012b, Gaudin et al. 2014, 2015, 2016, 2017, 2018).

The core of this technology is the biochip (9mm x 9mm), a solid support in which multiple ligands are precisely dispensed, immobilised, and stabilised in predefined x, y coordinates, creating ordered discrete test regions (DTRs) forming arrays. The biochip is also used as the reaction platform to perform simultaneous immunoassays. Nine biochips are held in a carrier (3×3 biochips per carrier = 9 reaction vessels per carrier). A handling tray with the capacity to accommodate 6 carriers (54 biochips) is also provided with the system (Fig. 1) and helps biochip carriers being processed through the immunoassay steps (Fig. 2). The immunoassay steps (i.e. reagent loading and washing) are manually performed under controlled incubation conditions as a customised thermoshaker unit is provided. Reagents, multi-analyte calibrators and/ or multi-analyte controls information is entered with the aid of a handheld barcode scanner. Biochips are applied to one single platform: the semi-automated biochip analyser Evidence Investigator (Fig. 1). Once the biochip carrier is inserted in the image station of the analyser, the chemiluminescent reactions produced at the different DTRs on the surface of the nine biochips contained in a carrier are simultaneously detected and recorded by a cooled charge coupled device (CCD) camera in the Evidence Investigator. The CCD camera has a sensor that converts incident photons produced in the chemiluminescent reaction into electrons and the light output generated is quantified. The dedicated software processes, reports and archives the data generated for retrospective access. The system is Laboratory Information Management System (LIMS) integrated for convenient reporting, with multi-format options for the review of the results e.g. by array, by users, by date or sample code. The data are traceable, and the system is password protected for various user levels. This simple easy to apply and operate biochip platform generates an increased number of screening test results from a single sample previously extracted by provided extraction protocol.

Thorough quality control, monitoring is implemented through every single step, from the fabrication of the biochips to the management of data in the system software. Post-process checks for biochip functionality, integrity and stability have been introduced to ensure product quality. System checks are incorporated in the hardware and software to alert the user when the quality of the results may be compromised (Molloy et al. 2005).

The technology is very flexible, implements both competitive and sandwich immunoassays as well as monoclonal and polyclonal antibodies, of which the majority are developed in house. The developed arrays might be either qualitative or quantitative, where number of assays and time of analysis

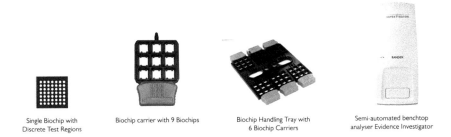

Single Biochip with Discrete Test Regions

Biochip carrier with 9 Biochips

Biochip Handling Tray with 6 Biochip Carriers

Semi-automated benchtop analyser Evidence Investigator

Figure 1: Biochip (9mm × 9mm) with DTRs, biochip carrier (9 biochips), biochip handling tray (54 biochips) and the semi-automated benchtop analyser Evidence Investigator.

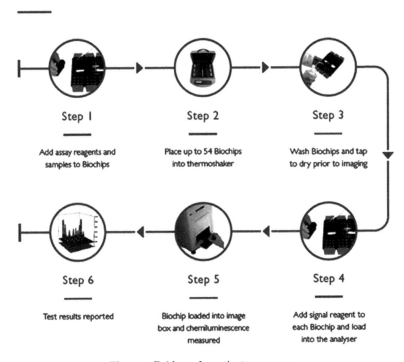

Evidence Investigator Process

Step 1

Add assay reagents and samples to Biochips

Step 2

Place up to 54 Biochips into thermoshaker

Step 3

Wash Biochips and tap to dry prior to imaging

Step 6

Test results reported

Step 5

Biochip loaded into image box and chemiluminescence measured

Step 4

Add signal reagent to each Biochip and load into the analyser

Figure 2: Evidence Investigator process

depends on the specific biochip array. Biochips are customised depending on their application, either for mycotoxins or other food and feed contaminants, like antibiotic residues, antiparasitic residues, unauthorised substances and pathogens residues as well as for toxicology and clinical applications.

4. Application of BAT to the Screening of Mycotoxins

BAT is employed for the semi-quantitative detection of mycotoxins from cereals, cereals milling products as well as cereal based feed implementing simple, single step, generic liquid/liquid extraction. Extraction involves 10 min mixing previously milled, homogeneous sample with extraction solution consisting of acetonitrile, methanol and water followed by 2 min centrifugation and dilution in the kit buffer prior to application to the biochip. Myco arrays have been developed to accommodate 5 (Myco 5, EV4137A/B), 6 (Myco 6, EV4348), 7 (Myco7, EV4065) or 9 (Myco 9, EV3941) competitive immunoassays on the biochip surface for the simultaneous detection of predominant and modified mycotoxins. Flexibility of the technology was already utilised to customise the arrays, to accommodate the mycotoxins to be screened according to various jurisdictions or prevalence in a particular geographical region or other customised requirements.

Additionally, profile of matrices considered for the screening destined to human consumption may be also extended. Biochip technology can be improved following future incoming jurisdictions and adjusted to be fully automated (with the exception of sample preparation) to fit the end user requirements.

4.1 Detection of Predominant Mycotoxins and Modified Forms from a Single Sample

The biochip arrays show broad specificity profile (Table 1), which allows the detection of a broad range of mycotoxins from a single sample. With the biochip arrays, not only predominant mycotoxins of interest for food safety and regulatory purposes are detected but also modified and masked forms from a single sample.

The significant interest except of main, legislated and already well know mycotoxins is associated with their masked forms. Masked toxins are metabolites of the parent mycotoxin formed in the plant or fungus, e.g. by conjugation with polar compounds. Mycotoxins can also be modified by bacteria, mammals and by processing of the edible plants.

However, is the presence in food and feed of these masked forms a cause of concern? Scientific Opinion publications report on the risk for animal and public health related to the presence in food and feed of some "masked mycotoxins", including deoxynivalenol and its acetylated and modified forms (EFSA Panel on Contaminants in the Food Chain (CONTAM 2017a), zearalenone and its modified forms (EFSA Panel on Contaminants in the Food Chain (CONTAM 2017b). As the consequence of EFSA reports, which are invaluable tools for the global discussions, some modifications of current legislations or new jurisdictions have already been implemented [T-2/HT-2 – 2014]. Moreover, more are currently being considered by the European Commission, including legislation covering not only deoxynivalenol but also its acetylated and modified forms. It is also highly probable that with better understanding by means of new scientific reports, especially covering the

Table 1: Mycotoxins detected with the biochip arrays

Mycotoxins detected with biochip arrays	
Immunoassay	*Mycotoxins detected*
Aflatoxin B1	Aflatoxin B1
	Aflatoxin B2
Aflatoxin G1	Aflatoxin G1
	Aflatoxin G2
Deoxynivalenol	Deoxynivalenol
	3-acetyl-deoxynivalenol
	15-acetyl-deoxynivalenol
	Deoxynivalenol-3-glucoside
Diacetoxyscirpenol	Diacetoxyscirpenol
Fumonisin B1	Fumonisin B1
	Fumonisin B2
	Fumonisin B3
Ochratoxin A	Ochratoxin A
Paxilline	Paxilline
T-2 toxin	T-2 toxin
	HT-2 toxin
Zearalenone	Zearalenone
	α-Zearalenol
	β-Zearalenol
	Zearalanone
	α-Zearalanol
	β-Zearalanol

metabolomic pathways these forms of toxins interact with, numerous global jurisdictions will be further adjusted.

Another point of discussion covers the occurrence of masked mycotoxins. The zearalenone 14-β-D-glucopyranoside and deoxynivalenol 3-β-D-glucopyranoside have been proven to occur in naturally infected cereals. (Berthiller 2013, 2015). The existence of deoxynivalenol-3-glucoside in cereal grain and beer has been reported (Berthiller et al. 2005, Lancova et al. 2008). The acetylated derivatives of deoxynivalenol, 3-acetyl-deoxynivalenol and 15-acetyl-deoxynivalenol were found in wheat and maize (Perkowski et al. 1990, Boutigny et al. 2012). The metabolite of 15-acetyl-deoxynivalenol, 15-acetyl-deoxynivalenol-3-glucoside was detected in Dried Distillers Grains (DDGS) from Ontario, Canada (Renaud et al 2019). In humans, the significant metabolism and excretion of deoxynivalenol and its modified form deoxynivalenol-3-glucoside within 24 hours by the analysis of urine has been reported (Vidal 2018). These are examples of already discovered masked mycotoxins, however there may be many more naturally occurring. Unfortunately, it takes considerable time from the discovery of a new mycotoxin, through to assessment of its health risk, implementation in legislation to routine testing availability. The unprecedented advantage of mycotoxins biochip array technology is the possibility of implementation of additional assays to the current screening portfolio, once more masked mycotoxins of health concern will be established and their analytical standards will be commercially available.

4.2 Detection of Multi-mycotoxin Contamination

Mycotoxins co-occurrence is an indisputable fact proved in numerous reports which raises even more concerns of their detrimental effect. Cereals, directly used for consumption or as an ingredient of various feed and food commodities, may be naturally contaminated by one or several fungi simultaneously. Each fungal strain can produce either single or multiple mycotoxins. Animal diets or food made up of multiple grains may be a source of multiple mycotoxins for humans and animals (Smith et al. 2016). A three-year (2009–2011) survey on the worldwide occurrence of mycotoxins in feedstuffs and feed, samples (n=7049) showed that 48% presented two or more of the tested mycotoxins (aflatoxins, deoxynivalenol, fumonisins, ochratoxin A, and zearalenone) (Rodrigues and Naehrer 2012). In an extensive survey of samples (finished feed, maize and maize silage provided by the Biomin Mycotoxin Survey), collected worldwide for the years 2012 to 2015 (n=1926), it was reported that of 162 compounds detected, up to 68 metabolites were found in a single sample. In the same survey, a total of 57 regulated mycotoxins and mycotoxins with guidance levels, masked mycotoxins and emerging mycotoxins were investigated in a subset of 1113 samples from 46 countries. In this sample subset, typically, between 50% and (in some cases) almost all samples contained 3 or more (out of 6) regulated toxins or toxins with guidance levels, both masked toxins (deoxynivalenol-3-glucoside, zearalenone-14-sulfate) and 3 or more (out of 5) emerging mycotoxins including beauvericin and enniatins (Kovalsky et al. 2016). In a study performed in cereal crops and processed products from Nigeria (n=363) 43% of the samples were contaminated with more than one toxin (Chilaka et al. 2016). The reported surveys confirm the importance and relevance of multiplex testing, which is the future of cost effective mycotoxin analysis.

The risk analysis of mycotoxins is performed considering individual mycotoxin contamination at the investigated daily intake rate. Nevertheless, mycotoxin co-occurrence presents a health risk, the interactions between various mycotoxins and between mycotoxins and other food and feed contaminants are being widely investigated and show additive, synergistic or antagonistic effect. The biological activity of mycotoxins in terms of dose-activity relationships, may significantly change depending on the concomitance of other bioactive compounds such as other mycotoxins and other food components (Dellafiora and Dall'Asta 2017). A study regarding the assessment of the cytotoxic effects on the cell line Caco-2 of the mycotoxins deoxynivalenol and ochratoxin A (alone or combined), indicated a higher toxic effect of the mycotoxins when they were co-exposed. In the same study, the addition of the antioxidant resveratrol to the combined mycotoxins, resulted in increased cytotoxicity (Cano-Sancho et al. 2015). In human cell lines from the kidney, intestine, blood and liver the consequences of combined exposure to the environmental contaminant cadmium and to the food contaminant deoxynivalenol are specific to the target organ (Le et al. 2018).

There is still much more research required to fully understand all the interactions between mycotoxins, but one thing is acknowledged, the future of mycotoxins testing is multiplexity. That is exactly what the BAT platform can offer, co-occurrence detection of mycotoxins from a single food/feed sample.

4.3 Aspects of the Analytical Performance

Nine point calibration curves for each of the mycotoxins assays are generated simultaneously when the multi-analyte calibrators are applied to the biochips in the biochip carrier (9 biochips). The detection range can be selected depending on the end-user requirements using alternative sample dilutions. Either, contamination levels comparable to available confirmatory limits of detection or regulatory limits can be detected. The available risk assessment together with recent scientific reports underline the significance of predominant mycotoxins detection not only at established legislation but also at lower levels, which may have a potential health risk to humans and animals due to their natural co-occurrence and subsequently, potential of additive or synergistic effect.

Optimal analytical performance of Myco 7 Array, which includes immunoassays for aflatoxin B1, aflatoxin G1, deoxynivalenol, fumonisin B1, ochratoxin A, T-2 toxin and zearalenone has been reported in the screening of cereals and their milling products as well as cereal based feed (Plotan et al. 2016, Freitas et al. 2019). The analytical evaluation of Myco 7, based on European Commission Decision 2002/657/EC, where the cereal based food and feed samples were assessed showed low matrix effect, high sensitivity comparable to confirmatory methods (screening decision level for aflatoxin B1 and ochratoxin A was 0.25 ppb; for aflatoxin G1, deoxynivalenol, zearalenone, T2-toxin, fumonisin B1 was 0.5 ppb, 100 ppb, 2.5 ppb, 5 ppb and 10 ppb, respectively), good precision (repeatability and within-laboratory reproducibility showed low overall CVs, 10.6 and 11.6%, respectively), good recovery (overall average 104%) and excellent correlation with assigned values for samples within the Food Analysis Performance Assessment Scheme (Fapas®) programme (all samples were within the schemes' z-score ±2 range) and samples assessed by LC-MS/MS (Plotan et al. 2016). This rigorous validation was reviewed with the conclusion that the study was a refreshing reminder that the end point for assay development is not proof of concept, but rather functional assays that can be widely used (Berthiller et al. 2018). The study of Myco 7 Array for semi-quantitative screening of several mycotoxins in maize also indicated optimal analytical performance with precision (CVs <10.1%, except for the sum fumonisin B1 and fumonisin B2: 21.2%) and recovery values (within 73.6% for aflatoxin G1 and 108.4% for deoxynivalenol) meeting performance criteria for mycotoxin analysis according to Regulation EC no. 401/2006. When certified reference material (Test Veritas) was analysed, the results obtained with Myco 7 were in agreement with the established data (Freitas et al. 2019). Further assessment of the precision and reliability of Myco 7 under inter-laboratory repeatability conditions was carried out in accordance with the Association of American Feed Control Officials (AAFCO) Method Performance Criteria, proficiency testing samples (AAFCO and Fapas®) were analysed and the results indicated that Myco 7 fulfilled the performance criteria. Myco 7 is therefore fit for purpose for use in the official food control program for mycotoxins.

Myco 9 includes two more mycotoxin assays: diacetoxyscirpenol and paxilline. The analytical parameters validation of Myco 9 followed Commission Regulation 519/2014 for cereals (corn, barley, buckwheat, sugar beet, rye, wheat, oat, rapeseed and soya) and cereal based feed. When both fortified and naturally contaminated proficiency testing samples from Fapas® (maize, cereal based animal feed, wheat flour, oat flour) and UKGTN (barley and wheat) were analysed, all the samples were within the accepted $-2 \leq$ Z-score ≤ 2. The assessment of Quality Control (QC) samples from Fapas® (maize, cereal based animal feed, wheat flour and infant food cereal based) and Trilogy (maize) also showed all the samples within the acceptance criteria ($-2 \leq$ Z-score ≤ 2). The majority of the proficiency testing and QC samples presented multi-contamination, 14% samples were contaminated with single toxin and 86% were multi-contaminated with 2 or more toxins.

5. Conclusions

Food and feed safety is considered an essential element inherent in global food security, the latter specifically means that food as normally consumed should not pose health risks to the consumer (Eisenbrand 2015). Toxicity assessment for all mycotoxin derivatives that occur in food is important for the estimation of the health risk posed by the sum of different forms of a given mycotoxin. Mycotoxins pose a major health concern basically because they may elicit a number of adverse effects, and they are practically unavoidable contaminants of food and feed (Dellafiora and Dall'Asta 2017). It should be a high priority for research to extend current multi-toxin methods to include newly discovered transformation products of mycotoxins. The recognition of the toxicological relevance of masked mycotoxins in food commodities provides a new impetus for the establishment of overall toxicity estimates to be used by regulatory bodies, food manufacturers and monitoring authorities to protect consumers' health. (Berthiller 2013). Exposure to mixtures of mycotoxins is common as mycotoxins typically co-occur in agricultural crops and through diversified diets (Eskola et al. 2019). The progression of the instrument intensive techniques such as mass spectrometry and screening techniques such as immunoassays, with time has led to the proliferation of techniques capable of detecting groups of mycotoxins. BAT, the innovative immunological strategy presented in this chapter, allows semi-quantitative multi-analytical detection of predominant mycotoxins and their modified forms from a single sample of cereals, cereals milling products or cereal based feed, using single step extraction protocol, which leads to test consolidation, an increased result output and a reduction of samples to be assessed by confirmatory methods. Consolidation of tests using BAT, significantly reduces labour cost and analysis time, consumables and wastes. Simultaneous miniaturised immunoassays, defining arrays of DTRs, take place on the biochip surface. Myco arrays have been developed to accommodate 5 (Myco 5), 6 (Myco 6), 7 (Myco 7) or 9 (Myco 9) immunoassays on the biochip surface with the capacity up to 44 immunoassays. The analytical performance complies with current European regulatory requirements (Plotan et al. 2016, Freitas et al. 2019). The application of the Myco arrays to the semi-automated benchtop analyser Evidence Investigator allows the handling up to 54 biochips/samples at a time and the dedicated software processes, reports and archives the data generated for retrospective access.

One of the multiple advantages of the presented Myco arrays is the possibility of mycotoxin detection at the levels below established global regulations, which still might cause or enhance adverse health effects to humans and animals. The Myco arrays are then fit for purpose to facilitate both the sensitive detection and the monitoring of mycotoxins at regulatory levels. This ability maximises the screening of these undesirable contaminants at various levels, which all can cause adverse health effects in humans and animals and can be of huge benefit considering the consequences of climate change and expected increase of mycotoxin occurrence. The current trend of mycotoxins analysis shows requirement of more sensitive screening of mycotoxins in various feed and food commodities due to new scientific reports and new risk assessment reports being generated by EFSA. This could inevitably lead to more restrictive and extended global legislations in the future. It is also probable that the importance of multi-mycotoxin detection at low levels will become even more critical from novel 'omic' research reports. The outcome of research projects may also enforce a need of routine screening not only predominant mycotoxins, but also their masked forms. Currently, research facilities using developed methods can detect over 100 mycotoxins and new mycotoxins are being continuously discovered (Malachova et al. 2014, Renaud et al. 2019). However, these could not be implemented within routine testing techniques and commercially available immunoassay platforms due to unavailability of reference standards and certified reference materials obligatory for development. The future of mycotoxin control lies in continued productive collaboration between academia, all establishments involved in the safe production of food and feed and the regulatory authorities. In the EU, this collaborative approach is reflected for instance by the Research and Innovation programme Horizon 2020 for Safe Food and Feed through an Integrated ToolBox for Mycotoxin Management. The use of reliable, multiplex detection technologies have relevant application assuring safety of food and feed commodities regarding mycotoxin co-occurrence and their undesired presence.

References

Adekoya, I., Njobeh, P., Obadina, A., Landschoot, S., Audenaert, K., Okoth, S. et al. 2019. Investigation of the metabolic profile and toxigenic variability of fungal species occurring in fermented foods and beverage for Nigeria and South Africa using UPLC-MS/MS. Toxins 11: E85.

Al-Rubaye, A.G., Nabok, A., Catanante, G., Marty, J-L., Takács, E. and Székács, A. 2018. Label-free optical detection of mycotoxins using specific aptamers immobilized on gold nanostructures. Toxins 10: 291.

Berardo, N., Pisacane, V., Battilani, P., Scandolara, A., Pietri, A. and Marocco, A. 2005. Rapid detection of kernel rots and mycotoxins in maize by near-infrared reflectance spectroscopy. Journal of Agricultural and Food Chemistry 53: 8128–8134.

Berthiller, F., Dall'Asta, C., Schuhmacher, R., Lemmens, M., Adam, G. and KrsKa, R. 2005. Masked mycotoxins determination of deoxynivalenol glucoside in artificially and naturally contaminated wheat by liquid chromatography-tandem mass spectrometry. Journal of Agricultural and Food Chemistry 53: 3421–3425.

Berthiller, F., Crews, C., Dall'Asta, C., De Saeger, S., Haesaert, G., Karlovsky, P. et al. 2013. Masked mycotoxins: A review. Molecular Nutrition and Food Research 57: 165–186.

Berthiller, F., Maragos, C.M. and Dall'Asta, C. 2015. Introduction to masked mycotoxins. pp. 1–13. *In:* Dall'Asta, C., Berthiller, F. (eds.). Masked Mycotoxins in Food: Formation, Occurrence and Toxicological Relevance, RSC Publishing. United Kingdom.

Berthiller F., Cramer B., Iha M.H., Krska R., Lattanzio V.M.T., MacDonald S. et al. 2018. Developments in mycotoxin analysis an update for 2016-2017. World Mycotoxins Journal 11: 5–31.

Boutigny, A.L., Beukes, I., Small, I., Zühlke, S., Spiteller, M., Van Rensburg, B.J. et al. 2012. Quantitative detection of *Fusarium* pathogens and their mycotoxins in South Africa maize. Plant Pathology 61: 522–531.

Boudra, H., Barnouin, J., Dragacci, S. and Morgavi, D.P. 2007. Aflatoxin M1 and ochratoxin A in raw bulk milk from French dairy herds. Journal of Dairy Science 90: 3197–3201.

Bryden, W.L. 2012. Mycotoxin contamination of the feed supply chain: Implications for animal productivity and feed security. Animal Feed Science and Technology 173: 134–158.

Cano-Sancho, G., González-Arias, C.A., Ramos, A.J., Sanchis, V. and Fernández-Cruz, M.L. 2015. Cytotoxicity of the mycotoxins deoxynivalenol and ochratoxin A on Caco-2 cell line in the presence of resveratrol. Toxicology In Vitro 29: 1639–1646.

Chilaka, C.A., De Boevre, M., Atanda, O.O. and De Saeger, S. 2016. Occurrence of *Fusarium* mycotoxins in cereal crops and processed products (Ogi) from Nigeria. Toxins 8: pii: E342.

Codex Alimentarius. 1995. *Codex Alimentarius* international food standards, General standard for contaminants and toxins in food and feed. CXS 193–1995.

Commission Directive 2003/100/EC (2003) Official Journal of the European Union, L285: 33–37.

Commission Recommendation 2006/576/EC (2006) Official Journal of the European Union L229: 7–9.

Commission Recommendation 2013/165/EU (2013) Official Journal of the European Union L91, 12–15.

Commission Regulation (EC) No 1881/2006 of 19 December 2006. Official Journal of the European Union L364: 5–24.

Commission Regulation (EU) No 519/2014 of 16 May 2014. Official Journal of the European Union L147: 29–43.

Dänicke, S. and Winkler, J. 2015. Invited review: Diagnosis of zearalenone (ZEN) exposure of farm animals and transfer of its residues into edible tissues (carryover). Food and Chemical Toxicology 84: 225–249.

Dellafiora, L. and Dall'Asta, C. 2011. Forthcoming challenges in mycotoxins toxicology research for safer food – A need for multi-omics approach. Toxins 9: 18.

EFSA, Scientific Report of EFSA. 2013. Deoxynivalenol in food and feed: Occurrence and exposure. EFSA Journal 11: 3379.

EFSA Panel on Contaminants in the Food Chain (CONTAM). 2014a. Scientific opinion on the risks for human and animal health related to the presence of modified forms of certain mycotoxins in food and feed. EFSA Journal 12: 3916.

EFSA Panel on Contaminants in the Food Chain (CONTAM). 2014b. Scientific Opinion on the risks for human and animal health related to the presence of beauvericin and enniatins in food and feed. EFSA Journal 12: 3802.

EFSA Panel on Contaminants in the Food Chain (CONTAM). 2017a. Risks to human and animal health related to the presence of deoxynivalenol and its acetylated and modified forms in food and feed. EFSA Journal 15: e04718.

EFSA Panel on Contaminants in the Food Chain (CONTAM). 2017b. Risks for animal health related to the presence of zearalenone and its modified forms in feed. EFSA Journal 15: e04851.

Eisenbrand, G. 2015. Current issues and perspectives in food safety and risk assessment. Human and Experimental Toxicology 34: 1286–1290.

Ekins, R. 1990a. Measurement of free hormones in blood. Endocrinology Reviews 11: 5–46.

Ekins, R., Chu, F. and Biggart, E. 1990b. Multispot, multianalyte, immunoassay. Annales de Biologie Clinique 48: 655–666.

Ekins, R. 1998. Ligand assays: From electrophoresis to miniaturized microarrays. Clinical Chemistry 44: 2015–2030.

Eskola, M., Kos, G., Elliott, C.T., Hajšlová, J., Mayar, S. and Krska, R. 2019. Worldwide contamination of food-crops with mycotoxins: Validity of the widely cited 'FAO estimate' of 25. Critical Reviews in Food Science and Nutrition 1–17.

FitzGerald, S.P., Lamont, J.V., McConnell, R.I. and Benchikh, E.O. 2005. Development of a high-throughput automated analyzer using biochip array technology. Clinical Chemistry 51: 1165–1176.

Freitas A., Barros, S., Brites, C., Barbosa, J. and Sanches Silva, A. 2019. Validation of a biochip chemiluminescent immunoassay for multi-mycotoxins screening in maize (Zea mays L.). Food Analytical Methods 12: 2675–2684.

Gaudin, V., Hedou, C., Soumet, C. and Verdon, E. 2014. Evaluation and validation of a biochip multi-array technology for the screening of 6 families of antibiotics in honey according to the European guideline for the validation of screening methods for residues of veterinary medicines. Food Additives and Contaminants Part A 31: 1699–1711.

Gaudin, V., Hedou, C., Soumet, C. and Verdon, E. 2015. Evaluation and validation of a biochip multi-array technology for the screening of 14 sulphonamides and trimethoprim residues in honey according to the European guideline for the validation of screening methods for residues for veterinary medicines. Food and Agricultural Immunology 26: 477–495.

Gaudin, V., Hedou, C., Soumet, C. and Verdon, E. 2016. Evaluation and validation of a multi-residue method based on biochip technology for the simultaneous screening of six families of antibiotics in muscle and aquaculture products. Food Additives and Contaminants Part A 33: 403–419.

Gaudin, V., Rault, A., Hedou, C., Soumet, C. and Verdon, E. 2017. Strategies for the screening of antibiotic residues in eggs: Comparison of the validation of the classical microbiological method with an immunobiosensor method. Food Additives and Contaminants Part A 34: 1510–1527.

Gaudin, V., Hedou, C., Soumet, C. and Verdon, E. 2018. Multiplex immunoassay based on biochip technology for the screening of antibiotic residues in milk: Validation according to the European guideline. Food Additives and Contaminants Part A 35: 2348–2365.

Gizachew, D., Szonyi, B., Tegegne, A., Hanson, J. and Grace, D. 2016. Aflatoxin contamination of milk and dairy feeds in the Greater Addis Ababa milk shed, Ethiopia. Food Control 59: 773–779.

Habler, K. and Rychlik, M. 2016. Multi-mycotoxin stable isotope dilution LC-MS/MS method for Fusarium toxins in cereals. Analytical and Bioanalytical Chemistry 408: 307–317.

Habler, K., Gotthardt, M., Schlüler, J. and Rychlik, M. 2017. Multi-mycotoxin stable isotope dilution LC-MS/MS method for Fusarium toxins in beer. Food Chemistry 218: 447–454.

Hossain, M.Z. and Goto, T. 2015. Determination of sterigmatocystin in grain using gas chromatography-mass spectrometry with an on-column injector. Mycotoxin Research 31: 17–22.

Kovalsky, P., Kos, G., Nährer, K., Schwab, C., Jenkins, T., Schatzmayr, G. et al. 2016. Co-occurrence of regulated, masked, and emerging mycotoxins and secondary metabolites in finished feed and maize – An extensive survey. Toxins 8: 363.

Lancova, K., Hajslova, J., Poustka, J., Krplova, A., Zachariasova, M., Dostalek, P. et al. 2008. Transfer of Fusarium mycotoxins and "masked" deoxynivalenol (deoxynivalenol-3-glucoside) from field barley through malt to beer. Food Additives and Contaminants Part A 25: 732–744.

Lattanzio, V.M., Visconti, A., Haidukowski, M. and Pascale, M. 2012a. Identification and characterization of new Fusarium masked mycotoxins, T2 and HT2 glycosyl derivatives, in naturally contaminated wheat and oats by liquid chromatography-high-resolution mass spectrometry. Journal of Mass Spectrometrometry 47: 466–475.

Lattanzio, V.M., Nivarlet, N., Lippolis, V., Della Gatta, S., Huet, A.C., Delahaut, P. et al. 2012b. Multiplex dipstick immunoassay for semi-quantitative determination of Fusarium mycotoxins in cereals. Analytical Chemical Acta 718: 99–108.

Lattanzio, V.M., von Holst, C. and Visconti, A. 2013. Experimental design for in-house validation of a screening immunoassay kit. The case of a multiplex dipstick for Fusarium mycotoxins in cereals. Analytical and Bioanalytical Chemistry 405: 7773–7782.

Le, T.H., Alassane-Kpembi, I., Oswald, I.P. and Pinton, P. 2018. Analysis of the interactions between environmental and food contaminants, cadmium and deoxynivalenol, in different target organs. Science of the Total Environment 622-623: 841–848.

Magan, N., Medina, A. and Aldred, D. 2011. Possible climate-change effects on mycotoxin contamination of food crops pre- and postharvest. Plant Pathology 60: 150–163.

Malachová, A., Sulyok, M., Beltrán, E., Berthiller, F. and Krska, R. 2014. Optimization and validation of a quantitative liquid chromatography-tandem mass spectrometric method covering 295 bacterial and fungal metabolites including all regulated mycotoxins in four model food matrices. Journal of Chromatography 1362: 145–156.

Malachová, A., Stránská, M., Václavíková, M., Elliott, C.T., Black, C., Meneely, J. et al. 2018. Advanced LC-MS-based methods to study co-occurrence and metabolization of multiple mycotoxins in cereals and cereal-based food. Analytical and Bioanalytical Chemistry 410: 801–825.

Milićević D. Jurić, V., Stefanović, S., Jovanović, M. and Janković, S. 2008. Survey of slaughtered pigs for occurrence of ochratoxin A and porcine nephropathy in Serbia. International Journal of Molecular Sciences 9: 2169–2183.

Molloy, R.M., McConnell, R.I., Lamont, J.V. and FitzGerald, S.P. 2005. Automation of biochip array technology for quality results. Clinical Chemistry Laboratory Medicine 43: 1303–1313.

Moretti, A., Pascale, M. and Logrieco, A.F. 2019. Mycotoxin risks under climate change scenario in Europe. Trends in Food Science and Technology 84: 38–40.

O'Mahony, J., Moloney, M., McConnell, R.I., Benchikh, E.O., Lowry, P., Furey, A. et al. 2011. Simultaneous detection of four nitrofuran metabolites in honey using multiplexing biochip screening assay. Biosensors and Bioelectronics 26: 4076–4081.

Oruc, H.H. 2012. Simultaneous detection of six different groups of antimicrobial drugs in milk, meat, urine and feed matrices. Uludağ Üniversitesi Journal Fakültesi Veteriner Medicine 31: 29–33.

Oruc, H.H., Rumbeiha, W.K., Ensley, S., Olsen, C. and Schrunk, D.E. 2013. Simultaneous detection of six different groups of antimicrobial drugs in porcine oral fluids using biochip array-based immunoassay. Kafkas Üniversitesi Veteriner Fakültesi Dergisi 19: 407–412.

Paterson, R.R.M. and Lima, N. 2010. How will climate change affect mycotoxins in food? Food Research International 43: 1902–1914.

Paterson, R.R.M. and Lima, N. 2017. Thermophilic fungi to dominate aflatoxigenic/mycotoxigenic fungi on food under global warming. International Journal of Environmental Research and Public Health 14(2): 199–210.

Payros, D., Dobrindt, U., Martin, P., Secher, T., Bracarense, A.P., Boury, M. et al. 2017. The food contaminant deoxynivalenol exacerbates the genotoxicity of gut microbiota. MBio 8: pii: e00007–17.

Perkowski, J., Plattner, R.D., Goliński, P., Vesonder, R.F. and Chelkowski, J. 1990. Natural occurrence of deoxynivalenol, 3-acetyl-deoxynivalenol, 15-acetyl-deoxynivalenol, nivalenol, 4, 7-dideoxynivalenol and zearalenone in Polish wheat. Mycotoxin Research 6: 7–12.

Pleadin, J., Perši, N., Zadravec, M., Sokolović, M., Vulić, A., Jaki, V. et al. 2012. Correlation of deoxynivalenol and fumonisin concentration determined in maize by ELISA methods. Journal of Immunoassay and Immunochemistry 33: 414–421.

Plotan, M., Devlin, R., Porter, J., Benchikh, M.E., Rodríguez, M.L., McConnell, R.I. et al. 2016. The use of biochip array technology for rapid multimycotoxin screening. Journal of AOAC International 99: 878–889.

Popa, I.D., Schiriac, E.C., Matiut, S. and Cuciureanu, R. 2012a. Method validation for simultaneous determination of 12 sulfonamides in honey using biochip array technology. Farmacia. 60: 143–154.

Popa, I.D., Schiriac, E-C. and Cuciureanu, R. 2012b. Multi-analytic detection of antibiotic residues in honey using a multiplexing biochip assay. Revista Medico-Chirurgicala a Societatii de Medici si Naturalisti din Iasi. 116: 324–329.

Porter, J., O'Loan, N., Bell, B., Mahoney, J., McGarrity, M., McConnell, R.I. et al. 2012. Development of an evidence biochip array kit for the multiplex screening of more than 20 anthelmintic drugs. Analytical and Bioanalytical Chemistry 403: 3051–3056.

Renaud, J.B., Miller, J.D. and Sumarah, M.W. 2019. Mycotoxin testing paradigm: Challenges and opportunities for the future. Journal of AOAC International 102: 1681–1688.

Rodrigues, I. and Naehrer, K. 2012. A three-year survey on the worldwide occurrence of mycotoxins in feedstuffs and feed. Toxins 4: 663–675.

Schaafsma, A.W., Nicol, M.E., Savard, M.E., Sinha, R.C., Reid, L.M. and Rottinghaus, G. 1998. Analysis of Fusarium toxins in maize and wheat using thin layer chromatography. Mycopathologia 142: 107–113.

Scott, P.M., Lawrence, J.W. and van Walbeek, W. 1970. Detection of mycotoxins by thin-layer chromatography: Application to screening of fungal extracts. Applied Microbiology 20: 839–842.

Schuhmacher, R., Berthiller, F., Buttinger, G. and Krska, R. 2005. Simultaneous determination of type A-&B trichothecenes and zearalenone in cereals by high performance liquid chromatography-tandem mass spectrometry. Mycotoxin Research 21: 237–240.

Smith, M.-C., Madec, S., Coton, E. and Hymery, N. 2016. Natural co-occurrence of mycotoxins in foods and feeds and their in vitro combined toxicological effects. Toxins 8: 94.

Soleimany, F., Jinap, S., Rahmani, A. and Khatib, A. 2011. Simultaneous detection of 12 mycotoxins in cereals using RP-HPLC-PDA-FLD with PHRED and a post-column derivatization system. Food Additives and Contaminants Part A Chem. Anal. Control Expo. Risk Assess 28: 494–501.

Squire, R.A. 1981. Ranking animal carcinogenesis: A proposed regulatory approach. Science 214: 877–880.

Tala, M. and Kebede. B. 2016. Occurrence, importance and control of mycotoxins: A review. Cogent Food and Agriculture 2: 1191103.

Vidal, A., Claeys, L., Mengelers, M., Vanhoorne, V., Vervaet, C., Huybrechts, B. et al. 2018. Humans significantly metabolize and excrete the mycotoxin deoxynivalenol and its modified form deoxynivalenol-3-glucoside within 24 hours. Science Reports 8: 5255.

World Health Organisation and International Agency for Research on Cancer. 2002. International Agency for Research on Cancer IARC Monographs on the Evaluation of Carcinogenic Risks to Humans. IARC Press: Lyon, France 1–601.

Zhao, Z., Liu, N., Yang, L., Deng, Y., Wang, J., Song, S. et al. 2015. Multi-mycotoxin analysis of animal feed and animal-derived food using LC-MS/MS system with timed and highly selective reaction monitoring. Analytical and Bioanalytical Chemistry 407: 7359–7368.

Biosensor and Aptamer: New in Mycotoxins Detection

Elsaadani Moez[1,3]*, Sorli Brice[2] and Montet Didier[3,4]

[1] Faculty of Biotechnology, Misr University for Science and Technology (MUST), 6th October City, Egypt
[2] UM IES, UMR CNRS 5214, Montpellier University, France
[3] Cirad, UMR QualiSud, 73 rue Jean-François Breton, 34398 Montpellier Cedex 5, France
[4] UMR QualiSud, QualiSud, Université Montpellier, CIRAD, Montpellier SupAgro, Université d'Avignon, Université de La Réunion, Montpellier, France

1. Introduction

The main problem with the traditional mycotoxin analysis methods is the duration of analysis, the non-portability of very expensive equipment, the need for skilled operators and the pre-treatment or pre-concentration of samples. The availability of rapid, sensitive, simple, portable and inexpensive methods for the rapid determination of food contaminants constitutes a growing need for human safety. Consequently, the use of analytical procedures based on affinity biosensors has recently aroused great interest, mainly due to their ability to solve a large number of analytical problems and challenges in a wide variety of fields.

The main advantages of biosensors compared to traditional analytical methods are summarized by rapid detection, high sensitivity, easy preparation, reusability and low costs. In recent years, there has been growing interest in the development of biosensors. Such rapid growth is driven by several factors, including medical and health issues, the increasing incidence of chronic diseases and of course, serious security challenges and military and agricultural/food security applications (Bhalla et al. 2016).

In recent years, the importance of monitoring and controlling many critical parameters in areas such as clinical diagnostics, the food industry, the environment, forensics and drug development has increased. It is therefore necessary to have reliable analysis devices capable of carrying out rapid and precise analyzes. Conventional methods offer high sensitivity and selectivity,

*Corresponding author: moez.elsaadani@yahoo.fr

but they are expensive, time consuming, require highly qualified personnel and pre-processing or pre-concentration of the sample. One way to overcome the many disadvantages of these methods is to develop biosensors.

There are mainly three generations of biosensors: in the first generation biosensors, the product of the reaction diffuses towards the transducer and provokes the electrical response. In the second generation, specific mediators between the reaction and the transducer generate an improved response, and in the third generation, the reaction itself elicits the response and no diffusion of product or mediator is directly involved.

It all begins when Cremer (1906) demonstrated that the concentration of an acid in a liquid is proportional to the electrical potential that is created between parts of the fluid located on either side of a glass membrane. In 1916, Griffin and Nelson demonstrated for the first time the immobilization of the enzyme invertase on aluminum hydroxide and carbon (Griffin and Nelson 1916). The first biosensor was developed by Leland in 1962 for the detection of glucose. He was known as the "father of biosensors"(Clark and Lyons 1962). After Leland published his article on enzyme electrodes, Guilbault and Montalvo (1969) discovered in 1969 the first potentiometric biosensor for the detection of urea. Then, in 1975, the first commercial biosensor was developed by Yellow Spring Instruments (YSI) (Suzuki et al. 1975). Finally, Cammann (1977) introduced the term "Biosensor", but it was not until 1997 that the IUPAC (International Union of Pure and Applied Chemistry) introduced the definition of biosensors for the first time. Today, work on biosensors and published articles continue to progress using different biological elements in combination with various types of transducers (Fig. 1).

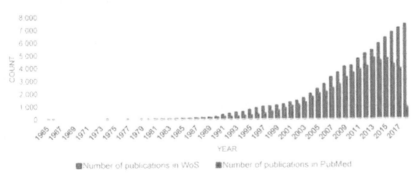

Figure 1: Annual number of publications on biosensors
(Web of Science and PubMed, May 2019).

2. Definitions

Several definitions have been proposed to define biosensors, but according to IUPAC, a biosensor is an integrated receptor transducer device, capable of measuring biological or chemical reactions using a biological recognition element by generating proportional signals at the concentration of a target.

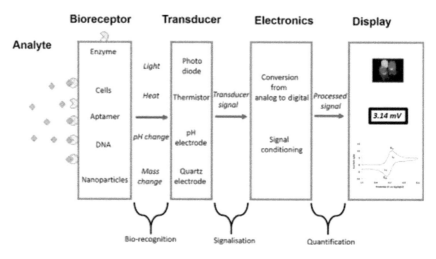

Figure 2: Schematic representation of a biosensor (Bhalla et al. 2016)

A typical biosensor is shown in Fig. 2, it includes the following components:

The target is a substance of interest that needs to be detected. For example, Ochratoxin A is a target for a biosensor designed to detect Ochratoxin A.

The bioreceptor is a molecule that specifically recognizes the target. For example, enzymes, cells, aptamers, deoxyribonucleic acid (DNA), ribonucleic acid (RNA) and antibodies.

The transducer is an element that converts one form of energy into another. The role of the transducer is to convert the biorecognition event into a measurable signal. Most transducers produce optical or electrical signals generally proportional to the concentration of target-bioreceptor interactions.

The electronic is the part of the biosensor which is responsible for processing the signal and preparing it for display. It consists of complex electronic circuits that condition signals, such as amplifying and converting signals from analog to digital format.

The display consists of a user interpretation system, generating figures or curves understandable by the user. It often consists of a combination of hardware and software that generates user-friendly results for the biosensor. The displayed signal can be digital, graphic, tabular or image, depending on the needs of the end user.

3. Biosensor Characteristics

Each biosensor has certain static and dynamic attributes. These properties must be optimized because they reflect the performance of the biosensor.

Selectivity is the ability of a bioreceptor to detect a specific target in a sample containing other adjuvants and contaminants. Selectivity is also the ability of the sensor to respond only to the target.

Sensitivity or limit of detection (LOD) is the minimum quantity of the target that can be detected by a biosensor. In several applications, a biosensor is required to detect a target concentration as low as ng/mL, in order to confirm the presence of traces of the target in a sample.

Stability is the degree of sensitivity to disturbances in and around the biosensor. These disturbances can cause a signal drift of a biosensor during measurement, which in turn can cause an error in the measured concentration and can affect the precision of the biosensor. There are factors that can affect stability, for example, temperature, affinity, and degradation of the bioreceptor.

Linearity (the range) is the attribute that shows the accuracy of the measured response (relative to a straight line), mathematically represented as $y = m \times c$, where c is the concentration of the analyte, y is the output signal, and m is the sensitivity of the biosensor. The resolution of the biosensor and the linear range of the target are associated with the linearity of the biosensor. The resolution of the biosensor is defined as the necessary change in the concentration of a target to modify the response of the biosensor. The linear range is defined as the range of target concentrations for which the response of the biosensor changes linearly with the concentration. On the other hand, the sensor can also follow curves which are not linear.

Repeatability is the ability of the biosensor to generate identical responses for a duplicated experimental configuration. Reproducibility is characterized by the precision and accuracy of the transducer. Accuracy is the ability of the sensor to provide identical results each time a sample is measured or an average value close to the actual value.

Regeneration time is the time required to return the reversible biosensor to working condition after interaction with the sample. In case of irreversible biosensor, regeneration is not possible.

4. Development of a Biosensor

After the identification of the target, the development of a biosensor consists in:

- Select an appropriate bioreceptor for the target,
- Select an immobilization method,
- Select a transducer which translates the link reaction into a measurable signal,
- Select a biosensor taking into account the measurement range, linearity and minimization of interference and improvement of sensitivity,
- Packaging of the biosensor in a complete device.

Biosensors can be classified according to the type of biorecognition elements or the type of transduction they use. Most forms of transduction can be classified into one of the five main classes: Optical, piezoelectric, thermal, electrochemical and electrical detection. In addition, this group can be divided into two categories: labeled biosensor and unlabeled biosensor.

4.1 Types of Transduction

4.1.1 Optical Biosensor

The transduction process induces a change in phase, amplitude, polarization and frequency of the input light in response to the physical or chemical change produced by the biorecognition process (Bosch et al. 2007).

The main components of an optical biosensor are a light source, an optical transmission medium, immobilized biological recognition elements and an optical detection system. The reaction causes a change in absorption or fluorescence resulting from a change in the refractive index of the surface between two media with different densities. According to Lambert-Beer's law (1852), the absorbance is proportional to the concentrations of the materials in the sample. These types of biosensors are a powerful alternative to traditional analytical techniques with their high specificity and sensitivity, as well as their small size and cost-effectiveness.

4.1.1.1 Biosensors based on energy transfer by fluorescence resonance (FRET)

Biosensors based on energy transfer by fluorescence resonance are devices that allow energy transfer without radiation from the donor to the receiver (Fig. 3). FRET is commonly used to demonstrate communication or interaction between two molecules. FRET amplifies conformational changes by emitting light, which can be captured by a sensor. An excited fluorochrome emits an essentially virtual photon, which is then absorbed by a fluorochrome receptor. The distance between the donor fluorochrome and the acceptor must be between 1 and 10 nm to allow energy transfer.

4.2.1.2 Biosensors based on surface plasmon resonance (SPR)

Surface plasmon resonance is a powerful tool capable of measuring the binding kinetics of two molecules without a fluorescent marker. SPR occurs when light is reflected off the internal surface of a material of various refractive indexes. Between two layers, a thin layer of a good conductor, such as gold or silver, with specific energy to raise the surface plasmon, is placed (Cooper 2002) (Fig. 4).

4.1.1.3 Colorimetric biosensors

The colorimetric method is an interesting optical method, which allows rapid identification of pathogens in the sample by color change. Response signals can be observed and resolved with the naked eye without the need for an analytical tool (Rubab et al. 2018).

4.1.2 Mass-sensitive Biosensors

These biosensors are considered to be biosensors based on the transduction method, which contains minor changes in the mass of the biosensor. They are

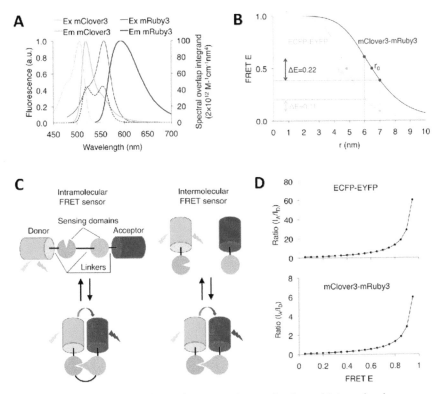

Figure 3: Two types of FRET biosensors: intramolecular and intermolecular FRET biosensors (Bajar et al. 2016).

often used as piezoelectric crystals to accurately determine small variations in mass, which is why they are also called piezoelectric biosensors (Fig. 5).

The surface of the piezoelectric sensor is covered with a selective bioreceptor in which the solution containing the targets is placed. Targets bind to bioreceptors that reduce the frequency of oscillation as the crystal mass increases and produce an electrical signal when mechanical force is applied. Piezoelectric biosensors are based on the principle of acoustics (sound vibrations), which is why they are also called acoustic biosensors. There are two types of piezoelectric transducer: Quartz crystal microbalance (QCM) and Micro-cantilever.

The QCM uses detection molecules (quartz, tourmaline, lithium, oriented zinc oxide or aluminum nitride) which are attached to a piezoelectric surface in which the interactions between the target and the detection molecules generate a change of mass on the crystal sensitive surface resulting in the decrease of the resonant frequency, this electrical signal is proportional to the quantity of target (Janshoff et al. 2000, Cooper 2003).

The Micro cantilever detects changes in the vibration frequency. The principle of this detection is based on the transduction of molecular

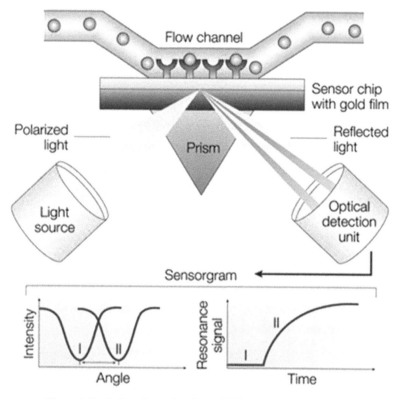

Figure 4: Typical configuration for an SPR biosensor (Cooper 2002).

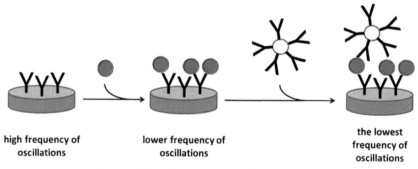

high frequency of oscillations

lower frequency of oscillations

the lowest frequency of oscillations

Figure 5: Piezoelectric biosensors (Pohanka 2018)

adsorption and specific molecular interactions on a cantilever surface in modification of the mechanical response of a cantilever. A micro-cantilever biosensor works by transducing the molecular interaction between the target and the capture molecule, which is immobilized as a layer on a surface of a cantilever. Biomolecular interactions occurring on a solid state interface lead to an increase in mass (Backmann et al. 2005). Viscosity and density can be measured by detecting changes in vibration frequency (Vashist 2007).

4.1.3 Thermometric Biosensor

These biosensors are constructed by immobilizing the bioreceptor on temperature sensors. The temperature is measured via a sensitive thermistor once the target meets the bioreceptor. The total heat produced or absorbed is proportional to the molar mass and the total number of molecules in the reaction. Thermal biosensors do not require frequent recalibration and are insensitive to the optical and electrochemical properties of the sample (Mohanty and Kougianos 2006).

4.1.4 Electrochemical Biosensor

According to the 1999 recommendation of the IUPAC, an electrochemical biosensor is an integrated autonomous device capable of providing specific quantitative or semi-quantitative analytical information using a biological recognition element which is retained in direct contact with an electrochemical transduction (Thevenot et al. 1999). The electrochemical measurement is based on the measurement of the current produced by the oxidation and reduction reactions. Electrochemical sensors directly convert the existing electrical signal into the electronic field and allow the development of compact system designs with simple instrumentation. The resulting electrical signal is related to the recognition process and proportional to the concentration of the target. Electrochemical biosensors can be classified into amperometric, potentiometric, impedance meter and conductimetric biosensors (Mortari and Lorenzelli 2014, Zhao et al. 2014).

4.1.4.1 Amperometric biosensors

These biosensors are used to examine electrochemical reactions while measuring the variation of a constant potential between two electrodes. They are based on the measurement of current as a function of time resulting from the oxidation and reduction of an electroactive species in a biochemical reaction which mainly depends on the concentration of a target with fixed potential. In this kind of biosensor, there are three types of electrodes that are usually used. The working electrode generally, which can generally be gold, carbon or platinum. A reference electrode usually made of silver or silver chloride which controls the potential of the working electrode and the third electrode called the counter which is used to measure the potential (Wang et al. 1996, Wang 2008). When the oxidation or reduction of certain molecules happened at the metal electrodes, electrons are transferred from the target to the working electrode or vice versa (Banica 2012). The direction of the electron flow depends on the properties of the target which can be controlled by the electrical potential applied to the working electrode. If the working electrode is brought to a positive potential, an oxidation reaction occurs. Likewise, if the working electrode is brought to a negative potential, a reduction reaction occurs. A third electrode called a counter electrode is often used to help measure the flow of current. The concentration of the target in a solution is proportional to the response of the biosensors (Heller 1996). Also, they use mediators for their biochemical reactions, which can participate in the redox reaction with the biological component and help to accelerate the transfer of electrons (Chaubey and Malhotra 2002).

4.1.4.2 Potentiometric biosensors

This type of transducer measures the different generated potential through an ion selective membrane which separates two solutions with a practically zero current flow (Knocki 2007). Potentiometric biosensors are based on the measurement of the oxidation and reduction potential of an electrochemical reaction. In addition, a pH meter consists of an immobilized enzyme membrane, where hydrogen ions are produced or absorbed by the catalyzed reaction. In this sensor, the biorecognition element converts the recognition process into a potential signal to provide an analytical signal (Fig. 6).

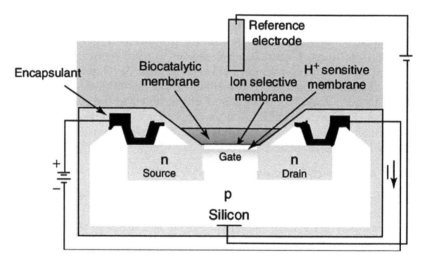

Figure 6: Potentiometric biosensor based on FET sensitive to ions (Takhistov 2005).

The potentiometric biosensor consists of two electrodes, the working electrode develops a variable potential from the recognition process and the reference electrode which provides constant potential. The operating principle of potentiometric transduction depends on the potential difference between the two electrodes (working and reference) which accumulate during the recognition process.

4.1.4.3 Impedance spectroscopy biosensors

Electrochemical impedance spectroscopy (EIS) combines the analysis of the resistive and capacitive properties of materials. Mainly, the sample is placed on the detection device, a controlled alternating voltage is applied to the electrode with monitoring of the current flowing in the sample. The results of the electrical impedance of the sample are calculated by the voltage/current ratio. That is to say, with the voltage variation applied to the sample, the resulting current can be in phase or out of phase with the applied voltage (resistive or capacitive). The EIS system includes three electrodes system (working electrode, counter electrode and reference electrode), a potentiostat and a frequency response analyzer (FRA). The working electrode provides current measurement, the counter electrode provides current to the target and the reference electrode provides voltage measurement.

4.1.5 Electric Biosensor

4.1.5.1 Conductimetric biosensors (metric impedance)

This kind of biosensor provides information on the ability of an electrolytic solution to conduct an electric current between the electrodes. The conductimetric biosensor is composed of two electrodes: a reference and a working electrode which are separated by a certain distance or a medium. Conductivity measurement is based on the biocatalytic reaction of the sample on an electrode. The reaction will produce ions which will result in the change of conductivity/impedance. For the measurement of the conductivity, an alternating current distributor is used to supply the electrode. Thus, the ionic composition changes and provides conductance. (Adley and Ryan 2015).

4.1.5.2 Capacitive biosensors

Capacitive biosensors belong to the sub-category of metric impedance biosensors. When a target interacts with the biorecognition elements, which are immobilized on the insulating dielectric layer, the capacitive biosensor measures the evolution of the dielectric properties, the charge distribution, the size, the shape and the conductivity of the dielectric layer at the electrolyte-electrode interface (Berggren and Johansson 2001). Consequently, a change in the dielectric properties of the material between a pair of electrodes leads

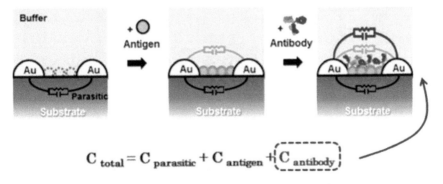

$$C_{total} = C_{parasitic} + C_{antigen} + C_{antibody}$$

Figure 7: Capacitive biosensors (Jung et al. 2014).

to a change in the capacitance, which is correlated to the bound molecules and to the quantity captured by biorecognition elements on the surface of the electrode.

4.2 Classification Depending on Label and Label-free Biosensor

Another way to classify biosensors is the labeled or unmarked biosensor (Fig. 8). Generally, labeled detection requires a secondary molecule and an amplification. The nature of the molecule used may be a radioactive dye, fluorescent, an enzyme, a metal complex, nanoparticles (Schmitteckert and Schlicht 1999, Daniels and Pourmand 2007, Wu et al. 2011, Duan and Wang 2011, Yang et al. 2011, Wang et al. 2011). This marker is attached to the target molecule or to the bioreceptor. The activity of the marker or the modification of the chemical or physical properties of the surface of the transducer are measured to carry out the desired analysis. Conjugation and easy detection are characteristics necessary to ensure the amplification of the signal in order to improve the sensitivity and selectivity of the method. However, the labeling and immobilization stages are very long and costly. Also, nonspecific adsorptions could be observed and consequently they can reduce the reproducibility, the sensitivity and the selectivity of the biosensor (Daniels and Pourmand 2007).

On the other hand, detection without labeling allows the sensors to directly detect the molecule of interest and allows the study of biomolecular interactions in real time. In this type of biosensors, the target and the bioreceptor are used in their natural form, they are not modified. The bioreceptor is immobilized on the surface of the transducer and the sample containing the target is directly incubated with the functionalized surface. The analysis is then carried out by studying the evolution of the electrical or physical properties of the surface,

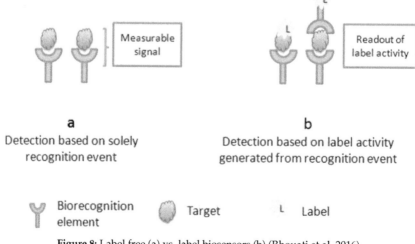

Figure 8: Label free (a) vs. label biosensors (b) (Rhouati et al. 2016).

which depend on the affinity of the interaction between the target and its receptor and therefore on the concentration of the target in the sample. The use of biosensor without labeling increases the affinity and decreases non-specific adsorptions (Li et al. 2010).

Generally, biosensors can be classified according to the type of bioreceptor they use. It is essential for a bioreceptor to be selective and sensitive to the specific target analyte in order to prevent interference from other substances in the sample matrix. The bioreceptor used by biosensors can be divided into five main mechanisms: Enzymes (Clarkand Lyons 1962), Antibodies (Tromberg et al. 1987), Nucleic acids (Cagnin et al. 2009, Lazerges et al. 2012), Cells (Vo-Dinh and Cullum 2000) and Aptamers.

5. Aptamers

These kinds of biosensors are called Aptasensor because they use aptamer DNA or RNA as bioreceptors. Aptamers are short, simple strands of artificial DNA or RNA oligonucleotides. They are artificial single-stranded oligonucleotides of DNA or RNA, which are selected from a combinatorial sequence library based on their ability to recognize a target with high affinity and specificity (Hermann and Patel 2000).

The word aptamer is derived from the Latin "aptus" "to adapt" which defines a polymer which "corresponds" to its target (Ellington and Szostak 1990). They take various forms due to their tendency to form single-stranded helices and loops. They are selected from a combinatorial sequence library based on their ability to recognize a target with high affinity and specificity. The aptamers have been studied in bio-recognition in numerous studies as a diagnostic and treatment tool, as well as in the development of new drugs and drug delivery systems. They can be generated against various targets such as proteins, drugs, organic or inorganic molecules (Tombelli et al. 2005, Bonel et al. 2011, Rhouati et al. 2013a).

Over the past 3 decades, antibodies have been widely used as a molecular recognition platform (Song et al. 2012). After their discovery, aptamers were compared to antibodies because of their similar ability to bind to specific targets. Aptamers have been successfully developed for viruses, proteins, cells and small molecules (Binning et al. 2013). However, aptamers offer several advantages over antibodies as molecular recognition molecules (Ellington and Szostak 1990). Now aptamers are widely known as substitutes for antibodies, as these molecules overcome weaknesses in antibodies.

5.1 Comparison between Aptamers and Antibodies

The identification and production of antibodies are very expensive in vivo processes involving the screening of a large number of colonies. In addition, their production requires animal immunization, inducing an immune response against a target in a biological system. The clinical commercial success of antibodies has led to the need for very large-scale production in mammalian cell culture (Birch and Racher 2006). On the contrary, the aptamers selected can be synthesized with great precision and reproducibility via chemical reactions. In addition, aptamers can be easily modified by various chemical reactions to increase their stability and resistance to nucleases (Birch and Racher 2006, FerreiraI and Missailidis 2007).

Antibodies are very sensitive to temperature and pH variations and undergo irreversible denaturation. Unlike antibodies, aptamers are stable at elevated temperatures. Even if they undergo reversible denaturation, they can be re-naturated in a few minutes (Jayasena 1999).

In the case of toxins or molecules that do not cause strong immune responses, it is difficult to identify and produce antibodies, but aptamers can be generated against these molecules. In addition, aptamers have a strong affinity and specificity for certain ligands which cannot be recognized by antibodies, such as ions or small molecules (Jayasena 1999).

Thanks to their advantages, many different targets have been explored. This research has made it possible to develop numerous applications based on aptamers. Aptamers have been used as an alternative to antibodies in many bioanalytical methods such as the enzyme linked immunosorbent assay, affinity columns and biosensors.

Table 1: Details of the differences between aptamers and antibodies

Descriptions	Aptamers	Antibodies
Size	Small (<30 kDa)	Large (>75 kDa)
Molecular structure	Nucleic Acid	Protein
Target molecules	All molecules and cells	Limited (Immunogenic targets)
Stability	Reversible even after denaturing	Loses functions easily after denaturing
Affinity	Strong	weak
Chemical modification	Easy to modify	Difficult to modify
Immunogenicity	Non-immunogenic	immunogenic
Lifetime	Several years	~6 months
Storage	Stable at room temperature	Must be frozen
Fabrication	*In vitro*	*In vivo*
Cost	Low cost	High cost

5.2 In Vitro Selection of Aptamers

The aptamers are generated by an *in vitro* selection procedure called SELEX (systematic evolution of ligands by exponential enrichment). This technique makes it possible to isolate aptamers of high affinity for a given target from approximately 1012 to 1015 combinatorial oligonucleotide banks.

In general, the SELEX process consists of three steps which are repeated in order to search for a sequence of nucleotides which are capable to bind to the target (Tuerk and Gold 1990). During the first step, a combinatorial library of oligonucleotides is synthesized, each oligonucleotide contains a central random region of 20 to 80 nucleotides, flanked by a binding region at each end. In this step, the target molecules are incubated for a planned period with the library in an appropriate buffer and under certain pH and temperature conditions.

During the second step, the free oligonucleotides are separated, and the linked oligonucleotides are eluted. Generally, this step is combined with several other methods to make the selection quick and easy.

During the third step, the selected sequences are amplified by PCR (Polymerase Chain Reaction) using primers corresponding to the fixed regions of the library.

The aptamers are continuously developed through this continuous process, and their characteristics are identified using various biological tests. However, it is difficult to identify optimal sequences using traditional cloning and sequencing approaches. Recently, a few studies have shown that the use of high throughput sequencing in the screening of aptamers constitutes a powerful tool for identifying aptamers (Cho et al. 2010). Sequence alignments, secondary structure analysis and linkage studies are necessary to identify the final sequence and characteristics of the aptamer identified (Stoltenburg et al. 2007).

5.3 Biorecognition of the Target by Aptamers

The formation of the aptamer-target complex is induced by different types of recognition:

5.3.1 Direct Format

The direct format consists of a direct or simple interaction of the aptamer with its target (Fig. 9). This type of recognition is very effective in detecting small molecules because they are often covered by aptamer loops. On the other hand, large molecules are structurally more complex than smaller targets and allow the interaction of several discriminating contacts such as hydrogen bonds, electrostatic interactions, and the complementary form.

5.3.2 Sandwich Format

The sandwich format uses two aptamers or one aptamer and an antibody for the recognition of the target molecule. This type of format is based on the immobilization of the first aptamer on the surface of the electrode (Fig. 10) and the aptamer or secondary antibody are functionalized with signaling fragments linked to the captured target to generate signals. Generally, the first and second aptamers have different nucleic acid sequences. However, there are target molecules that contain two identical binding sites, thus allowing the use of a single aptamer on the electrode. The use of the sandwich format allows

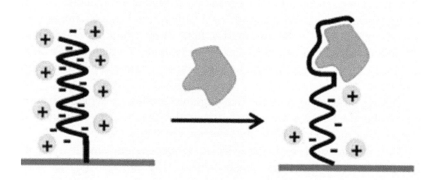

Figure 9: Aptamer-target direct format.

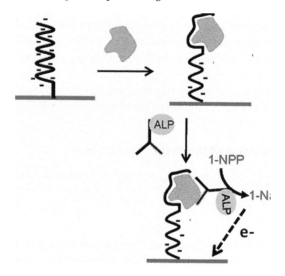

Figure 10: Aptamer-target sandwich format.

the detection of the target with a very high sensitivity and selectivity. However, several incubation steps are necessary, which is very time consuming.

5.3.3 Competitive Format

The competitive format is based on immobilizing the target on the surface of the electrode, followed by exposure to different concentrations of the target and a fixed concentration of the aptamer, which leads to competition for aptamer (Fig. 11).

5.4 Techniques for Aptamers Immobilization

The first step in the protocol for building aptasensor is immobilizing the

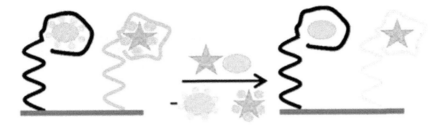

Figure 11: Aptamer-target competitive format.

aptamers on the surface of the electrode. To guarantee high orientation, reactivity, accessibility and stability of the aptasensor, it is essential to carry out a control step (Nimse et al. 2014). At the same time, before immobilization of the aptamer, pretreatment of the aptamer solution is necessary to avoid deformation of the aptamer.

The aptamer can be physically adsorbed on the surface, covalently linked with functional groups or coupled to "self-assembled monolayer" (SAM) (Table 2).

- Physical adsorption

  This immobilization technique is based on direct adsorption of the aptamer via weak types of bonds like; Van der Waals, hydrogen bonds, charge transfers or even hydrophilic interactions between the functional groups of the active biomolecule and the surface of the support. It is a non-denaturing technique which preserves the structure of the adsorbed molecule and which allows the regeneration of the biosensor (Meng et al. 2012, Wei et al. 2015). However, the orientation of the aptamer is random and the attachment of the aptamer to the surface of the electrode is weak.

Table 2: Different types of techniques for aptamers immobilization

Techniques	Principe	Advantages	Limitations
Physical adsorption	Electrostatique forces Van der Waals Interaction	Simple and fast	Weak bond Random orientation of the Aptamer
Covalent bond	Interactions between surface functional groups and aptamer chemical groups	Wide range of group flexibility	Several stages of conjugation Non-specific binding
Self-assembled Monolayer (SAMs)	Hydrophilic and hydrophobic groups with respective affinity for the transducer and the aptamer	Oriented recognition Stability	More suitable for silicon and gold surfaces

- Covalent bond

  This immobilization technique is based on the formation of a covalent bond between the aptamer and the surface of the electrode by the activation of a carboxylic acid (Danping et al. 2014). However, this method requires chemical changes to the electrode surface and the aptamers which must be changed, making the test more expensive than other immobilization techniques.

- Self-Assembled Monolayer (SAMs)

  This technique consists of spontaneous formation of monolayers by adsorption or by chemical bonding of molecules. (Zhang and Yadavalli 2011). The SAM method is one of the simplest techniques to provide a reproducible, ultrafine surface suitable for changes in aptamers, which improves sensitivity, speed and reproducibility. However, the use of binding molecules makes the test expensive.

6. Aptasensors for Mycotoxins Analysis

Mycotoxins are toxic secondary metabolites formed by certain fungi that can contaminate a wide range of food products around the world, in the field or after harvest in the stages of food production chains. They are capable of causing disease and death in humans and animals through the ingestion of food derived from contaminated cultures.

The most common mycotoxin-producing molds belong to the genera Fusarium, Aspergillus and Penicillium (Moss 1992, Sweeney and Dobson 1999, Medinaet al. 2006, Dzuman et al. 2014). These types of mold produce one or more of the six major classes of mycotoxins: aflatoxins, trichothecenes, fumonisins, zearalenone, ochratoxins and sterigmatocystins (Richard 2007).

In addition to the effects on human and animal health, mycotoxins also have multiple economic impacts on society. The economic impacts of mycotoxins on human society can be viewed in two ways: (i) direct market costs associated with the loss of trade or reduced income due to often contaminated food or feed due to international standards in force, and (ii) losses in human health due to adverse effects linked to the consumption of mycotoxins.

Foods with mycotoxin levels above a maximum level authorized by international standards are rejected and must be destroyed (Wu 2006). The Food Organization of the United Nations (FAO) has estimated that mycotoxins affect 25% of the world's crops each year (Rice and Ross 1994), causing annual losses of around 1 billion tons of food (Trail et al. 1995). On a larger scale, this limit or prevents export and global trade and therefore affects the global economy.

Health losses occur when food contains mycotoxins at levels that can cause disease. It is clear that mycotoxins affect human health. (Bennett and Klich 2003, Richard 2007). In 1996, in many West African countries, more than 98% of people tested were positive for mycotoxins in the blood indicating exposure of the population. Consumption of disease-causing mycotoxins adds an economic burden to the health system, which is difficult to quantify, with a need for health professionals and pharmaceuticals (Pitt et al. 2012). In addition, the additional costs associated with mycotoxins include management costs at all levels: prevention, sampling, mitigation, litigation and research costs (Cleveland et al. 2003).

6.1 Mycotoxins Analysis

The detection and quantification of Mycotoxins in food is very important to prevent the consumption of contaminated products, preserving and protecting the health of humans and animals. Generally, most of the chemical methods of analysis of mycotoxins comprise several stages such as: extraction, cleaning, separation, detection and quantification (van Egmond 1991).

The methods commonly used today for mycotoxins are mainly based on high performance liquid chromatography with fluorescence detection (HPLC/FLD) (Schweighardt et al. 1980), thin layer chromatography (TLC) (Nesheim et al. 1973) and more recently HPLC/MS-MS.

Alternatively and more recently, other methods have been used such as separation/detection by liquid chromatography-mass spectrometry LC-MS (Abramson 1987), LC-MS/MS (Becker et al. 1998), ELISA (Pestka et al. 1981) and chromatography in gas phase-mass spectrometry (GC-MS) (Jiao et al. 1992).

More recent work reports methods using Aptamers. Aptamers are single-stranded oligonucleotides (DNA or RNA) selected in vitro to bind with high affinity and specificity to targets (Cruz-Aguado and Penner 2008). The applications of Aptamers are known and developed in chromatography, electrophoresis, mass spectrometry and biosensors (Tombelli et al. 2005, Bonel et al. 2011, Rhouati et al. 2013a).

6.2 Aptasensor for Mycotoxins Determination

Aptasensors are increasingly used in bioanalysis of mycotoxins due to their higher stability, lower cost, ease of production, and labelling compared to other bioanalysis methods. However, the selection of aptamers for small molecules is still challenging (Ruscito and De Rosa 2016). Up to date, aptamers have been selected for ochratoxin A (OTA), aflatoxin B1 (AFB1), aflatoxin M1 (AFM1), fumonisin B1 (FB1), zearalenone (ZEA), deoxyvalenol (DON), the T-2 toxin, and patulin. The examples are summarized in Table 3.

Because of that, the development of aptasensor for mycotoxins has risen sharply in the last few years with many different aptasensing technologies and by far, most of these studies took as target analyte OTA, because of its widespread occurrence and hazardous effects on animal and human health (Rhouati et al. 2013). In this section, we will report some examples of the main aptasensors created for mycotoxin analysis included their sensibility.

Table 3: Aptamers selected against various mycotoxins (KD – dissociation constant of the aptamer-mycotoxin complex, LOD – limit of detection)

Mycotoxins	Aptamer structure	Sensitivity	References
Ochratoxin A	5'-GAT CGG GTG TGG GTG GCG TAA AGG GAG CAT CGG ACA-3'	LOD 0.01 ng/L	Chrouda et al. 2015
Aflatoxin B1	5'-GTT GGG CAC GTG TT GTC TCT CTG TGT CTC GTG C CCT TCG CTA GGC CCA C-3'	LOD 0.1 ng/mL	Seok et al. 2015
Aflatoxin B2	5'-AGC AGC ACA GAG GTC AGA TGC TGA CAC CCT GGA CCT TGG GAT TCC GGA AGT TTT CCG GTA CCT ATG CGT GCT ACC GTG AA-3'	KD 9.8 nM,	Ma et al. 2015
Aflatoxin M1	5'-GTT GGG CAC GTG TTG TCT CTC TGT GTC TCG TGC CCT TCG CTA GGC CCA CA-3'	KD 10 nM	Hamula et al. 2006
Fumonisin B1	50-ATA CCA GCT TAT TCA ATT AAT CGC ATT ACC TTA TAC CAG CTT ATT CAA TTA CGT CTG CAC ATA CCA GCT TAT TCA ATT AGA TAG TAA GTG CAA TCT-3'	LOD 33 ng/mL	Chen et al. 2015
Zearalenone	5'-AGC AGC ACA GAG GTC AGA TGT CAT CTA TCT ATG GTA CAT TAC TAT CTG TAA TGT GAT ATG CCT ATG CGT GCT ACC GTG AA-3'	LOD 0.5 ng/mL	Yugender et al. 2017
T-2 toxin	5'-CAG CTC AGA AGC TTG ATC CTG TAT ATC AAG CAT CGC GTG TTT ACA CAT GCG AGA GGT GAA GAC TCG AAG TCG TGC ATC TG-3'	LOD 0.93 pg/mL	Khan et al. 2018

6.3   Aptasensors for Ochratoxin A Detection

The first mycotoxin that has been targeted by aptasensors was ochratoxin A (OTA) which is one of the most toxic and widespread ochratoxin in food. Cruz-Aguado and Penner (2008) selected the first mycotoxin aptamer for OTA. After this selection, various aptasensing strategies (labeled and unlabeled) have been reported (Table 4). OTA determination using unlabeled aptasensors is still a challenge, despite the recent advances in unlabeled aptasensors.

Barthelmebs et al. (2011) selected OTA-specific aptamers (H8 and H12) to develop the first assay by enzyme-linked aptamers (ELAA). Two different competitive assays (indirect and direct) were studied and compared to the traditional ELISA test for the detection of OTA in wine samples. The first strategy was the indirect competition between the immobilized OTA-BSA (bovine serum albumin) conjugate complex and the free OTA present in the sample which binds to the free labeled aptamer in the solution. The second strategy was based on direct competition between free OTA and the OTA – HRP conjugate (protein called horseradish peroxidase) which binds to the immobilized labeled aptamer. This test was successfully applied to the detection of OTA in wines with a good sensitivity of 1 ng/mL in 125 min of analysis. By avoiding the limitations due to unstable and expansive antibodies, the ELAA method presented characteristics similar to those obtained by ELISA.

Yang et al. (2011) used gold nanoparticles (AuNPs) as a colorimetric indicator to describe another OTA aptasensor detection system. The principle of this method was based on the stability of AuNP against salt-induced aggregation in the presence of aptamer. After binding of the aptamer with OTA, the conformation of the aptamer changes, resulting in the aggregation of AuNP, which changes the color from red to blue. This experiment took 5 min, with a detection limit of 8 ng/mL, but a pre-concentration step was necessary.

Wang et al. (2011) used a fluorescent aptasensor for the detection of OTA in red wine samples with a detection limit of 1.9 ng/mL. The aptamer was labeled with FAM (fluorescent dye), while graphene was used as an extinguishing agent. In the absence of OTA, the aptamer was adsorbed on graphene, quenching the fluorescence of the FAM. After binding to OTA, the aptamer changed its conformation and formed the G-quadruplex structure which was resistant to adsorption. Therefore, the intensity of the fluorescence was proportional to the concentration of OTA in the sample.

Duan and Wang (2011) reported a fluorescence aptasensor based on the conformation change of the aptamer upon binding to OTA. The aptamer was immobilized on a microplate and its complementary oligonucleotide was labeled with FAM (Carboxyfluorescein). The formation of the OTA-Aptamer complex decreased the intensity of the fluorescence. A very low detection limit of 0.01 ng/mL was obtained when using this method on samples of corn flour.

Zhang et al. (2013) reported a new aptasensor based on the improvement of the emission of terbium ion in solution in presence of ssDNA. The aptasensor was constructed by immobilizing the aptamer on magnetic beads. In absence of OTA, the aptamer hybridized with two ssDNA probes present in the solution, blocking thus the terbium ion emission. After the addition of OTA, the aptamer structure switched to G-quadruplex and released the ssDNA probes resulting in an enhancement in terbium ion fluorescence in solution proportional to the amount of OTA with a limit of detection as low as 0.02 μg/L.

Lu et al. (2015) used luminescent metal complexes (octahedral Iridium) selective for G-quadruplex structures to construct the aptasensor. They used an aptamer hybridized with a cDNA strand. OTA binding induced the disassembly of the duplex, allowing the aptamer to fold into a quadruplex structure and decrease of luminescence.

Park et al. (2014) developed a label-free localized surface plasmon resonance (LSPR) aptasensor based on the strategies of aptamer switching and gold nanorods. The detection of OTA was achieved by monitoring the change in the magnitude of the LSPR wavelength which depended on the location of the analyte relative to the surface of the nanoparticle and the degree of alteration of the refractive index.

Prabhakar et al. (2011) described the first impedimetric label-free aptasensor for OTA detection by the immobilization of the aptamer onto Langmuir–Blodgett (polyaniline–stearic acid) film deposited onto indium tin-oxide coated glass plates. The change in the magnitude of transfer resistance due to OTA binding is observed at the sensor surface with a limit of detection as low as 0.1 µg/L.

Table 4: Different types of aptasensor used for the OTA determination

Detection methods	Samples	Detection limit (ng/mL)	References
Colorimetry	Wine	1	Yang et al. 2012
Fluorescence	Wine	0.2	Zhao et al. 2013
	Beer	9.64	Guo et al. 2011
	Wheat	0.02	Zhang et al. 2013
Luminescence	Wheat	0.0003	Hun et al. 2013
	Corn	0.0001	Wu et al. 2011
Electrochemical (Labeled)	Wheat	0.07	Bonel et al. 2011
	Beer	0.05	Rhouati et al. 2013b
	Wine	0.11	Barthelmebs 2011
Electrochemical (Label free)	Wheat	0.001	Tong et al. 2011
	Beer	0.00012	Hayat et al. 2013
	Coffee	0.048	Castillo et al. 2012

6.4 Aptasensors for Aflatoxins Detection

Le et al. (2010) have selected the first aptamer against Aflatoxin B1. Then, Ma et al. (2014, 2015) identified specific aptamers for Aflatoxin B1 and B2, while Aflatoxin M1 aptamer has been selected by Malhotra et al. (2014). All these aptamers have been used in many reports as biorecognition elements in aptasensing strategies.

Luan et al. (2015) described a colorimetric label-free aptasensor for Aflatoxin B1 and Aflatoxin B2 detection. In the absence of Aflatoxin, the nanoparticles were stabilized and dispersed by the aptamer leaving the solution red under high NaCl conditions. The target-induced conformational change exposes the AuNPs to NaCl-induced aggregation leading to a color change. The limit of detection was found to be 0.025 µg/L.

Seok et al. (2015) designed an ultra-sensitive probe that was capable of detecting aflatoxin B1 by the use of colorimetric assay based on Aflatoxin B1-induced DNA structural changes and peroxidase mimicking DNAzyme. The probe consisted of two split DNAzyme halves, as well as an aptamer. AFB1 binding induced a structural deformation of the aptamer-DNAzyme complex, which caused splitting of the DNAzyme halves thus decreasing peroxidase mimicking activity.

Ma et al. (2016) prepared an Aflatoxin B1-reponsive hydrogel for the colorimetric determination with high sensitivity. This hydrogel was composed of a recognition unit which was the aptamer, and a cross linkers composed of a pair of complementary DNA chains. Gold nanoparticles were added into the hydrogel, in order to determine Aflatoxin. In the presence of the Aflatoxin, the destruction of the hydrogel was noted as a result of the formation of Aflatoxin–aptamer complex. Meanwhile, the gold nanoparticles were subsequently released, resulting in a transition from colorless to red. The limit of detection of Aflatoxin was estimated to be as low as 5.5×10^{-10} g/L.

Nguyen et al. (2013) have applied an electrochemical aptasensor to aflatoxins detection using Cyclic Voltammetry and square wave voltammetry to monitor the biomolecular interaction between Aflatoxin M1 and aptamer. The aptamer has been immobilized on Fe_3O_4 incorporated polyaniline film polymerized on interdigitated electrode (IDE). The binding of the Aflatoxin M1 resulting a blockage of the charge transfer to the electrode surface, which influenced the switching rate of polyaniline film and decreased the current, which is inversely proportional to the analyte concentration. This aptasensor has shown high sensitivity of 0.00198 µg/L, excellent stability and reproducibility.

In the last year, Xia et al. (2019) have developed an enzyme-free, amplified, and ultra-fast aptasensor capable of one-test-tube homogeneous determination of aflatoxin B1. This aptasensor was composed of a dual-terminal proximity structure, for that one molecule could switch up to two fluorophores. Consequently, the signal was amplified without using enzymes. The aptasensor could detect the presence of Aflatoxin B1 within 1 min.

Table 5: Different types of aptasensor used for the aflatoxin determination

Detection methods	Samples	Detection limit (ng/mL)	References
Colorimetry	–	0.025	Luan et al. 2015
Fluorescence	Corn, Rice and Wheat	0.0002	Chen et al. 2014
Chemiluminescence	Corn	0.11	Shim et al. 2014
Electrochemical	Milk	0.00115	Istamboulie et al. 2016
	Beer and Wine	0.12	Yugender et al. 2016

6.5 Aptasensors for Other Mycotoxins

After OTA and aflatoxins, the aptamers recognizing fumonisin B1 (FB1) and zearalenone (ZEN) have been selected (Chen et al. 2013, Chen et al. 2014b).

Chen et al. (2015) functionalized the mass sensitive aptasensor cantilevers in the array with self-assembled monolayers (SAMs) of thiolated FB1 aptamer. Reference cantilevers were modified with 6-mercapto-1-hexanol SAMs to avoid interferences in the environment. Then, the analyte concentration was proportional to the differential deflection amplitude between sensing and reference cantilevers with a Limit of detection of 33 µg/L.

Wu et al. (2017) explored the fluorescence of up conversion nanoparticles and conjugated with the complementary oligonucleotide of ZEN aptamer for use as signal probes to develop a highly sensitive ZEN aptasensor. In the presence of ZEN, the aptamer binds to its target and releases the cDNA resulting in a decrease of the fluorescence intensity. The monitoring of luminescence allowed the quantitative detection of ZEN with a low limit of detection (0.007 µg/L).

7. Conclusion and Perspectives

Mycotoxin contamination is a threat to the health and life of Humans and animals. The detection and quantification of mycotoxin in food is very important to prevent the consumption of contaminated products, preserving and protecting the health of humans and animals. Generally, most of the chemical methods of analysis of the mycotoxins comprise several stages such as: extraction, cleaning, separation, detection, quantification.

The main problem with mycotoxin analysis methods is the duration of the analysis, the non-portability of equipment which is very expensive and the need to have qualified operators. For that reason, biosensor have been widely used for the determination of mycotoxins. On account of the high selectivity of aptamers toward specific targets, aptamers have been used to replace antibodies for the quantification of mycotoxins in all the analytical methods including biosensor.

Aptasensors are increasingly used in bioanalysis of mycotoxins due to their higher stability, lower cost, ease of production, and labelling compared to other bioanalysis methods. Although good results are achieved using these aptasensors for mycotoxin determination, some problems and drawbacks remain:

- The number of aptamers used in aptasensor for the determination of mycotoxin are limited. Additional efforts are demanded for searching new structures and increase the selectivity of the binding.
- To date, aptamers are not selected against all type of mycotoxins.
- The simultaneous determination of several mycotoxins is highly demanded to discover the products of biochemical conversion of primary contaminants which often present in the agro-food objects together.

References

Abramson, D. 1987. Measurement of ochratoxin A in barley extracts by liquid chromatography-mass spectrometry. Journal of Chromatography 391: 315–320. DOI: 10.1016/S0021-9673(01)94330-4.

Adley, C. and Ryan, M. 2015. Conductometric biosensors for high throughput screening of pathogens in food. Woodhead Publishing Series in Food Science, Technology and Nutrition. 315–326. DOI: 10.1016/B978-0-85709-801-6.00014-9.

Backmann, N., Zahnd, C., Huber, F., Bietsch, A., Pluckthun, A., Lang, H.P. et al. 2005. A label-free immunosensor array using single-chain antibody fragments. Proceedings of the National Academy of Sciences of the United States of America 102(41): 14587–14592.DOI: 10.1073/pnas.0504917102.

Bajar, B., Wang, E., Zhang, S., Lin, M. and Chu, J. 2016. A guide to fluorescent protein FRET Pairs. Sensors (Basel) 16(9). DOI: 10.3390/s16091488.

Banica, F.G. 2012. Chemical Sensors and Biosensors: Fundamentals and Applications. John Wiley & Sons. DOI: 10.1002/9781118354162.

Barthelmebs, L. 2011. Electrochemical DNA aptamer-based biosensor for OTA detection, using superparamagnetic nanoparticles. Sensors and Actuators B: Chemical 156: 932–937. DOI: 10.1016/j.snb.2011.03.008.

Barthelmebs, L., Hayat, A., Prieto-Simon, B. and Marty, J.L. 2011. Enzyme-linked aptamer assays (elaas), based on a competition format for a rapid and sensitive detection of ochratoxin A in wine. Food Control 22: 737–743. DOI: 10.1016/j.foodcont.2010.11.005

Becker, M., Herderich, M., Schreier, P. and Humpf, H.U. 1998. Column liquid chromatography electrospray ionization-tandem mass spectrometry for the analysis of Ochratoxin. Journal of Chromatography 818: 260–264. DOI: 10.1016/S0021-9673(98)00594-9.

Bennett, J.W. and Klich, M. 2003. Mycotoxins. Clinical Microbiology Reviews 16(3): 497–516. DOI: 10.1128/CMR.16.3.497–516.2003.

Berggren, C.B. and Johansson, G. 2001. Capacitive biosensors. Electroanalysis 13: 173–180. DOI: 10.1002/1521-4109(200103).

Bhalla, N., Jolly, P., Formisano, N. and Estrela, P. 2016. Introduction to biosensors. Essays in Biochemistry 60(1): 1–8. DOI: 10.1042/EBC20150001.

Binning, J.M., Wang, T., Luthra, P., Shabman, R., Borek, D., Liu, G. et al. 2013. Development of RNA aptamers targeting Ebola virus VP35. Journal of Biochemistry 52(47): 8406–8419. DOI: 10.1021/bi400704d.

Birch, J.R. and Racher, A.J. 2006. Antibody production. Advanced Drug Delivery Reviews 58(5–6): 671–685. DOI: 10.1016/j.addr.2005.12.006.

Bonel, L., Vidal, J.C., Duato, P. and Castillo, J.R. 2011. An electrochemical competitive biosensor for ochratoxin A based on a DNA biotinylated aptamer. Biosensors and Bioelectronics 26(7): 3254–3259. DOI: 10.1016/j.bios.2010.12.036.

Bosch, M., Rojas, F. and Ojeda, C. 2007. Recent development in optical fiber biosensors. Sensors 7: 797–859. DOI: 10.3390/s7060797.

Cagnin, S., Caraballo, M., Guiducci, C., Martini, P., Ross, M., Santaana, M. et al. 2009. Overview of electrochemical DNA biosensors: New approaches to detect the expression of life. Sensors (Basel) 9(4): 3122–3148. DOI: 10.3390/s90403122.

Castillo, G., Mosiello, L. and Hianik, T. 2012. Impedimetric DNA aptasensor for sensitive detection of ochratoxin A in food. Electroanalysis 24: 512–520. DOI: 10.1002/elan.201100485.

Chaubey, A. and Malhotra, B. 2002. Mediated biosensors. Biosensors and Bioelectronics 17(6–7): 441–456. DOI: 10.1016/s0956-5663(01)00313-x.

Chen, J., Zhang, X., Cai, S., Wu, D., Chen, M., Wang, S. et al. 2014a. A fluorescent aptasensor based on DNA-scaffolded silver-nanocluster for ochratoxin A detection. Biosensors and Bioelectronics 57: 226–231. DOI: 10.1016/j. bios.2014.02.001.

Chen, X., Huang, Y., Duan, N., Wu, S., Ma, X., Xia, Y. et al. 2013. Selection and identification of ssDNA aptamers recognizing zearalenone. Analytical and Bioanalytical Chemistry 405(20): 6573–6581. DOI: 10.1007/s00216-013-7085-9.

Chen, X., Huang, Y., Duan, N., Wu, S., Xia, Y., Ma, X. et al. 2014b. Selection and characterization of single stranded DNA aptamers recognizing fumonisin B1. Microchimica Acta 181(11): 1317–1324. DOI: 10.1016/j.foodchem.2014.06.039.

Chen, X., Bai, X., Li, H. and Zhang, B. 2015. Aptamer-based microcantilever array biosensor for detection of fumonisin B-1. Royal Society of Chemistry Advances 5(45): 35448–35452. DOI: 10.1039/C5RA04278J.

Cho, M., Xiao, Y., Nie, J., Stewart, R., Csordas, A., Thomson, J. et al. 2010. Quantitative selection of DNA aptamers through microfluidic selection and high-throughput sequencing. Proceedings of the National Academy of Sciences of the United States of America 107(35): 15373-15378. DOI: 10.1073/pnas.1009331107.

Chrouda, A., Sbartai, A., Baraket, A., Renaud, L., Maaref, A. and Jaffrezic-Renault, N. 2015. An aptasensor for ochratoxin A based on grafting of polyethylene glycol on a boron-doped diamond microcell. Analytical Biochemistry 488: 36–44. DOI: 10.1016/j.ab.2015.07.012.

Clark, L. and Lyons, C. 1962. Electrode systems for continuous monitoring in cardiovascular surgery. Annals of the New York Academy of Sciences 102: 29-45. DOI: 10.1111/j.1749–6632.1962.tb13623.x.

Cleveland, T.E., Dowd, P., Desjardins, A., Bhatnagar, D. and Cotty, P. 2003. United States Department of Agriculture-Agricultural Research Service research on pre-harvest prevention of mycotoxins and mycotoxigenic fungi in US crops. Pest Management Science 59(6-7): 629–642.DOI: 10.1002/ps.724.

Cooper, M. 2002. Optical biosensors in drug discovery. Nature Reviews Drug Discovery 1(7): 515–528. DOI: 10.1038/nrd838.

Cooper, M. 2003. Label-free screening of bio-molecular interactions. Analytical and Bioanalytical Chemistry 377(5): 834–842. DOI: 10.1007/s00216-003-2111-y.

Cremer, M. 1906. The cause of the electromotive properties of the tissue, at the same time a contribution to the teaching of polyphasic electrolyte chains. Journal of Biology 47: 562–608.

Cruz-Aguado, J. and Penner, G. 2008. Determination of ochratoxin a with a DNA aptamer. Journal of Agriculture Food Chemistry 56(22): 10456–10461. DOI: 10.1021/jf801957h.

Danping, X., Shangguan, L., Qi, H., Xue, D., Gao, Q. and Zhang, C. 2014. Click chemistry-assisted self-assembly of DNA aptamer on gold nanoparticales-modified screen-printed carbon electrodes for label-free electrochemical aptasensor. Sensors and Actuators B: Chemical 192: 558–564. DOI: 10.1016/j. snb.2013.11.038.

Daniels, J. and Pourmand, N. 2007. Label-free impedance biosensors: Opportunities and challenges. Electroanalysis 19(12): 1239–1257. DOI: 10.1002/elan.200603855.

Duan, N. and Wang, Z. 2011. An aptamer-based fluorescence assay for ochratoxin A. Analytical Chemistry 39: 300–304. DOI: 10.1016/S1872-2040(10)60423-9.

Dzuman, Z., Zachariasova, M., Lacina, O., Veprikova, Z., Slavikova, P. and Hajslova, J. 2014. A rugged high-throughput analytical approach for the determination and quantification of multiple mycotoxins in complex feed matrices. Talanta 121: 263–272. DOI: 10.1016/j.talanta.2013.12.064.

Ellington, A. and Szostak, J. 1990. In vitro selection of RNA molecules that bind specific ligands. Nature 346(6287): 818-822. DOI: 10.1038/346818a0.

FerreiraI, C. and Missailidis, S. 2007. Aptamer-based therapeutics and their potential in radiopharmaceutical design. Brazilian archives of biology and technology 50: 63–76. DOI: 10.1590/S1516-89132007000600008.

Griffin, E. and Nelson, J. 1916. The influence of certain substances on the activity of invertase. Journal of the American Chemical Society 38(3): 722–730. DOI: 10.1021/ja02260a027.

Guilbault, G. and Montalvo, G. 1969. Urea-specific enzyme electrode. Journal of the American Chemical Society 91: 2164. https://doi.org/10.1021/ac50159a062.

Guo, Z., Ren, J., Wang, J. and Wang, E. 2011. Single-walled carbon nanotubes based quenching of free FAM-aptamer for selective determination of ochratoxin A. Talanta 85(5): 2517–2521. DOI: 10.1016/j.talanta.2011.08.015.

Hamula, C., Zhang, H., Li, X. and Le, X. 2006. Selection and analytical applications of aptamers. Trends Analytical Chemistry 25: 681–691. DOI: 10.1016/j.trac.2006.05.007.

Hayat, A., Andreescu, S. and Marty, J.L. 2013. Design of PEG-aptamer two piece macromolecules as convenient and integrated sensing platform: Application to the label free detection of small size molecules. Biosensors and Bioelectronics 45: 168–173. DOI: 10.1016/j.bios.2013.01.059.

Heller, A. 1996. Amperometric biosensors. Current Opinion in Biotechnology 7(1): 50–54. DOI: 10.1016/S0958-1669(96)80094-2.

Hermann, T. and Patel, D. 2000. Adaptive recognition by nucleic acid aptamers. Science 287(5454): 820–825. DOI: 10.1126/science.287.5454.820.

Hun, X., Liu, F., Mei, Z., Ma, L., Wang, Z. and Luo, X. 2013. Signal amplified strategy based on target-induced strand release coupling cleavage of nicking endonuclease for the ultrasensitive detection of ochratoxin A. Biosensors and Bioelectronics 39(1): 145–151. DOI: 10.1016/j.bios.2012.07.005.

International Agency for Research on Cancer. 1992. IARC working group on the evaluation of carcinogenic risks to humans. IARC Monograph on the Evaluation of Carcinogenic Risks Human 57: 7–398.

Istamboulie, G., Paniel, N., Zara, L., Reguillo, L., Barthelmebs, L. and Noguer, T. 2016. Development of an impedimetric aptasensor for the determination of aflatoxin M1 in milk. Talanta 146: 464–469. DOI: 10.1016/j.talanta.2015.09.012.

Janshoff, A., Galla, H. and Steinem, C. 2000. Piezoelectric mass-sensing devices as biosensors – An alternative to optical biosensors. Angewandte Chemie International Edition in English 39(22): 4004–4032. DOI: 10.1002/1521-3773(20001117)39:22<4004::aid-anie4004>3.0.co;2-2.

Jayasena, S.D. 1999. Aptamers: An emerging class of molecules that rival antibodies in diagnostics. Clinical Chemistry 45(9): 1628–1650. PMID: 10471678.

Jiao, Y., Blaas, W., Ruhl, C. and Weber, R. 1992. Identification of ochratoxin A in food samples by chemical derivatization and gas chromatography-mass spectrometry. Journal of Chromatography 595(1–2): 364–367. DOI: 10.1016/0021-9673(92)85183-t.

Khan, I., Zhao, S., Niazi, S., Mohsin, A., Shoaib, M., Duan, N. et al. 2018. Silver nanoclusters based FRET aptasensor for sensitive and selective fluorescent detection of T-2 toxin. Sensors and Actuators B: Chemical 277: 328–335. DOI: 10.1016/j.snb.2018.09.021.

Knocki, R. 2007. Recent developments in potentiometric biosensors for biomedical analysis. Analytica Chimica Acta 599(1): 7–15. DOI: 10.1016/j.aca.2007.08.003.

Lazerges, M., Perrot, H., Rabehagasoa, N. and Compere, C. 2012. Thiol- and biotin-labeled probes for oligonucleotide quartz crystal microbalance biosensors of microalga alexandrium minutum. Biosensors (Basel) 2(3): 245–254. DOI: 10.3390/bios2030245.

Le, L.C., Cruz-Aguado, J.A. and Penner, G.A. (2010). DNA Ligands for Aflatoxin and Zearalenone. US 13/391426. https://patents.google.com/patent/WO2011020198A1/en

Li, B., Dong, S. and Wang, E. 2010. Homogeneous analysis: Label-free and substrate-free aptasensors. Chemistry an Asian Journal 5(6): 1262–1272. DOI: 10.1002/asia.200900660.

Lu, L., Wang, M., Liu, L., Leung, C. and Ma, D. 2015. Label-free luminescent switch-on probe for ochratoxin A detection using a G-Quadruplex-Selective Iridium(III) complex. ACS Applied Materials and Interfaces 7(15): 8313–8318. DOI: 10.1021/acsami.5b01702.

Luan, Y., Chen, Z., Xie, G., Chen, J., Lu, A., Li, C. et al. 2015. Rapid visual detection of aflatoxin B1 by label-free aptasensor using unmodified gold nanoparticles. Journal of Nanoscience and Nanotechnology 15(2): 1357–1361. DOI: 10.1166/jnn.2015.9225.

Ma, X., Wang, W., Chen, X., Xia, Y., Wu, S., Duan, N. et al. (2014). Selection, identification, and application of Aflatoxin B1 aptamer. Eur. Food Res. Technol. 238: 919–925. DOI: 10.1007/s00217-014-2176-1.

Ma, X., Wang, W., Chen, X., Xia, Y., Duan, N., Wu S. et al. (2015). Selection, characterization and application of aptamers targeted to Aflatoxin B2. Food Control 47: 545–551. DOI: 10.1016/j.foodcont.2014.07.037

Ma, Y., Mao, Y., Huang, D., He, Z., Yan, J., Tian, T. et al. 2016. Portable visual quantitative detection of aflatoxin B1 using a target-responsive hydrogel and a distance-readout microfluidic chip. Lab on a Chip 16(16): 3097–3104. DOI: 10.1039/C6LC00474A.

Malhotra, S., Pandey, A.K., Rajput, Y. and Sharma, R. 2014. Selection of aptamers for aflatoxin M1 and their characterization. Journal of Molecular Recognition 27(8): 493–500. DOI: 10.1002/jmr.2370.

Medina, A., Valle-Algarra, F., Mateo, R., Gimeno-Adelantado, J., Mateo, F. and Jimenez, M. 2006. Survey of the mycobiota of Spanish malting barley and evaluation of the mycotoxin producing potential of species of Alternaria, Aspergillus and Fusarium. International Journal of Food Microbiology 108(2): 196–203. DOI: 10.1016/j.ijfoodmicro.2005.12.003.

Meng, D., Yang, T., Zhao, C. and Jiao, K. 2012. Electrochemical logic aptasensor based on graphene. Sensors and Actuators B: Chemical 169: 255–260. DOI: 10.1016/j.snb.2012.04.078.

Mohanty, S. and Kougianos, E. 2006. Biosensosrs: A tutorial review. Open Journal of Applied Biosensors 25(2): 35–40. DOI: 10.1109/MP.2006.1649009.

Mortari, A. and Lorenzelli, L. 2014. Recent sensing technologies for pathogen detection in milk: A review. Biosensors and Bioelectronics 60: 8–21. DOI: 10.1016/j.bios.2014.03.063.

Moss, M.O. 1992. Secondary metabolism and food intoxication-moulds. Society of Applied Bacteriology Symposium Series 21: 80S–88S. DOI: 10.1111/j.1365-2672.1992.tb03627.x.

Nesheim, S., Francis, O. and Langham, W. 1973. Analysis of Ochratoxins A and B and their esters in barley, using partition and thin-layer chromatography. I. Development of the method. Journal of the Association of Official Analytical Chemists 56: 817–821. DOI: 10.1093/jaoac/56.4.817.

Nguyen, B., Tran, L., Do, Q., Nguyen, H., Tran, N. and Nguyen, P. 2013. Label-free detection of aflatoxin M1 with electrochemical Fe3O4/polyaniline-based aptasensor. Materials Science and Engineering C: Materials for Biological Applications 33(4): 2229–2234. DOI: 10.1016/j.msec.2013.01.044.

Nimse, S., Song, K., Sonawane, M., Sayyed, D. and Kim, T. 2014. Immobilization techniques for microarray: Challenges and applications. Sensors (Basel) 14(12): 22208–22229. DOI: 10.3390/s141222208.

Park, J., Byun, J., Mun, H., Shim, W., Shin, Y., Li, T. et al. 2014. A regeneratable, label-free, localized surface plasmon resonance (LSPR) aptasensor for the detection of ochratoxin A. Biosensors and Bioelectronics 59: 321–327. DOI: 10.1016/j.bios.2014.03.059.

Pestka, J., Steinert, B. and Chu, F. 1981. Enzyme-linked immunosorbent assay for detection of ochratoxin A. Applied and Environmental Microbiology 41(6): 1472–1474. PMID: 7247398.

Pitt, J., Wild, C., Baan, R., Gelderblom, W., Miller, J., Riley, R. et al. 2012. Improving public health through mycotoxin control. International Agency for Research on Cancer, World Health Organization.

Pohanka, M. 2018. Overview of piezoelectric biosensors, immunosensors and DNA sensors and their applications. Materials (Basel) 11(3): 448. DOI: 10.3390/ma11030448.

Prabhakar, N., Matharu, Z. and Malhotra, B. 2011. Polyaniline Langmuir-Blodgett film based aptasensor for ochratoxin A detection. Biosensors and Bioelectronics 26(10): 4006–4011. DOI: 10.1016/j.bios.2011.03.014.

Rhouati, A., Yang, C., Hayat, A. and Marty, J.L. 2013a. Aptamers: A promising tool for ochratoxin A detection in food analysis. Toxins (Basel) 5(11): 1988–2008. DOI: 10.3390/toxins5111988.

Rhouati, A., Hernandez, D., Meraihi, Z., Munoz, R. and Marty, J.L. 2013b. Development of an automated flow-based electrochemical aptasensor for on-line detection of ochratoxin A. Sensors and Actuators B: Chemical 176: 1160–1166. DOI: 10.1016/j.snb.2012.09.111.

Rhouati, A., Gilvanda, N., Hayat, A. and Marty, J.L. 2016. Label-free aptasensors for the detection of mycotoxins. Sensors 16. 2178. DOI: 10.3390/s16122178.

Rice, L. and Ross, P. 1994. Methods for detection and quantitation of fumonisins in corn, cereal products and animal excreta. Journal of Food Protection 57(6): 536–540. DOI: 10.4315/0362-028X-57.6.536.

Richard, J. 2007. Some major mycotoxins and their mycotoxicoses – An overview. International Journal of Food Microbiology 119(1–2): 3–10. DOI: 10.1016/j.ijfoodmicro.2007.07.019.

Rubab, M., Shahbaz, H.M., Olaimat, A.N. and Oh, D.H. 2018. Biosensors for rapid and sensitive detection of Staphylococcus aureus in food. Biosens Bioelectron 105: 49–57. DOI: 10.1016/j.bios.2018.01.023.

Ruscito, A. and DeRosa, M. 2016. Small-molecule binding aptamers: Selection strategies, characterization, and applications. Frontiers in Chemistry 4: 14. DOI: 10.3389/fchem.2016.00014.

Schmitteckert, E. and Schlicht, H. 1999. Detection of the human hepatitis B virus X-protein in transgenic mice after radioactive labelling at a newly introduced phosphorylation site. Journal of General Virology 80: 2501–2509. DOI: 10.1099/0022-1317-80-9-2501.

Schweighardt, H., Abdelhamid, A., Böhm, J. and Leibetseder, J. 1980. Method for quantitative determination of Ochratoxin A in foods and feeds by high-pressure liquid chromatography (HPLC). Zeitschrift fur Lebensmittel Untersuchung und Forschung 170: 355–359. DOI: 10.1007/BF01042973.

Seok, Y., Byun, J., Shim, W. and Kim, M. 2015. A structure-switchable aptasensor for aflatoxin B1 detection based on assembly of an aptamer/split DNAzyme. Analytica Chimica Acta 886: 182–187. DOI: 10.1016/j.aca.2015.05.041.

Shim, W., Kim, M., Mun, H. and Kim, M. 2014. An aptamer-based dipstick assay for the rapid and simple detection of aflatoxin B1. Biosensors and Bioelectronics 62: 288–294. DOI: 10.1016/j.bios.2014.06.059.

Song, K., Lee, S. and Ban, C. 2012. Aptamers and their biological applications. Sensors (Basel) 12(1): 612–631. DOI: 10.3390/s120100612.

Stoltenburg, R., Reinemann, C. and Strehlitz, B. 2007. SELEX – A (r)evolutionary method to generate high-affinity nucleic acid ligands. Biomolecular Engineering 24(4): 381–403. DOI: 10.1016/j.bioeng.2007.06.001.

Suzuki, S., Ikuo, S. and Nobuyuki, S. 1975. Ethanol and lactic acid sensors using electrodes coated with dehydrogenase—Collagen membranes. Bulletin of the Chemical Society of Japan 48(11): 3246–3249. DOI: 10.1246/bcsj.48.3246.

Sweeney, M. and Dobson, A. 1999. Molecular biology of mycotoxin biosynthesis. FEMS Microbiology Letters 175(2): 149–163. DOI: 10.1111/j.1574-6968.1999.tb13614.x.

Takhistov, P. 2005. Biosensor technology for food processing, safety and packaging. Handbook of Food Science, Technology and Engineering 3: 128(1)–128(20). DOI: 10.1201/b15995-143.

Thevenot D., Durst, R. and Wilson, G. 1999. Electrochemical biosensors: Recommended definitions and classification. Pure and Applied Chemistry 71(12): 2333–2348. DOI: 10.1016/s0956-5663(01)00115-4.

Tombelli, S., Minunni, M. and Mascini, M. 2005. Analytical applications of aptamers. Biosensors and Bioelectronics 20(12): 2424–2434. DOI: 10.1016/j.bios.2004.11.006.

Tong, P., Zhang, L., Xu, J. and Chen, H. 2011. Simply amplified electrochemical aptasensor of ochratoxin A based on exonuclease-catalyzed target recycling. Biosensors and Bioelectronics 29(1): 97–101. DOI: 10.1016/j.bios.2011.07.075.

Trail, F., Mahanti, N. and Linz, J. 1995. Molecular biology of aflatoxin biosynthesis. Microbiology 141 (Pt 4): 755–765. DOI: 10.1099/13500872-141-4-755.

Tromberg, B., Sepaniak, M., Vo-Dinh, T. and Griffin, G. 1987. Fiber-optic chemical sensors for competitive binding fluoroimmunoassay. Analytical Chemistry 59(8): 1226–1230. DOI: 10.1021/ac00135a033.

Tuerk, C. and Gold, L. 1990. Systematic evolution of ligands by exponential enrichment: RNA ligands to bacteriophage T4 DNA polymerase. Science 249(4968): 505–510. DOI: 10.1126/science.2200121.

Vashist, S. 2007. A review of microcantilevers for sensing applications. Journal of Nanotechnology 3: 1–15. DOI: 10.2240/azojono0115.

van Egmond, H. 1991. Methods for determining ochratoxin A and other nephrotoxic mycotoxins. International Agency for Reasearch of Cancer Scientific Publication 115: 57–70. PMID: 1820355.

Vo-Dinh, T. and Cullum, B. 2000. Biosensors and biochips: Advances in biological and medical diagnostics. Fresenius Journal of Analytical Chemistry 366(6–7): 540–551. DOI: 10.1007/s002160051549.

Wang, J., Cai, X., Rivas, G., Shiraishi, H., Farias, P. and Dontha, N. 1996. DNA electrochemical biosensor for the detection of short DNA sequences related to the human immunodeficiency virus. Analytical Chemistry 68(15): 2629–2634. DOI: 10.1021/ac9602433.

Wang, W., Wu, W., Zhong, X., Wang, W., Miao, Q. and Zhu, J. 2011. Aptamer based PDMS-Gold nanoparticle composite as a platform for visual detection of biomolecules with silver enhancement. Biosensors and Bioelectronics. 26(7): 3110–3114. https://doi.org/10.1016/j.bios.2010.10.034.

Wang, Y., Zhang, J. and Li, G. 2008. Electrochemical sensors for clinic analysis. Sensors 8: 2043–2081. DOI: 10.3390/s8042043.

Wei, Y., Zhang, J., Wang, X. and Duan, Y. 2015. Amplified fluorescent aptasensor through catalytic recycling for highly sensitive detection of ochratoxin A. Biosensors and Bioelectronics 65: 16–22. DOI: 10.1016/j.bios.2014.09.100.

Wu, F. 2006. Mycotoxin reduction in Bt corn: Potential economic, health, and regulatory impacts. Transgenic Research 15(3): 277–289. DOI: 10.1007/s11248-005-5237-1.

Wu, S., Duan, N., Wang, Z. and Wang, H. 2011. Aptamer-functionalized magnetic nanoparticle-based bioassay for the detection of ochratoxin A using upconversion nanoparticles as labels. Analyst 136(11): 2306–2314. DOI: 10.1039/c0an00735h.

Wu, Z., Xu, E., Chughtai, M., Jin, Z. and Irudayaraj, J. 2017. Highly sensitive fluorescence sensing of zearalenone using a novel aptasensor based on upconverting nanoparticles. Food Chemistry 230: 673–680. DOI: 10.1016/j.foodchem.2017.03.100.

Xia, X., Wang, Y., Yang, H., Dong, Y., Zhang, K., Lu, Y. et al. 2019. Enzyme-free amplified and ultrafast detection of aflatoxin B1 using dual-terminal proximity aptamer probes. Food Chemistry 283: 32–38. DOI: 10.1016/j.foodchem.2018.12.117.

Yang, C., Lates, V., Prieto-Simon, B., Marty, J.L. and Yang, X. 2012. Aptamer-DNAzyme hairpins for biosensing of Ochratoxin A. Biosensors and Bioelectronics 32(1): 208–212. DOI: 10.1016/j.bios.2011.12.011.

Yang, C., Wang, Y., Marty, J.L. and Yang, X. 2011. Aptamer-based colorimetric biosensing of Ochratoxin A using unmodified gold nanoparticles indicator. Biosensors and Bioelectronics 26(5): 2724–2727. DOI: 10.1016/j.bios.2010.09.032.

Yugender, K., Catanante, G., Hayat, A., Vengatajalabathy, K. and Marty, J.L. 2016. Disposable and portable electrochemical aptasensor for label free detection of aflatoxin B1 in alcoholic beverages. Sensors and Actuators B: Chemical 235: 466–473. DOI: 10.1016/j.snb.2016.05.112.

Yugender, K., Hayat, A., Satyanarayana, M., Sunil, V., Catanante, G., Vengatajalabathy, K. et al. 2017. Aptamer-based zearalenone assay based on the use of a fluorescein label and a functional graphene oxide as a quencher. Microchimica Acta 184(11): 4401–4408. DOI: 10.1007/s00604-017-2487-6.

Zhang, J., Zhang, X., Yang, G., Chen, J. and Wang, S. 2013. A signal-on fluorescent aptasensor based on Tb^{3+} and structure-switching aptamer for label-free detection of Ochratoxin A in wheat. Biosensors and Bioelectronics 41: 704–709. DOI: 10.1016/j.bios.2012.09.053.

Zhang, X. and Yadavalli, V. 2011. Surface immobilization of DNA aptamers for biosensing and protein interaction analysis. Biosensors and Bioelectronics 26(7): 3142–3147. DOI: 10.1016/j.bios.2010.12.012.

Zhao, Q., Geng, X. and Wang, H. 2013. Fluorescent sensing ochratoxin A with single fluorophore-labeled aptamer. Analytical and Bioanalytical Chemistry 405(19): 6281–6286. DOI: 10.1007/s00216-013-7047-2.

Zhao, X., Lin, C., Wang, J. and Oh, D. 2014. Advances in rapid detection methods for foodborne pathogens. Journal of Microbiology and Biotechnology 24(3): 297–312. DOI: 10.4014/jmb.1310.10013.

Index